The Scientific Sublime in Imperial Rome

The Scientific Sublime in Imperial Rome

Manilius, Seneca, Lucan, and the *Aetna*

PATRICK GLAUTHIER

OXFORD
UNIVERSITY PRESS

OXFORD
UNIVERSITY PRESS

Oxford University Press is a department of the University of Oxford.
It furthers the University's objective of excellence in research, scholarship,
and education by publishing worldwide. Oxford is a registered trade mark of
Oxford University Press in the UK and in certain other countries.

Published in the United States of America by Oxford University Press
198 Madison Avenue, New York, NY 10016, United States of America

CIP data is on file at the Library of Congress.

ISBN 9780197787557

DOI: 10.1093/9780197787588.001.0001

Integrated Books International, United States of America

MIX
Paper
FSC FSC® C183721

The manufacturer's authorised representative in the EU for product safety is
Oxford University Press España S.A. of el Parque Empresarial San Fernando
de Henares, Avenida de Castilla, 2 - 28830 Madrid (www.oup.es/en)

To Katherine and Oliver

Contents

Acknowledgments ix
Texts and Abbreviations xi

Introduction 1

1. Locating the Scientific Sublime 12
   1.1. Introduction 12
   1.2. The Dual Logic of the Sublime 16
   1.3. Encoding the Sublime 19
   1.4. Sublime Spectacles and Subjects 22
   1.5. The Fall 29
   1.6. The Scientific Sublime: Theoretical Perspectives from Plato to Hermogenes 40

2. Manilius and the Aesthetics of the Infinite 56
   2.1. Introduction 56
   2.2. The Milky Way 59
   2.3. Phaethon and the Celestial Daredevils 67
   2.4. Mathematical Poetry and the Revelation of God 74
   2.5. Calculating the Horoscope 83
   2.6. Infinity and Empire 91

3. Seneca's *Natural Questions*: The Promise and Danger of the Sublime 100
   3.1. Introduction 100
   3.2. Progressing toward Transcendence 103
   3.3. Working with Others 112
   3.4. Comets and *magnitudo mundi* 119
   3.5. A Vital Calmness 130
   3.6. Delusions of Grandeur 137

4. Lucan: Natural Inquiry in a Time of Civil War 145
   4.1. Introduction 145
   4.2. Sublime Distraction 148
   4.3. Knowing Too Much 160
   4.4. Spectacular Irrelevance 169
   4.5. Cato and the Snakes 178
   4.6. Acoreus and the Nile 187

5. *Aetna* and the Wonders of the Earth 199
5.1. Introduction 199
5.2. A Turn toward the Earth 203
5.3. An Epiphany in the Amphitheater 209
5.4. The Mechanical Sublime 216
5.5. From Causal Analysis to Paradoxography 226
Epilogue: The Scientific Sublime and the Case of the Elder Pliny 242

Bibliography 259
Index Locorum 277
General Index 291

Acknowledgments

In writing this book, I have benefitted greatly from the knowledge and generosity of multiple friends and colleagues, and it is a pleasure to thank them here. Gareth Williams introduced me to Seneca and supervised the dissertation out of which this study has grown; his criticism and advice have improved the final product immeasurably. Katharina Volk dared to offer a graduate seminar on Manilius and has always been a kind and insightful reader of my work; her feedback on this project has been invaluable. Margaret Graver, Håkan Tell, and Bryce Walker commented on early versions of several chapters. Joseph Farrell, Kirk Freudenburg, Margaret Graver, and Håkan Tell read a draft of the entire manuscript and spent an afternoon discussing it with me courtesy of Dartmouth's Leslie Center for the Humanities and its director, Rebecca Biron. Participants at a Dartmouth conference on the sublime helped me sharpen my thoughts on the paradoxographical tradition and the Elder Pliny. As readers for Oxford University Press, Philip Hardie and an anonymous colleague asked probing questions and pushed me to clarify important points. A Dartmouth Junior Faculty Fellowship and sabbatical provided essential research leave, which was spent in West Philly. During that period, Peter Struck facilitated access to the Penn Libraries system, and Sheila Murnaghan kindly extended the courtesy later down the road. Stefan Vranka encouraged the project from an early date and expertly guided the manuscript through the press. Rada Radojicic and Deva Thomas managed the production process with diligence and exactitude, answering numerous questions along the way. Keith Cline provided excellent copyediting, and Elizabeth Hadley and Jennifer Lynn gave crucial assistance with proofreading. For all remaining errors and any misguided lines of interpretation, I take full responsibility.

This book is dedicated with gratitude and lots of love to Katherine, who has been exceedingly patient for a very long time, and to Oliver, who is always excited to read.

Texts and Abbreviations

Latin quotations come from the following main texts, unless otherwise indicated: Goold (1998) for Manilius's *Astronomica*, Hine (1996a) for Seneca's *Natural Questions*, Shackleton Bailey (1997) for Lucan's *Civil War*, and Goodyear (1966) for the *Aetna*. In addition, I follow Bailey (1947) for Lucretius's *De rerum natura*, Mayhoff (1892–1909) for Pliny's *Natural History*, and Halliwell (2022) for Longinus's *On the Sublime*. Except where noted, translations are my own.

In general, abbreviations for ancient texts and authors follow S. Hornblower, A. Spawforth, and E. Eidinow, eds., 2012, *The Oxford Classical Dictionary*, 4th ed., Oxford, Oxford University Press. The following abbreviations are used throughout:

CHHP	Algra, K., J. Barnes, J. Mansfeld, and M. Schofield, eds. 1999. *The Cambridge History of Hellenistic Philosophy*. Cambridge: Cambridge University Press.
D-K	Diels, H., and W. Kranz, eds. 1960. *Die Fragmente der Vorsokratiker*. 3 vols. 9th ed. Berlin: Weidmannsche Verlagsbuchhandlung.
EGP	Laks, A., and G.W. Most, eds. 2016. *Early Greek Philosophy*. 9 vols. Cambridge, MA: Harvard University Press.
FLP	Courtney, E., ed. 1993. *The Fragmentary Latin Poets*. Oxford: Oxford University Press.
L-S	Long, A.A., and D.N. Sedley, eds. 1987. *The Hellenistic Philosophers*. 2 vols. Cambridge: Cambridge University Press.
OLD	Glare, P.W., ed. 2012. *Oxford Latin Dictionary*. 2 vols. 2nd ed. Oxford: Oxford University Press.
SVF	von Arnim, H., ed. 1903–1924. *Stoicorum Veterum Fragmenta*. 4 vols. Leipzig: B.G. Teubner.

Introduction

The sublime refers to both an experience and a discourse, the theory of sublimity (what it is, how it operates, what sets it in motion, what it *means*).[1] The two are not necessarily separate, especially in the ancient world; a description of the experience can imply a theory, and the explication of a theory can trigger the experience. The experience is one of transcendence, broadly construed, and it is often described as wonder, awe, stupefaction, ecstasy, or even terror. What I call the scientific sublime is the experience of sublimity through or in connection with the study of the natural world, from its origins to its constituent elements, from the celestial sphere up above to the subterranean spaces down below, and everything in between. The scientific sublime renders such study an exhilarating adventure or an intoxicating journey, a dangerous expedition through which our consciousness expands and our perspective shifts as we leave behind the trivialities of everyday life and approach something that appears to be truer and more real. Given the stakes, the very act of asking questions or daring to hazard an explanation of this or that phenomenon can become more important than finding correct answers that accurately correspond to reality or that fit seamlessly within the intellectual system of a particular philosophical school; simply to come close to catching a glimpse of the truth is well worth the effort.

The present study reads four Latin texts through the lens of the scientific sublime: Manilius's *Astronomica*, Seneca's *Natural Questions*, Lucan's *Civil War*, and the anonymous *Aetna*. (In the Epilogue, I briefly apply this critical filter to the Elder Pliny's *Natural History*.) In these four works, the ethos of the scientific sublime shapes and informs the representation of the study of nature, but as Manilius, Seneca, Lucan, and the *Aetna* poet repeatedly seek to energize their readers with the electric vitality of the study of the physical world, ideological differences come into view. Reading their respective

[1] For the history of the discourse, see Kirwan (2005), Doran (2015), and Shaw (2017). Day (2013, 30–71) and Porter (2016, 1–56) integrate recent approaches to Greek/Latin material into the broader story. I offer my own succinct overview of the classical sublime in Glauthier (2023).

deployments of the scientific sublime against one another and considering how those deployments interact with and respond to other modes of sublimity allow us to differentiate the authors' value systems and worldviews more clearly and simultaneously shine a spotlight on a multifaceted argument about how and why we ought to study the natural world in the first place—the contours of this argument are the subject of this book. By approaching these texts through the scientific sublime, then, a major thread of literary and intellectual dialogue and debate becomes apparent, one that cuts across genres, moving from the late Augustan Principate through Neronian Rome and into the Flavian period. The scientific sublime thereby emerges as a central feature of Latin literary production and frames a new story about the development and dynamics of Latin literature.

Some scholars have argued that employing the terms "science," "scientist," and "scientific" in ancient contexts is anachronistic. The methods, goals, assumptions, and institutions of the modern sciences, the argument goes, differ from those of any ancient intellectual pursuit, and the application of modern intellectual categories to Greco-Roman activities runs the risk of misrepresenting the past. This is a crucial insight.[2] From an ancient perspective, most of the fields of inquiry that figure in the present study constituted or could be claimed to constitute subcategories of philosophy—these include natural philosophy, physics, and meteorology.[3] Others had developed into or simply were autonomous fields, but philosophically minded individuals readily repurposed, elaborated on, or argued with their premises and claims, thereby drawing them back into the ambit of philosophy itself—such fields include mathematics, astronomy, astrology, and medicine. I do not use the terms "science" or "scientist," nor do I claim that Manilius, Seneca, Lucan, or the *Aetna* poet engaged in "science" or were "scientists." Instead, I employ the ancient terminology, which should be taken to refer to ancient intellectual pursuits. In the following pages, for instance, "meteorology" names the branch of ancient philosophy that

[2] French (1994, ix–xxii) and Rihll (1999, 1–23) are good introductions to the various problems. Keyser (2013, esp. 17–18, 29–32), however, argues forcefully in favor of using the term "science" in ancient contexts. A twenty-first-century reluctance to write of "science" in connection with Latin literature should not be conflated with an older view that denigrates Roman intellectual pursuits by comparison with those of the Greeks. This is the perspective from which Stahl (1962, vi), for instance, can write of his own *Roman Science* that "it may be said to be a book on a nonexistent subject."

[3] I use "natural philosophy" in the sense of "speculation about the ultimate principles and constituents of the natural world" (French 1994, x).

studied not only the weather but such diverse as phenomena as the Milky Way, earthquakes, rivers, and the sea. Where appropriate, I also use phrases such as "natural inquiry," "study of nature," and "investigator of nature," all of which have ancient precedents. However, I have named the central phenomenon in which I am interested the "scientific sublime." In this formulation, "scientific" is a convenient shorthand that refers to the various subcategories of philosophy, broadly construed, that studied and explained the nature, the mechanics, and especially the causes of different aspects of the material world.

Nor should "the scientific sublime" be taken to imply uniformity. Ancient investigators of nature adopted various principles, engaged with multiple discourses, and often had strong religious and ethical convictions that motivated their work and connected it to broader social contexts. Within a single philosophical school, variety and multiplicity could be the norm, both from one generation to the next and at roughly the same time. Stoicism was particularly flexible, and some writers freely borrowed from more than one tradition, including those that might not have qualified as philosophical to many ancient readers. In the *Natural Questions*, for instance, Seneca argues with his Stoic predecessors, makes extensive use of Lucretius's Epicurean masterpiece, and thinks in and through the language and imagery of epic and tragedy. The result is an intellectual hybrid of natural philosophy, physics, and meteorology that castigates first-century mores, contemplates Rome's political horizons, operates via poetic citation and intertext, and seeks to ennoble the reader's anxious mind by bringing it into contact with the divine. It is a singular creation, but its expansive vision of the study of nature is hardly an outlier.

At the same time, the sublime itself refuses to be confined within neat and tidy boundaries. The scientific sublime is a mode of sublimity that is grounded in one area of experience (the study of nature), but it is inflected differently in different authors and can even appear in contexts that probe the limits or question the very possibility of knowledge about the natural world; such contexts appear repeatedly in the present study. Moreover, any given mode of sublimity tends to respond to or interact with other modes. In a prominent simile in the first book of his *Civil War*, Lucan compares Caesar to a thunderbolt that shatters the heavens, terrifies those on earth with its blinding light and deafening crash, and rages unstoppably, causing destruction far and wide (1.151–157). The passage vividly depicts an awesome natural phenomenon, draws on the physics of Lucretius and Seneca, exploits longstanding connections between political speech and thunderbolts, and alludes to natural disasters that occur in *De rerum natura*, the *Georgics*, and the *Aeneid*. In just a few lines, then, we find multiples modes of sublimity, all reinforcing one another: the sublimity of nature, of natural inquiry, of imperial power, of rhetorical excellence, of epic poetry, etc.[4] In other instances, we find different modes of sublimity set in opposition to one another. Prominent passages in the *Natural Questions* suggestively contrast the ecstatic rush of studying nature with the seductive thrill of political domination and the delusional ecstasy of unrestrained eros. Similarly, the author of the *Aetna* juxtaposes the ennobling pleasure of celestial inquiry with the explosive power of studying the earth and finds the former wanting. Although my primary focus in this study is the scientific sublime, no mode of sublimity exists in a vacuum, and we will often see competing modes mirroring one another or jockeying for position. From one end of classical antiquity to the other, the sublime remains multiple, contested, and dynamic.

Classical scholars tend to approach the sublime in literature through the critical frameworks of one or more major theorists of the discourse on sublimity. Although the historical details are sketchy, the discourse itself began to emerge in fifth-century BCE Greece.[5] While its technical terminology was fluid, the images and metaphors through which it described the experience remained remarkably stable and made their presence felt in a variety of genres. To return to an example that we have already encountered, on the comic stage, Pericles can be imagined hurling rhetorical thunderbolts from the heights of Olympus centuries before the author of *On the Sublime* compares the power of sublime speech to lightning.[6]

In a sense, it is with this fragmentary treatise that any discussion of the sublime in antiquity must begin. *On the Sublime* (*Subl.*) is the only ancient work on the discourse of sublimity to have survived; it likely dates to the mid-first century CE, and for the sake of convenience, I refer to its author as Longinus.[7] The range of objects and phenomena that Longinus deems sublime is expansive, from individual words to moments of silence, from specific ideas to certain rhythms, from poetry and statues and paintings to rivers, oceans, mountains, and stars. The simple fact that such objects and phenomena have the capacity to unleash feelings of wonder or awe, that we

[4] For the allusions, see the notes in Roche (2009). I return to Lucan's simile in Ch. 4.3.

[5] For details and discussion, see Ch. 1.2.

[6] Ar. *Ach.* 530–531 and Long. *Subl.* 1.4, 12.4, and 34.4.

[7] For discussion of date and authorship, see most recently Halliwell (2022, x–xix). A prominent proponent of a third-century CE date is Heath (2012; 2013, 169–179).

can experience them in a state of quasi-religious ecstasy or madness, proves that "nature planted in our souls an irresistible passion for everything great and, from our perspective, more than human" (35.2); the sublime is our shared vocation, and we cannot resist its awesome power. Although numerous points of Longinus's theory of sublimity remain contested, many of his formulations have provided critics with powerful heuristic tools for reading a wide range of ancient texts ("the sublime is a kind of pinnacle and preeminence of language," 1.3; "the sublime does not persuade listeners but drives them out of themselves," 1.4; "the sublime, applied at the right moment, shatters everything like a bolt of lightning," 1.4; "the sublime is an echo of greatness of mind," 9.2; "the sublime lifts those who employ it near to the greatness of mind of god," 36.1). To read an ancient work of literature through Longinus—to identify similar expressions, arguments, or imagery, and to interpret them through Longinus's discourse on sublimity—is an accepted and exciting literary critical methodology.

Besides Longinus, the two most significant theorists of sublimity for classical scholars have been Edmund Burke (*A Philosophical Enquiry into the Sublime and the Beautiful*, 1757, revised 1759) and Immanuel Kant (*Critique of the Power of Judgment*, 1790, revised 1793). Their analyses of how the sublime operates and what sets it in motion have exercised the greatest influence on classicists. Burke (1998, 86, from *Enquiry* 1.7) analyzes the sublime as a combination of pain and pleasure:

> Whatever is fitted in any sort to excite the ideas of pain, and danger, that is to say, whatever is in any sort terrible, or is conversant about terrible objects, or operates in a manner analogous to terror, is a source of the *sublime*; that is, it is productive of the strongest emotion which the mind is capable of feeling. . . . When danger or pain press too nearly, they are incapable of giving any delight, and are simply terrible; but at certain distances, and with certain modifications, they may be, and they are delightful, as we every day experience.

The sublime is an experience of terror at a distance, an idealized terror that is recuperated by pleasure, and it is the most intense emotional experience possible.

Subsequently, Burke divides the experience into two halves, one negative, the other positive. It is an experience of the soul (Burke 1998, 101, from *Enquiry* 2.1)

> in which all its motions are suspended, with some degree of horror. In this case the mind is so entirely filled with its object, that it cannot entertain any other, nor by consequence reason on that object which employs it. Hence arises the great power of the sublime, that far from being produced by them, it anticipates our reasonings, and hurries us on by an irresistible force.

The sublime consists in an encounter with an overwhelming object that paralyzes the soul, incapacitating our ability to reason, and the astonishing ability of the soul to overcome that obstacle and to reassert its independence and power, its rational capabilities. Throughout the treatise, Burke analyzes the causes of this experience, which include obscurity, power, forms of privation (vacuity, darkness, solitude, and silence), vastness, infinity, magnitude, difficulty, magnificence, and quick transitions from darkness to light and from light to darkness. He adduces examples from numerous Greek and Latin texts and is familiar with Longinus.

Kant distinguishes between two kinds of sublimity, the mathematically sublime and the dynamically sublime. Despite significant differences between the views of Burke and Kant, both of Kant's categories show the same combination of contraries as Burke's conception and operate dialectically. The mathematically sublime names the experience of an object that appears "absolutely great," something "in comparison with which everything else is small" (Kant 2000, 131, 134, from *Critique* §25). It is something even to be able to think of which "demonstrates a faculty of the mind that surpasses every measure of the senses" (Kant 2000, 134, from *Critique* §25). In trying to comprehend such objects, the mind falters and then triumphs. In the case of a visitor to the Vatican entering St. Peter's Basilica, Kant (2000, 136, from *Critique* §26) writes, "there is a feeling of the inadequacy of his imagination for presenting the ideas of a whole, in which the imagination reaches its maximum and, in the effort to extend it, sinks back into itself, but is thereby transported into an emotionally moving satisfaction." As in Burke's *Enquiry*, there is a twofold process at play: the mind is paralyzed and then liberated, stunned and then exalted. It should be kept in mind that the mathematically sublime is not limited to encounters with manmade objects. Kant (2000, 140, from *Critique* §26), for instance, discusses the experience of trying to fathom great magnitudes in nature, moving from a tree to a mountain to the earth to the Milky Way and "the immeasurable multitude of such Milky Way systems, called nebulae." This kind of endeavor activates the same dialectical process as setting foot into St. Peter's.

Kant's discussion of the dynamically sublime focuses on "nature as a power" (Kant 2000, 143, from *Critique* §28). When we recognize nature's fearful might, its ability to dominate and to crush us, without actually fearing it, we are in a position to experience the dynamically sublime (Kant 2000, 144–145, from *Critique* §28):

> Bold, overhanging, as it were threatening cliffs, thunder clouds towering up into the heavens, bringing with them flashes of lightning and crashes of thunder, volcanoes with their all-destroying violence, hurricanes with the devastation they leave behind, the boundless ocean set into a rage, a lofty waterfall on a mighty river, etc., make our capacity to resist into an insignificant trifle in comparison with their power. But the sight of them only becomes all the more attractive the more fearful it is, as long as we find ourselves in safety, and we gladly call these objects sublime because they elevate the strength of our soul above its usual level, and allow us to discover within ourselves a capacity for resistance of quite another kind, which gives us the courage to measure ourselves against the apparent all-powerfulness of nature.

Initially, nature's might terrifies the subject, making us aware of our human smallness and frailty. When viewed from a safe distance, however, the mind claims nature's grandeur for itself and proclaims its own superiority to that which had previously been threatening. The quickening thrill of the literary storm or the breathtaking view into the painted abyss immobilize and inspire; one might react to Kant's own rhetoric similarly.[8]

As is the case with Longinus, classicists have become used to reading ancient literature through the theories of Burke and Kant. In Kant's case, this maneuver is facilitated by the suspicion that he draws some of his arguments and imagery directly from Lucretius.[9]

[8] On this reading, the passage from Kant is a concrete example of what de Bolla (1989, 12) describes as "a natural tendency for the discourse on the sublime to produce the conditions necessary for the construction of the discourse of the sublime, a discourse which produces from within itself sublime experience."

[9] See Porter (2007, 177–184; 2016, 466–473). For possible Senecan influence on Kant, see Williams (2012, 224–225).

But this is not the primary methodology of the present work. As I explain in Chapter 1, I employ Robert Doran's (2015) analysis of the discourse of the sublime. Doran shows that all the major theorists of the sublime, from Longinus to Kant, construe the sublime as an experience of emotional intensity that exhibits a dual structure: that of being overwhelmed and uplifted, overpowered and ennobled. It is a form of secular transcendence for which Greek and Latin texts established a paradigm upon which the early modern world wrung changes. Adopting Doran's straightforward definition grounds the study of the classical sublime and simultaneously opens it up. The sublime does not belong to Longinus, Burke, Kant, or anyone else, and there is no reason that the circumstances and ideologies of one critic should set the agenda and establish the parameters for the study of ancient texts and traditions. The role of the sublime in the *Astronomica*, *Natural Questions*, *Civil War*, and *Aetna* has not escaped critical attention in recent years, but Doran's framework allows me to uncover experiences of sublimity in unexpected places, to argue that the sublime functions in unfamiliar ways, and to call attention to previously underappreciated traditions in which the sublime plays significant roles. The classical sublime is multifaceted and contentious, and the definition that I adopt deepens our understanding of its expansive and sometimes unpredictable dynamism. James Porter has demonstrated at length the rich possibilities that emerge when one decenters Longinus from the study of the sublime in the ancient world; the present study builds on Porter's example.[10]

The study of the sublime in antiquity, especially in the field of Latin literature, has developed in exciting ways during the first decades of the twenty-first century.[11] Instead of thinking about the sublime as a mere effect of language or style, critics have recognized that this experience plays a fundamental role in a wide range of ancient discourses and intellectual traditions. In particular, Porter and Philip Hardie have well demonstrated, in very different ways, that the experience of the sublime is central to numerous

[10] Porter (2016); see esp. "The Sublime without Longinus" (18–25).

[11] Especially relevant in the present context are Delarue (2006), Leigh (2006), Schrijvers (2006), Porter (2007), Barchiesi (2009), Hardie (2009, 67–228), Williams (2012, 113–116, 213–257), Day (2013), Chaudhuri (2014, 13–14, 62–63, 136–144, 181–192, 234–243), Schiesaro (2014), Gunderson (2015), Lowe (2015, esp. 193–216), Porter (2016), Williams (2016; 2017a), Lagière (2017, esp. 147–173), Zientek (2017), Lombardo (2018, esp. 14–20), Williams (2020), Garani (2020a), Hardie (2022a, 9–12, 30–105), Glauthier (2022, 89–101), Garani (2023), and Glauthier (2023, 20–27). Important twenty-first-century work on Long. *Subl.* includes Hunter (2009, 128–168), the second edition of Mazzucchi's commentary (2010), Halliwell (2011, 327–367), de Jonge (2012), Doran (2015, esp. 1–12, 15–19, 27–94), Porter (2016, 1–51, 57–177), de Jonge (2020), and Halliwell's editions (2021 in Italian, 2022 in English).

authors who write about the mechanics of the natural world and the causes of natural phenomena. Hardie focuses on Lucretius and his reception, especially by Virgil, Ovid, and Horace.[12] His powerful and wide-ranging readings have shown the centrality of Lucretius to the history of the sublime in the poetry of the late Republic and Augustan Principate. In many ways, Porter reinvigorated the study of the classical sublime with his own groundbreaking contribution to the study of Lucretius, and his subsequent book embedded that reading of *De rerum natura* within a long history of philosophers and investigators of nature.[13] It is no longer surprising to argue that Greek and Latin texts about nature's causes involve the sublime.

At the same time, Lucretius now dominates many discussions of the sublime in the Roman world. My first chapter turns repeatedly to Lucretius to illustrate concepts and patterns, and later chapters discuss or point to Lucretius often since *De rerum natura* is a core text for the four primary authors under consideration. And yet, while the histories of the sublime and the study of the natural world at Rome are inconceivable without Lucretius, I do not frame this investigation as one of Lucretian reception, nor do I attempt to read the individual texts through the Lucretian sublime. Collectively, the chapters that make up this book add new perspectives to the story of Lucretius's afterlife, and they simultaneously demonstrate that those histories include dimensions and narratives that are unrelated to or not primarily dependent on *De rerum natura*. The Lucretian sublime has received ample attention, but other points of departure are possible. Just as I decenter Longinus, I decenter Lucretius from the study of the sublime at Rome and allow a different narrative to develop.[14]

Each author at the heart of this study formulates his own version of the scientific sublime, elaborates it in contrast to competing versions, and situates it within the contemporary socio-political landscape. The scientific sublime is a medium through which these authors react to one another and define their own intellectual priorities and values. Manilius takes the reader

[12] Hardie (2009, 67–228). Hardie (2022b) incorporates Ovid into this narrative. Hardie (2013a) explores the sublime in Flavian epic.

[13] Porter (2007) on Lucretius. Porter's monumental *The Sublime in Antiquity* (2016) contains material from numerous previous studies; at 445–473, he expands on Lucretius. In contrast to Hardie and Porter, Most (2012, esp. 249–251) seeks to establish a fundamental opposition between the sublime in Longinus and Lucretius.

[14] Empedocles is another investigator of nature who has emerged as a central figure in the history of the sublime at Rome. Empedocles, however, does not play a prominent role in this study, and Ch. 1.1 deliberately opens with Parmenides, thereby demonstrating that it is possible to frame the story differently. For Empedocles, the sublime, and Latin literature, see Hardie (2009, 88–90, 99–109, 116–125, 136–152, 172, 197–198, 214–217).

on a wild rollercoaster ride through the heavens, reveling in the infinite dimensions of the cosmos, the unfailing reason that pervades the universe, and the astounding ability of his mathematical calculations to uncover the mind of god. Even if the numbers do not always add up, the mathematically grounded study of the stars is the one true way to come to know nature, and it brings the reader into contact with the divine. The supreme confidence, exuberant triumphalism, and sheer disregard for boundaries and limits mirror the indefatigable ethos of the Augustan regime. For Manilius, to study the celestial sphere through the verses of the *Astronomica* is to experience the thrilling power of Roman imperial rule.

In Nero's Rome, the study of the natural world is more fraught and less certain, a change in outlook that reflects the growing instability and paranoia of the age. The *Natural Questions* redirects the reader's field of vision from Manilius's heavens, which now appear to exist at an impossible epistemological distance, to the atmosphere and atmospheric phenomena, an inherently less predictable area of inquiry. Although Seneca insists that the study of nature leads to wisdom and virtue, the journey is full of false starts and dead ends, and we only ever experience a fleeting burst of transcendence. What's more, competing sources of sublimity, including Roman political power, threaten to derail the project and to sink the reader into vast seas of moral turpitude. Against such degradations of the sublime, Seneca presents the quickening thrill of meteorological study as an imperfect bulwark. Writing at roughly the same time, Lucan applies even greater pressure to the outlook and commitments of the scientific sublime, suggesting that the projects of Manilius and Seneca both are downright delusional. An almost pathological urge to narrate and to speculate about nature and natural phenomena, cut loose from any higher purpose, saturates the *Civil War*, potentially ennobling but ultimately crushing the reader with theoretical detail and open-ended questions. Moreover, those characters who have studied and understand the physics of the universe cease to inspire wonder—other sources of sublimity, including Caesar's military might, take precedence, relegating the promise of the scientific sublime to the naïveté of a bygone era.

In the wake of Nero's implosion, at the beginning of the Flavian period, the scientific sublime seeks to reassert its ability to expand the mind, ennoble the spirit, and bring the student of nature into contact with god. Unfortunately, Lucan's critique is too devastating—there is no going back. Rejecting the lofty aspirations of the *Astronomica* and *Natural Questions*, the *Aetna* poet formulates a dual imperative, to study the earth and to celebrate its inexplicable wonders, the handiwork of nature the demiurge. By the poem's end, however, it is uncomprehending awe in the face of nature's wonders, rather than the rush of causal analysis, that proves to be the surest way to spiritual elevation. As the author seeks to overcome the darkness and anxiety of the final years of the Julio-Claudian era, he exchanges the study of the natural world for a spectacular show at the center of which stands a personified and all-powerful nature. In the Epilogue, I suggest that a similar ambivalence permeates the Elder Pliny's *Natural History*, which dates to the same period as the *Aetna*. Pliny feels alienated from the values and aspirations of the scientific sublime and implies that the moral failings of his contemporaries have placed that experience out of reach. Consequently, he eschews causal analysis and finds transcendence in the contemplation of a seemingly unending parade of facts that demonstrate nature's power and grandeur. For both the *Aetna* poet and Pliny, the scientific sublime is an unattainable ideal that gradually fades into obsolescence. In its place, there emerges a spectacle fit for Rome's magnificent and soon-to-be-completed Flavian amphitheater, the spectacle of nature.

Such, in broad outline, is the story about the study of nature, the sublime, and Latin literature that develops over the course of this book. While the attitudes and values that motivate the individual authors reflect the evolving nature of Roman *imperium* and public life, it is important to remember that other perspectives will have been available to ancient readers at each stage along the way. In other words, this story does not constitute a comprehensive or exhaustive account of early imperial views of natural inquiry and the sublime. The texts on which I focus constitute one thread in a long-standing and multifaceted debate about the study of the heavens and the earth and how that study fits into the intellectual, political, and moral landscape of Roman society. By devoting sustained attention to this particular thread of that debate, I hope to shift the way we think about the Latin tradition that inquires into nature's mechanics and to enrich our understanding of the sublime at Rome. But before we can explore the details, we need to look more closely at the scientific sublime itself, its history and the images and narrative patterns that are central to it, and we need to think more carefully about how we locate the sublime in ancient texts in the first place.

The Scientific Sublime in Imperial Rome: Manilius, Seneca, Lucan, and the Aetna. Patrick Glauthier, Oxford University Press. © Oxford University Press 2025. DOI: 10.1093/9780197787588.003.0001

1

Locating the Scientific Sublime

1.1. Introduction

In the fifth century BCE, Parmenides composed a hexameter poem in which a goddess, possibly Night, revealed to the author the truth about a wide range of topics related to the natural world: the origins of the cosmos, the nature of the heavens and heavenly bodies, the mechanics of meteorological phenomena, like earthquakes, and biological processes, such as human reproduction.[1] The poem begins with a journey, a stupefying expedition through which human exploration and mystic initiation merge (D4 *EGP*, B1 D-K).[2] In these opening lines, Parmenides travels along the road of the goddess, guided by horses that can take him as far as he might wish. The mares strain at the chariot and the axle hisses in the naves, like a flute, set ablaze by the pressure of the fast whirling wheels. The scene is all energy and noise, a dazzling sight that overloads the senses. Maidens lead the way and persuade the gatekeeper to thrust the bolted bar from the gate with all haste, and then: "When [the gates] flew open, they made an infinite abyss of the doors" (ταὶ [sc. πύλαι] δὲ θυρέτρων / χάσμ' ἀχανὲς ποίησαν ἀναπτάμεναι, 17–18), spinning two bronze pivots in their sockets, fitted with pegs and rivets. The physical mechanisms of the gates implicitly creak and heave, echoing the tumultuous din of the chariot.[3] The jolting sound reverberates in stark contrast to the "infinite abyss" (χάσμ' ἀχανές) that momentarily floods the reader's field of vision and threatens to detain us indefinitely; the imagery recalls the "great abyss" that separates earth from Tartarus in Hesiod (χάσμα μέγ', *Th.* 740), confirming that we are on a downward

[1] Origins of the cosmos and celestial topics: D11–31 *EGP* (A37–43a, B9–15, 21 D-K). Earthquakes: D36 *EGP* (A44 D-K). Human reproduction: D44–50 *EGP*, with references to D-K.

[2] For the identity of the goddess, the importance of initiation, and parallels with Orphic literature, see Palmer (2009, 51–61).

[3] The repetition of ἄξων (the "axle" of the chariot at line 6, the "pivots" in the sockets of the doors at line 19) suggests the sensory parallel.

journey, an intellectual katabasis.[4] But Parmenides's abyss is not just "infinite" (ἀχανές)—it is "open mouthed, astonished, dumbstruck."[5] The epithet suggests the physical and psychological state of both author and reader, the appropriate response to the heroic adventure playing out before our eyes. Without hesitation, the maidens lead Parmenides into the unending darkness, and the goddess welcomes him warmly.

For Parmenides, the experience of philosophical inquiry, which includes the explanation of celestial and meteorological phenomena, is a mental journey that is astonishing, bewildering, and thrilling. To ask questions about the natural world and humankind's place in it is an exhilarating exercise that is deeply implicated in the inspired tradition of epic poetry and simultaneously infused with the ambitious energy of intellectual innovation, and it puts the investigator in direct communication with the divine. Later writers will call this kind of experience sublime, and from the fifth century to the end of antiquity, the study of the natural world is frequently presented in these terms.

The execution, however, is not always so dramatic. While Parmenides's chariot ride is the first in a long line of similarly symbolic excursions, several of which feature prominently in this study, it is also a precursor to the more general flight of the mind topos, a tradition that configures intellectual inquiry as a journey up to or through the heavens or around the earth.[6] An imperial-era epigram from the *Greek Anthology* succinctly demonstrates the trope's power and appeal (*AP* 9.577):[7]

> οἶδ' ὅτι θνατὸς ἐγὼ καὶ ἐφάμερος· ἀλλ' ὅταν ἄστρων
> μαστεύω πυκινὰς ἀμφιδρόμους ἕλικας,
> οὐκέτ' ἐπιψαύω γαίης ποσίν, ἀλλὰ παρ' αὐτῷ
> Ζανὶ θεοτρεφέος πίμπλαμαι ἀμβροσίης.

I know that I am mortal, a creature of a day, but when I investigate the multitudinous round-running spirals of the stars, I no longer touch the earth with my feet but, at the side of Zeus himself, I am filled with god-nourishing ambrosia.

[4] The house of Night (line 9) and the paths of Night and Day (line 11) look back to this same Hesiodic passage (*Th.* 744 and 748–757, respectively). Some read Parmenides's chariot ride as an ascent rather than a katabasis.

[5] Montanari (2015), s.v. ἀχανής 2A. The adjective, related to χάσκω ("yawn, gape"), can mean both "open mouthed, gaping" (alpha copulative) and "not opening the mouth, straight, narrow" (alpha privative). When it indicates emotion, the sense is simply "astonished, dumbstruck, speechless."

[6] See most recently Hardie (2022a, 30–105).

[7] The text is that of Page (1981, 113).

The swirling orbits of the stars catch the author up in their gyrations, lifting him off the ground. To study such phenomena is physically and spiritually elevating, both a trip to Olympus and an expansion of the mind's horizons. It is an activity that blurs the boundary between human and divine and infuses the human with the power and majesty of the gods, allowing the author to transcend, if only momentarily, his own mortality. The manuscripts attribute the epigram to Ptolemaeus, which scholars take to mean Claudius Ptolemaeus, the great second-century CE astronomer. Whether or not Ptolemy wrote the epigram, the poem presents the investigation of the natural world through mathematical astronomy as a breathtaking adventure beyond the limits of everyday experience and sense perception; in other words, it is sublime.

The study of nature can spark an electrifying thrill in almost any literary context. To take an important Latin example, in the sixth *Eclogue*, before we hear Virgil's report of Silenus's song (6.31–86), we witness the ecstatic reaction of the poem's embedded audience: "Then indeed you would have seen both Fauns and wild animals playing in time to the rhythm, then [you would have seen] hard oaks shake their tops; neither does the cliff of Parnassus exult (*gaudet*) so much in Apollo nor do Rhodope and Ismarus wonder (*miratur*) so much at Orpheus" (6.27–30). The song possesses a vital charge that brings inanimate nature to life. Silenus, in effect, becomes a god, and the reaction of the audience surpasses in affective intensity the exultation and wonder that Apollo and Orpheus inspire. An explanation for the impact of Silenus's song follows immediately: "For he was singing how, through the great void, atoms of earth and air and sea and, at the same time, liquid fire had been driven together" (*namque canebat uti magnum per inane coacta / semina terrarumque animaeque marisque fuissent / et liquidi simul ignis*, 6.31–33). Silenus's performance begins with natural philosophy—atoms and void, the four elements, the origins of the universe, the creation of the sun and the earth, the first clouds and rainfalls, and the origins of plant and animal life (6.31–40)—and that is what causes the emotional explosion (*namque*). What's more, the language is markedly Lucretian. Lucretius, for instance, writes *magnum per inane* five times, and *semina* is the commonest term for "atoms" in his poem.[8] The conspicuous clustering

[8] The phrase *magnum per inane* occurs at Lucr. 1.1018, 1103, 2.65, 105, and 109. For the Lucretian echoes, see Coleman (1977, 183–186) and Clausen (1994, 189–192).

of Lucretian vocabulary allows us to construe Silenus's song, at the outset, as a reading of Lucretius himself. Put differently, Silenus's power is Lucretius's power, and the experience of listening to Silenus is that of reading *De rerum natura*.[9]

Virgil's maneuver constitutes a reflection on the dynamics of Latin literary history. The song begins with the origins of the world, dwells on dark forms of mythological eros, contains proto-elegy, and both ends on and is set within a pastoral landscape—content, allusions, and named figures move the reader from Lucretius to Calvus to Gallus to Virgil himself.[10] As a whole, the poem stands in contrast to martial epic and poetry on contemporary political themes and aligns itself with Callimachus and Theocritus (*Ecl.* 6.1–12), but it is Silenus's *carmen* that tells a teleological story about Latin literature in which all strands converge on the *Eclogues*.[11] The experience of the sublime, activated initially through the study of the natural world, becomes foundational to the history of Latin poetry and is made to lead inexorably to Virgil's own bucolic creation.

In broad terms, Virgil frames the study of nature from the same vantage point as Parmenides and the author of the epigram ascribed to Ptolemy—it is an intoxicating adventure, an irresistible force, a pathway to knowledge and divinity that promises to ennoble both author and reader alike. This is the scientific sublime. In a sense, the present study takes the song of Silenus as a point of departure to argue that the connections between the study of nature and the sublime are indeed foundational to Latin literary history. Manilius, Seneca, Lucan, and the *Aetna* poet are heavily invested in the promise of this experience—the scientific sublime is a discourse through which they define their own value systems, promote their own intellectual projects, and critique their competitors, both past and present. But how exactly are we to define the sublime and locate it within actual texts?

[9] The song of Apollonius's Orpheus at *Argon.* 1.494–515 is another important background text; it covers cosmogonical and astronomical material, as well as the origins of mountains, rivers, and animal life, and similarly enthralls its own embedded audience. Stewart (1959, 184–185) points out the presence of Empedoclean reminiscences in Silenus's performance and concludes that this portion of the song represents "scientific" poetry in general.

[10] The *Io* of C. Licinius Calvus (fr. 9 *FLP*) is the source of *a, uirgo infelix* in Silenus's rendition of Pasiphae (*Ecl.* 6.47, 52). C. Cornelius Gallus is the subject of *Ecl.* 6.64–73; for fragments and background, see *FLP* 259–270.

[11] For Callimachus and Theocritus, see Clausen (1994, 174–178). Virgil's opening lines famously rework Apollo's admonition that Callimachus keep his sacrifices fat, his poetry lean (fr. 1.23–24 Pfeiffer).

1.2. The Dual Logic of the Sublime

In a comprehensive study of the major theorists of the sublime from Longinus to Burke and Kant, Robert Doran identifies a common perspective: despite differences of language, emphasis, and general philosophical orientation, the sublime is consistently construed as an experience of affective intensity that exhibits a dual structure.[12] At the affective level, the sublime is a feeling of unsurpassed power, something that is more than or beyond what is normal or even endurable. Its dual structure is paradoxical, for the subject feels simultaneously overwhelmed and elevated, overpowered and ennobled. It is a single experience that has two poles, that of being "at once below and above, inferior and superior, humbled and exalted," and this very tension "produces the special dynamism of the sublime, creating nexuses with diverse areas of human reality" (Doran 2015, 11). The contradictory tension that is the dual logic of the sublime plays out in a wide variety of ways, which means that the sublime cannot be reduced to a static thing or a single state. Experiences of sublimity range from wonder and exultation to terror and dread, but the nuance and complexity with which its dual logic operates make it impossible to draft a simple list of relevant emotions. Astonishment, ecstasy, immobilizing pleasure, exhilarating tranquility—it is always possible to imagine new formulations and to recalibrate the affective impact toward this or that end of the continuum.

As Doran demonstrates, the dual structure of the sublime is operative already in Longinus, through whose treatise it becomes normative for later theorists. In other words, Longinus's treatise has "a *structuring effect* on the modern discourse of sublimity insofar as it sets a basic pattern, which is then revised and developed by later writers" (Doran 2015, 9, emphasis original). In the present context, this does not mean that Longinus has surreptitiously reasserted his centrality to the study of the sublime. The ancient discourse on sublimity began in the fifth century BCE. A likely point of origin is Democritus's theorization of poetic inspiration. In one fragment, Democritus states: "On the one hand, whatever things a poet writes when god is in him and with holy breath are extremely noble" (ποιητὴς δὲ ἅσσα μὲν ἂν γράφῃ μετ' ἐνθουσιασμοῦ καὶ ἱεροῦ πνεύματος, καλὰ κάρτα ἐστίν).[13]

[12] Doran (2015); for his conclusions, see esp. 4 and 8–12.

[13] D217 *EGP* (B18 D-K). The other relevant fragments are D219 and 221 *EGP* (B17, 21 D-K). The classic discussions are Wehrli (1946, esp. 11–14) and Quadlbauer (1958, 56–57).

When the god physically invades and fills (= inspires) the poet, the poet is literally overpowered, and that is when he rises to the level of (artistic, social, political, spiritual) nobility (= he writes things that are "extremely noble"); this is the dual structure of the sublime. Throughout antiquity, the experience of the inspired poet is paradigmatic of sublimity, and the dual structure of the phenomenon is generally ready to hand.[14] But whatever the historical origins of the discourse, the paradoxical experience was very much in evidence in fifth-century literature and became a focus of heated debate.[15] It is apparent, for instance, in the opening of Parmenides's poem: the assault on the senses overwhelms, the goddess's greeting elevates; the subject's passivity is an index of submission, the revelation of knowledge empowers; and the infinite abyss both terrifies and draws us in, a charged distillation of the potency of sublimity. Longinus inherited the dual logic of the sublime.

In addition to underscoring the dual structure of the sublime, Doran defines the experience as one of "transcendence conceived aesthetically" (Doran 2015, 9). By this, he means an originally religious feeling reconfigured in a secular setting—for instance, a feeling of awe triggered by a work of art. In an ancient context, it is not so easy, and frequently it is not possible or appropriate, to try to separate the religious from the secular. Many systems, especially in the Roman period, promote a pantheistic worldview that suffuses the whole cosmos with the divine (see Ch. 1.6, "Stoics, Epicureans, and the *nil admirari* Tradition"). The study of nature thereby becomes a religious experience, a kind of mystic initiation that is extraordinarily intense precisely because it brings the student into direct contact with divinity.[16] This outlook is crucially important, in different ways, for understanding Manilius, Seneca, Lucan, and the *Aetna* poet, and it constitutes a way of thinking against which Lucretius passionately preaches. Whether or not specific ancient authors are operating within such a system, I treat the sublime as registering a moment of or aspiration toward transcendence, broadly construed. To pass beyond the limits of everyday sense perception, to be transported mentally through space or time, to become suddenly aware of nature's vastness or might, to apprehend in an instant a profound truth about the mechanics of the world or the presence of a divine force

[14] Two particularly influential treatments are Pl. *Ion* 533d–536d and *Phdr.* 245a–c. Note that at *Subl.* 8.4, Longinus combines madness, breath, and inspiration in one sentence; these are precisely the central concepts in the relevant fragments of Democritus.

[15] See Porter (2016, 319–350, 415–444).

[16] For "the extraordinary experience" of initiation into the mysteries, see Burkert (1987, 89–93).

within it—in such moments, ancient writers suggest, it may be possible to leave behind one's ordinary subjectivity and to become or to approximate something greater than that which is human. The dual logic of the sublime is its dual structure striving toward transcendence.

Doran's definition, as elaborated above, is helpful because it is concrete, succinct, and flexible. Currently, the field of classical studies stands in need of a definition that possesses just these very virtues.[17] What's more, a variety of critics take something akin to Doran's dual structure for granted. This makes sense since these critics tend to adopt the lenses of Burke and Kant and even our cursory examination of Burke's *Enquiry* and Kant's *Critique* revealed some of the structural similarities that Doran explores in depth. Gian Biaggio Conte (1994, 28), for one, in his classic essay on Lucretius's sublime reader, explains that the sublime operates by steps: "First it generates a sense of inadequacy [and] then it elevates the reader's mind to a higher level of understanding and fills him with enthusiasm." Similarly, Henry Day (2013, 61; cf. 48–52, 122, 203, 225–227) writes of the sublime's "double movement of rupture and recuperation," and Siobhan Chomse (2018, 390) characterizes the sublime as "a sensation at once fascinating and terrifying" that "causes the mind, almost simultaneously, to contract in terror and expand in awe." The sublime's dual structure is implicit in such formulations, but if we make that structure explicit, it becomes possible to create a scholarly dialogue that shares a basic definition and then branches out in different directions. In other words, foregrounding the dual logic of the sublime has the potential to open up new avenues of inquiry while simultaneously revealing connections between disparate authors and texts.

In adopting Doran's analysis, I do not imply that ancient writers defined the sublime in the same terms. Doran extrapolates the sublime's dual structure and its connection with transcendence from Longinus's arguments, imagery, and language. While Manilius, Seneca, Lucan, and the *Aetna* poet do not define the sublime, they create narratives that explore and exploit the experience of sublimity as I have construed it here. Doran's framework, then, offers a helpful way to begin to think about the similarities and differences between these narratives. At the outset, I will often make the dual structure of the experience explicit. In later chapters, this will become less necessary, although identifying the dual structure of the sublime will always remain possible.

[17] Relatively recent attempts at defining the sublime by classicists include Day (2013, 3), Gunderson (2015, 3–4), and Porter (2016, 5, 56, 393). Frequently, the sublime goes undefined.

1.3. Encoding the Sublime

James Porter has repeatedly emphasized that we can profitably discuss the sublime without limiting ourselves to passages where the traditional terms that designate sublimity appear—above all, *hupsos*, *sublimis*, and *sublimitas*.[18] Such terminology does not figure in the passages considered in Ch. 1.1 and does not play a significant role in those that lie at the heart of this study. If we do not restrict ourselves to occurrences of words like *hupsos* and *sublimis*, and if we resist the urge to allow the exempla of theorists such as Burke or Kant to delimit what qualifies as sublime, we run the risk of the sublime becoming everything and nothing at all. Both Porter and Philip Hardie have compiled taxonomies of subjects and ideas that indicate an experience of the sublime. Hardie (2009, 81–88, 93–116) writes of "markers of the sublime," such as thunderbolts, lightning, and Fame. Similarly, Porter (2016, 51–54) abstracts from his expansive analyses a collection of "logical and thematic markers of the sublime," such as immense heights, profound depths, and limits that are revealed in the very act of transgression. Such taxonomies will be helpful, but not every stroke of lightning or boundary crossed registers as sublime. To claim that a specific passage, image, or idea sparks an experience of sublimity is to judge that it operates at a particular pitch of emotion and that it both overwhelms and uplifts the reader. Instead of asserting that such texts simply are sublime, in some kind of objective way, we can think of them, in the right circumstances, as dramatizing the dual structure of the sublime or participating in narratives that operate according to the sublime's dual logic, moving the reader from low to high (or sometimes from high to low) and pointing toward a moment of transcendence. By focusing attention on how texts dramatize the sublime's dual logic, I hope to keep my analyses concrete and to foreground the deliberate staging of sublimity. The sublime does not exist in a vacuum; it is produced, and there is an audience.

Very often, that audience is embedded in the text, a figure who experiences the sublime's affective intensity or who feels wonder, amazement, awe, stupefaction, bewilderment, ecstasy, even erotic passion, terror, or dread. At one level, such figures focalize the experience of the sublime, offering a

[18] See esp. Porter (2016, 7–35). For *sublimis* and the sublime, see Hardie (2009, 197–202). Martin (2019) catalogs and discusses *sublimis* in late Republican and Augustan literature; his analyses, however, can be unnecessarily rigid (e.g., Germanicus and Manilius are omitted because they were supposedly obliged to use *sublimis* in a literal sense).

subjectivity for the reader to inhabit. A character in a text who feels the intense rush of the sublime, whether brought on by the sudden crash of a thunderbolt or something else entirely, extends an invitation to the reader to experience the world as wondrous, amazing, awesome, stupefying, etc. Such passages thematize the sublime. By making the experience itself a subject of interpretation, they pause the narrative and summon reflection on how the sublime operates, under what circumstances one can or ought to experience it, to what ends it leads or what it implies about the world around us. The use of characters in this manner is common—it occurs already in the Homeric poems—but has not received the attention it deserves. Throughout this study, I describe these kinds of figures as encoding the sublime.[19]

An important example comes from the opening of *De rerum natura* 3. While addressing Epicurus directly, Lucretius describes the experience of reading the master's texts. As soon as the master's voice—his *ratio*—begins to proclaim in prophetic fashion the nature of things (*simul ac ratio tua coepit uociferari / naturam rerum*, 3.14–15), Lucretius has a vision that dispels all those fears about the gods and the underworld that motivate so much of the poem (*diffugiunt animi terrores*, 3.16). The language of viewing saturates the passage: when the ramparts of the universe part, Lucretius sees (*uideo*, 3.17) the atoms moving through the void, the nature and home of the gods appear (*apparet*, 3.18), the quarters of Acheron appear nowhere (*nusquam apparent*, 3.25), and the earth withdraws, granting a panoramic view of all that moves in the void beneath our feet (*nec tellus obstat quin omnia dispiciantur, / sub pedibus quaecumque infra per inane geruntur*, 3.26–27). In a flash, Lucretius apprehends the truth and far-reaching ramifications of atomic theory, which both disorients and exalts him. In fact, he transcends human reality: he sees beyond the limits of our universe, all the way to the *intermundia*; he sees below the normal threshold of human vision, all the way down to the atomic level; he sees through the boundary of the earth, which circumscribes and defines ordinary human existence. The narrative constitutes a dramatization of the dual structure of the sublime.

[19] Cf. Hardie (2009, 209) on the "sublime audience" in an epic simile and Porter (2016, 370) on epic's "internal observers" or "narrative focalizers." A Homeric example occurs at *Il.* 8.555–559, which I discuss at Glauthier (2023, 18–19). In terms of narrative technique, these passages can resemble ekphrases; note Hardie (2009, 153) and see my discussion of ekphrasis and the sublime at Glauthier (2024, 215–217).

Then, however, Lucretius describes the emotional impact itself (3.28–30):

> his ibi me rebus quaedam diuina uoluptas
> percipit atque horror, quod sic natura tua ui
> tam manifesta patens ex omni parte retecta est.

At these things, then, some kind of divine *uoluptas* and *horror* seize me, because, thanks to your might, nature has been revealed so clearly and openly in every direction.

On their own, either *uoluptas* ("pleasure, joy") or *horror* ("shivering, trembling") might name the affective intensity of the sublime; they are the Empedoclean forces Love and Strife, the elemental qualities hot and cold,[20] the psychological manifestations of Venus and Mars, each one electrifying and ecstatic. If we take *uoluptas* and *horror* together as a hendiadys describing a mixture of simultaneous and contradictory experiences (*horror* as *uoluptas* and vice versa), their shock value becomes more readily apparent: "a sensuous shudder, an erotic earthquake, a catastrophic pleasure." The experience is simultaneously devastating and inspiring.[21] Significantly, the language of speech and vision encourages readers to hear and see along with Lucretius, to immerse themselves in the sounds and sights of the scene. By naming his psychological state, Lucretius enables readers more easily to inhabit his own subjectivity, rendering the narrative an invitation to experience the sublimity of Epicurus's teachings for oneself.

As is often the case, the sublime here is protreptic. To hear or see the truth about the natural world is an incomparable experience that pushes the reader toward philosophy and moral improvement.[22] At the same time, Lucretius's experience provides a model for his own readers. Lucretius is eager to imitate Epicurus (*te imitari aueo*, 3.6), and the vision itself is sparked by the master's oracular pronouncements on the nature of things (*uociferari / naturam rerum*, 3.14–15). As Lucretius and Epicurus converge, so do Lucretius and the reader. Put differently, Lucretius's reaction to Epicurus

[20] *Voluptas*, as sexual pleasure (*OLD* 5), is naturally hot. *Horror* can denote shivering from cold (*OLD* 5); cf. *frigidus horror* at Lucr. 6.1011.

[21] For epexegetic *atque*, see *OLD* 8. For *horror* as the seismic jolt of an earthquake, see Lucr. 6.591–595. Critics regularly treat *horror* as a negative experience that is followed by positive *uoluptas*, but such a reading implicitly takes Lucretius's expression as a hysteron proteron, rather than a hendiadys, and obscures the confounding nature of the experience.

[22] Cf. Conte (1994, 21–22, 31).

models the experience of reading Lucretius's own poem, and by encoding the sublime, Lucretius shows readers how the experience should play out. Readers who acknowledge the sublimity of the moment (re)affirm their membership in the community of the Epicurean elect and acknowledge the privileged status of Lucretius, who, like Epicurus before him, has become a kind of god. The sublime advertises and recruits, all while establishing hierarchies of power and knowledge. Freedom from disturbance may be the ostensible goal of Epicurean physics, and Lucretius certainly instructs readers not to wonder at various aspects of the natural world (see Ch. 1.6, "Stoics, Epicureans, and the *nil admirari* Tradition"), but the combination *diuina uoluptas atque horror* indicates a subjectivity that Lucretius values highly.

1.4. Sublime Spectacles and Subjects

It bears repeating that ancient authors stage the sublime and that the sublime's dual logic can be used to make any object or phenomenon appear sublime, to turn any object or phenomenon into a sublime spectacle. Such spectacles are evidence of a will to sublimity, a desire to dramatize the affective intensity of the sublime and to dazzle, stupefy, or astound. In the case of the scientific sublime, the object is an idea, a theory, a part of the natural world filtered through an idea or a theory. Alternatively, the phenomenon is the process of formulating or evaluating such ideas and theories, or an individual or a text that asks questions or speculates about the mechanics of nature. The multitudinous round-running spirals of the stars, the song of Silenus setting natural philosophy and meteorology to music, the *ratio* of Epicurus revealing the irrefutable truth of the atomic system—a will to sublimity configures each of these phenomena as a sublime spectacle.

The standard critical terms for sublimity mean "height, elevation" (*hupsos, sublimitas*) and "on high, elevated, lofty" (*sublimis*). "In Greek, as in many other languages," writes Stephen Halliwell (2022, xxxiii), "the vocabulary of 'height' . . . readily furnishes a conceptual metaphor, on a quasi-vertical scale, for various kinds of superiority, eminence, or exceptional status." From the Homeric poems to the end of antiquity, the dual logic of the sublime is regularly employed in the description of objects, phenomena, and experiences that are located or unfold high above the ground—a flash of lightning, a towering mountain, the confounding paths of the planets. Oftentimes, though, the scientific sublime elevates the mundane, transforming an ordinary object into an awe-inspiring spectacle that shifts the way we think about the world. Vitruvius highlights this process when he considers the nature of wind (1.6.2). According to Vitruvius, wind is produced when heat strikes moisture and the onrush of the force presses out a powerful blast of breath. To prove the point, he describes a contraption, akin to a kettle, that one fills with water and places over a fire. When the water boils, the contraption generates a strong blast of wind: "Thus it is possible to know and to make judgments, based on a small and very brief spectacle, about the great and vast principles of the nature of the heavens and the winds" (*ita scire et iudicare licet e paruo breuissimoque spectaculo de magnis et immanibus caeli uentorumque naturae rationibus*). The kettle contraption is the arena in which a magnificent show is staged, one that makes accessible the hidden vistas of the atmosphere, allowing us to reach beyond the limits of our senses. Indeed, water boiling in a kettle becomes a means "to express the truth of divinity concerning the hidden principles of the heavens" (*de latentibus caeli rationibus... diuinitatis exprimere ueritatem*).

When the dual logic of the sublime is used to describe a thinking subject, the subject becomes sublime and very often demonstrates or suggests the same will to sublimity that animates the authors themselves. In Greek, a subject who dynamically overwhelms and uplifts, whose greatness commands attention and inspires those around him, can be said to possess *megalopsuchia* ("high-mindedness, greatness or expansiveness of soul") and/or *megalophrosunē* ("greatness of spirit or mind, elevation or expansiveness of thought").[23] Such subjects are noble, aristocratic, free from all constraint, and axiomatically "better" than everyone else; they are Homeric heroes who move in our midst.

In Latin, this constellation of qualities becomes *magnanimitas* or *magnitudo animi*, and a sublime subject can be called *magnanimus*.[24] Using different language but appealing to the same ideal, Lucretius extols the "vehement potency of mind" (*acrem... animi uirtutem*, 1.69–70) that allowed Epicurus to triumph in his confrontation with religious superstition (*religio*)

[23] For *megalopsuchia* and the sublime, see Porter (2016, 195–201). Longinus can define the sublime as "an echo of greatness of mind" (ὕψος μεγαλοφροσύνης ἀπήχημα, 9.2), which is regularly rendered "an echo of a noble mind," and write that the sublime lifts us "near to the greatness of mind of god" (τὸ δ' ὕψος ἐγγὺς αἴρει μεγαλοφροσύνης θεοῦ, 36.1).

[24] For *magnitudo animi*, see Knoche (1935) and Cullyer (1999). Here and elsewhere, I am not primarily concerned with the place of *megalopsuchia*, *magnitudo animi*, and related terms in the schemata of philosophical ethics. The concepts are important insofar as they appear or are alluded to in texts that dramatize, encode, or analyze the sublime.

and to discover the truth about the structure and laws of nature (1.62–79). Stories about the gods and frightening natural phenomena did not physically or emotionally restrain Epicurus (*compressit*, 1.69): "But all the more they excited the vehement potency of his mind, so that he desired to be the first to break open the close-set bolts of nature's doors" (*eo magis acrem / irritat animi uirtutem, effringere ut arta / naturae primus portarum claustra cupiret*, 1.69–71). The overwhelming sights and sounds of nature energize the philosopher, and when the forces of oppression threaten to constrain him, he rises to the challenge, dramatically breaks free, and expands boundaries of human knowledge, making "us" equal to the heavens (*nos exaequat uictoria caelo*, 1.79). Lucretius's Epicurus possesses the greatness of soul and expansiveness of thought that characterize a Homeric hero or philosophical Giant, and his story stages the dual logic of the sublime, moving the reader from fear to exultation.[25]

This is not to say that only members of the elite can function as sublime subjects. Longinus, for one, makes greatness of soul (τὰ ψυχικὰ μεγέθη) a function of the soul's participation in immortality and the divine; without proper exercise, it withers and fades (*Subl.* 44.8). Whatever their identities, figures who are emboldened and empowered by their greatness of soul and expansiveness of thought generally encode the sublime. Their will to sublimity dares readers to activate that same mental and spiritual potency in themselves and thereby to exceed the limits of their everyday lives. Still, the sublime subject's clarion call for freedom from constraint does not necessarily constitute a challenge to the stratifications of one's social system. On the contrary, it can be an invitation to situate oneself at the top of that very system or, better still, to imagine that one has moved beyond it and that its rules no longer apply to oneself.[26]

The concept of the sublime subject is a helpful tool for thinking about how ancient authors represent and engage in dialogue about sublimity. Most of the sublime subjects who appear in the present study are paradigms for the philosopher or the investigator of nature, and their greatness of soul and expansiveness of thought become apparent in their intellectual

[25] Homeric hero: Conte (1994, 1–3), the classic treatment of Lucretius's Epicurus and the sublime. Philosophical Giant: Chaudhuri (2014, 58–63). For *megalopsuchia* in Epicurean ethics, see McConnell (2017b). For the possibility that readers are meant to picture Lucretius's Epicurus as riding a chariot in the manner of Parmenides, see Hardie (2022a, 38).

[26] For the sublime as an experience of liberation, see Hardie (2009, 182–184; 2022a, 40–44). For Longinus on slavery, political repression, and the sublime, see Ch. 1.6 ("Longinus and the Rhetorical Tradition").

adventures. We see this in the flight of the mind topos, as demonstrated by Parmenides, the author of the Ptolemy epigram, and Lucretius's Epicurus. Two additional topoi are relevant here, both of which are closely related to the flight of the mind and not always to be strictly demarcated from it or from one another—these are the ascent and the view from above.[27] In all three traditions, a subject's spatial location and sensory horizons relative to those of regular human beings on earth indicate or generate spiritual and intellectual elevation and expansion.

In the ascent, the subject moves mentally and/or physically upward in the pursuit of knowledge, virtue, and immortality. In Plato's *Symposium*, Diotima characterizes philosophy as an upward movement on the steps of the ladder of love that brings the student face to face with the form of the beautiful, and she holds out the possibility that a life lived through philosophy might allow us to transcend our own mortality (*Symp.* 211b–d, 212a). The ecstatic vision ends abruptly when Alcibiades and his companions barge in. From our perspective, the implication is that elevation of soul and expansion of thought are fleeting and that sublime subjectivity is aspirational at best since the chaos of everyday life is inimical to the goals of the philosopher.

The view from above allows readers to imagine themselves looking down at the earth from a great height, having made the ascent or launched their mind in flight. This elevated vantage point affords a more expansive view of the physical world and our place within it and demands that we set aside the biases and misperceptions of ordinary life, see reality clearly, recognize incontrovertible truths, and make unshakable value judgments. At the end of Cicero's *De re publica*, a dreaming Scipio finds himself in the heavens, and through conversations with his adoptive grandfather, P. Cornelius Scipio Africanus, and biological father, L. Aemilius Paulus, he learns about the structure of the cosmos, including the organization of the planets and the music of the spheres, and the fate of the soul after death. Initially, Scipio's state is one of fear; he trembles upon recognizing the soul of Africanus (*cohorrui*, 6.10), and Africanus's intimations about Scipio's demise terrify him (*perterritus*, 6.14). But once Paulus points to the radiant splendor of the Milky Way, revealing it to be the home of those souls who have lived virtuous lives in the service of the state, the dominant mode changes: "from that

[27] In general, see Rutherford (1989, 155–161), Hadot (1995, 238–250), and, most comprehensively, Hardie (2022a, 30–105).

point, as I contemplated everything, it all seemed magnificent and wondrous" (*praeclara cetera et mirabilia uidebantur*, 6.16); "I was considering these things in a state of stupefaction (*stupens*), but when I had composed myself, I spoke" (6.18); "while I wondered at these things (*haec ego admirans*), I repeatedly brought my eyes back to earth" (6.19). Scipio's wonder and stupefaction encode the affective intensity of the sublime, as he gazes at and for the first time truly understands the order and dimensions of a world that is now seen to be fundamentally divine.[28] Scipio thrills to the view from above and demonstrates the shift in values that it necessitates when he recognizes the relative smallness of earth within the cosmos and feels ashamed that the Roman Empire covers only a miniscule fraction of even that speck of dust (6.16). As in *Symposium*, the vision does not last; the philosopher must return to earth and try to save the Republic if he hopes for his soul to return to the heavens, liberated once and for all after his death.[29]

Sublime subjects appear to stand outside or beyond the categories and hierarchies of their societies, and this transgressive position can provoke a range of reactions. From our perspective, as inheritors of the intellectual traditions of ancient Greece and Rome, we are likely to view Plato's Diotima and Cicero's Scipio positively, but their contemporaries were under no obligation to reach a similar conclusion. In *Clouds*, Aristophanes's Socrates, suspended in a basket and swinging through the air, preposterously literalizes the basic premises of the flight of the mind, the ascent, the view from above, and the natural philosophers' claims to sublime subjectivity (*Nub.* 218–238). In his pursuit of celestial and meteorological knowledge, Socrates has exceeded the limitations of earthly life. From his elevated vantage point, the philosopher, like a god, looks down on the mortals who move below him ("Why do you summon me, creature of a day," τί με καλεῖς, ὦ 'φήμερε, 223). As he explains, he has physically suspended his mind so that it might commingle with the rarified air and the earth might not draw off the moisture of thought—which is precisely what happens to watercress (227–234). The rationale relies on the principles of Diogenes of Apollonia,[30] but the

[28] Paulus begins his speech by asserting that the world that Scipio sees is the precinct (*templum*) of a god (6.15).

[29] McConnell (2017a, 48) argues that throughout *Rep.* "Cicero actively seeks to develop and test *magnitudo animi* among his readers." The smallness of the earth relative to the cosmos is a common philosophical theme; see Arist. *Mete.* 1.14.352a, *Cael.* 2.14.297b–298a; Lucr. 6.647–679; Cic. *Tusc.* 1.40; Sen. *NQ* 4b.11.4, 1 praef. 7–10, *Dial.* 6.21.2; Plin. *HN* 2.174.

[30] See Dover (1968, 127).

effect is deflationary. Aristophanes deploys the patterns of thought and tropes common to the scientific sublime against Socrates to expose the absurdity of his project and his claims to greatness. High-mindedness and expansiveness of thought are revealed to be folly, pride, even a threat to the *polis*.[31] The passage does not encode the sublime, but it does contribute to the emerging discourse on sublimity and what it means to style oneself a sublime subject.

Clouds reveals that the sublime is a contested experience from an early date, especially in connection with the study of nature. To lay claim to sublime subjectivity is to assert privilege, which can constitute a threat to the status quo. Mockery and weaponization of the logic of the sublime and the pose of sublime subjectivity are two possible responses.[32]

In other circumstances, the sublime subject becomes a dark or even dangerous figure. The conflict between Hercules and Cacus in the eighth book of the *Aeneid* is an instructive example (8.184–279).[33] Cacus, the monstrous offspring of Vulcan, roams the Italian countryside, belching fire from his terrible jaws, the personification of a volcanic mountain in a perpetual state of eruption.[34] After Cacus steals Hercules's cattle, the hero is enraged (*furiis exarserat*, 8.219) and pursues the monster to his subterranean lair, where he has blockaded himself. A furious Hercules arrives on the scene (*ecce furens animis aderat Tirynthius*, 8.228) and, seething with anger (*feruidus ira*, 8.230), vainly tries to gain access. After much difficulty, he breaks open the cave, exposing to view the shadowy cavern (*umbrosae penitus patuere cauernae*, 8.242), with his opponent immobilized by the sudden appearance of unfamiliar light (*insperata deprensum* [sc. *Cacum*] *luce repente*, 8.247). Hercules pelts Cacus with random objects as the monster spews forth fire and smoke, enveloping the landscape in impenetrable darkness. Undeterred, Hercules leaps into the cave and brutally throttles the thief, squeezing his eyeballs from his head. He then reveals, once and for all, the interior of the

[31] Aristophanes's portrait of Socrates up among τὰ μετέωρα (*Nub.* 228) figured prominently in the charges brought against him, based, at least, on Plato's account (*Ap.* 19b–c, and cf. 26d–e). For discussion, see Porter (2016, 434–443).

[32] For Rome, see Hardie (2009, 180–228) on Horace's ironic/deflationary use of the Lucretian sublime.

[33] Note that philosophical ethics distinguishes between positive and negative expressions of greatness of soul; see McConnell (2017a, 58–59) on Cic. *Off.* 1.64–65. For further examples of dark or deranged sublime subjectivity, see Porter (2016, 198) on Pl. *Rep.* 6.494c–d and Glauthier (2023, 23–24) on Aeschylus's Clytemnestra and Lucan's Caesar.

[34] For parallels between Cacus and the volcanic monsters Typhoeus and Enceladus, see Hardie (1986, 110–118) and Johnston (1996, 60–61).

cave (*panditur*, 8.262), displays his stolen property to light of the upper world (*caelo ostenduntur*, 8.264), and drags the monster's corpse outdoors for the locals to gaze at indefinitely (*nequeunt expleri corda tuendo*, 8.265). This insatiable stare encodes the sublime, prompting the reader gaze in awe at the grim spectacle. Similarly, as the narrative moves from exhilarating terror to heroic triumph, it dramatizes the sublime's dual logic,[35] and Hercules himself emerges as a sublime subject, the virtual embodiment of the will to sublimity. While Cacus unleashes all the violent might of a natural disaster, Hercules presses forward, plunging headlong into danger, and emerges a god. And yet, his high-mindedness and greatness of spirit take the form of rage and fury, the cataclysmic anger that seethes at the heart of the *Aeneid*, always just below the surface. Hercules, as sublime subject, stands beyond ordinary categories of moral evaluation; he is both terrifying and salvific.[36]

Virgil's rendition of Hercules and Cacus is relevant for another reason. The contrast between darkness and light and the emphasis on revelation and exposure to view suggest narratives of philosophical enlightenment. From an allegorical perspective, Cacus's volcanic nature allows us to read Hercules's hard-won victory as a dramatization of the wild ride and heroic potential of natural philosophy—the mythological conquest of the monster figures the philosophical conquest of natural phenomena. More precisely, Virgil's imagery of light and darkness, revelation and exposure, comes from *De rerum natura*,[37] and Hercules's triumph over Cacus replays Epicurus's conquest of religious superstition.[38] This alignment makes Hercules a commentary on Lucretius's Epicurus and Lucretius himself. At one level, this is entirely unproblematic; if Lucretius makes Epicurus a Homeric hero, Virgil makes Hercules an epic philosopher. And yet, the mood seems to have shifted: Epicurus's vehement potency of mind is now enraged madness, and Lucretius's demolition of the mythological past looks like the brutal asphyxiation of tradition. The anonymous gaze of the local population suggests the interpretive dilemma (8.265–267):

[35] Gildenhard (2004, 35) compares *Subl.* 9.5–6 and writes that Virgil moves from the "repulsive" to the "awe-inspiring."

[36] Cf. Morgan (1998), who reads Hercules's violence as paradoxically constructive.

[37] See Hardie (2009, 171) on Virgil's use of *appareo*, *pateo*, *resero*, *recludo*. For light and darkness in Lucretius, see, e.g., 1.136–145, 2.59–61, and 3.1–3 with West (1969, 79–93).

[38] Gildenhard (2004, 37–38) and Hardie (2009, 171–173).

nequeunt expleri corda tuendo
terribilis oculos, uultum uillosaque saetis
pectora semiferi atque exstinctos faucibus ignis.

The people are unable to satisfy their hearts with gazing on the half-beast's terrible eyes, the face, the breast shaggy with bristles, and the fires extinguished in his/its jaws.

The spectacle enthralls the crowd, with no end in sight, thereby staging the infinite thrill of reading *De rerum natura*, but the inability of the spectators to find satisfaction suggests that conquered nature offers neither emotional nor spiritual closure. Although Lucretius can explain the physics of a volcano, his readers do not rest easy; the monster's eyes remain terrifying (*terribilis oculos*). These spectators, in other words, feel the paradoxical rush of the sensuous shudder, which prompts one to ask whether Lucretius's ostensible goal of *ataraxia* is attainable.[39]

The Hercules and Cacus episode nicely demonstrates the sophisticated complexity of the classical sublime and its ability to work on multiple levels at once. Hercules is not just, or even primarily, a stand-in for Lucretius or Lucretius's Epicurus. In the immediate context of *Aen.* 8, the narrative explains the origins of the cult of Hercules at Rome and the location of the Ara Maxima. Hercules also has deep ties to Augustus, and his victory over Cacus at the beginning of the book anticipates Augustus's triumph over Anthony at its end. Rome's Greek and Italian prehistory, the evolution of its religious, philosophical, and poetic traditions, its imperial present and inescapably Augustan future—all these threads overlap and branch out in different directions, and they all enact a sensuous shudder, a catastrophic pleasure, with no end in sight.

1.5. The Fall

The exploits of the sublime subject are axiomatically greater than the activities of everyday life because they are riskier and more dangerous. Frequently, narratives and/or images of falling underscore the hazardous nature of the

[39] Virgil's interrogation of the Lucretian outlook becomes more pointed if we recall that Lucretius makes Epicurus a god of far greater significance than Hercules (5.1–54).

adventure and enhance its potential rewards; we will encounter the fall repeatedly in subsequent chapters.

Acrobats, Heroes, and Flawed Sublimity

The Younger Pliny nicely captures the precarious nature of an act performed in a state of mental elevation (*Ep.* 9.26.2–4):

> An orator ought to be elevated and exalted, and from time to time glowing with heat and carried away, often drawing close to a headlong fall (*saepe accedere ad praeceps*), for usually lofty and exalted language approximates to the precipitous. The lower and more pedestrian path over level ground is safer. A slip occurs more frequently in the case of runners than those who creep along, but no praise accrues to crawlers who do not slip, whereas runners gain some glory even if they slip. Eloquence is like certain other skills in winning approval above all from its hazards. You observe what loud shouts are often aroused by men who strive for the heights on a tightrope, when they seem about to fall at any moment. For the most wondrous feats (*maxime mirabilia*) are those that are most unexpected and most hazardous (*maxime insperata, maxime periculosa*), which Greeks better express as παράβολα ["risky, daring"]. Hence the skill of a helmsman in sailing on calm waters is by no means commensurate with journeying on a stormy sea. As for the first, no one looks on in wonder as he enters the harbor, winning no praise or glory, but when the ropes are creaking, and the mast bends over, and the rudder groans, he then achieves fame and attains a stature closest to the gods of the sea. (Trans. based on Walsh 2006)

The orator in a state of sublimity—the orator as sublime subject—runs great risks, balancing precariously on a live wire high above the audience. While Pliny's ideal speaker does not lose his balance, he cannot win immortal fame without exposing himself to the possibility of a spectacular fall. Such performances elicit wonder and wild applause; they are overwhelming and inspiring, virtually transforming their subject into a god. It is against this backdrop that Pliny is eager to defend the sublimity of his own rhetorical acrobatics ("you seemed to me to have designated as inflated some passages in my writings that I thought sublime," *uisus es mihi in scriptis meis adnotasse quaedam ut tumida quae ego sublimia... arbitrabar*, 9.26.5).[40]

The basic premise of Pliny's letter, though, is that to fall in a sublime enterprise is no fault at all. The analysis of *Ep.* 9.26.2–4 expands upon a simple criticism that Pliny levels against a contemporary whose style is "sound and sensible" but lacking in grandeur: "He makes no errors, except for the fact that he makes no errors" (*nihil peccat, nisi quod nihil peccat*, 9.26.1). In other words, to achieve the heights of sublimity necessarily involves failure, which Pliny represents as a stupendous fall. Longinus expresses a similar preference for "greatness that is in some respects flawed" to "moderation in success that is entirely sound and free from falls" (πότερόν ποτε κρεῖττον ἐν ποιήμασι καὶ λόγοις μέγεθος ἐν ἐνίοις διημαρτημένον ἢ τὸ σύμμετρον μὲν ἐν τοῖς κατορθώμασιν, ὑγιὲς δὲ πάντη καὶ ἀδιάπτωτον, 33.1).[41] Balance and moderation are safe and secure, signs of a mundane outlook and restrained disposition. Subsequently, the sublime is defined as incomprehensible excess that makes total control both impossible and undesirable: "Accuracy in everything runs the risk of smallness, whereas in greatness, as in great wealth, it is necessary that something be overlooked" (τὸ γὰρ ἐν παντὶ ἀκριβὲς κίνδυνος μικρότητος, ἐν δὲ τοῖς μεγέθεσιν, ὥσπερ ἐν τοῖς ἄγαν πλούτοις, εἶναί τι χρὴ καὶ παρολιγωρούμενον, 33.2). The comparison of sublimity with extreme wealth is telling: it is precisely the wealthy who can afford to overlook the details of day-to-day life or simply delegate them to others who may lack the freedom to decline the command. Later, Longinus grows more heated: "What, then, did those divine men see, who reached for what was greatest in writing and looked down upon accuracy in the details?" (τί ποτ' οὖν εἶδον οἱ ἰσόθεοι ἐκεῖνοι καὶ τῶν μεγίστων ἐπορεξάμενοι τῆς συγγραφῆς, τῆς δ' ἐν ἅπασιν ἀκριβείας ὑπερφρονήσαντες, 35.2). The answer, to be considered more fully in Ch. 1.6 ("Longinus and the Rhetorical Tradition"), is that nature summons us to more magnificent things than *akribeia*. Those who are equal to the gods (οἱ ἰσόθεοι ἐκεῖνοι) see and think about the world from an elevated vantage point (ὑπερ-φρονήσαντες), which

[40] At Sen. *Dial.* 4.12.4, tightrope walking is an example of the mind's ability to overcome (*transcendit*) all impediments and to accomplish any task it sets itself; the funambulist is paradigmatically sublime. For more on sublime tightrope walking, see Ch. 2.3.

[41] The sense "free from falls" is surely felt in ἀδιάπτωτον (from πίπτω). Dion. Hal. *Lys.* 13 uses ἀσφαλής (from σφάλλω) similarly: Lysias's style, far from being sublime (ὑψηλή), is "free from slips" (ἀσφαλής), which is opposed to one that is "adventurous" or "full of risks" (παρακεκινδευμένη).

forces all the details of mundane life to fade away, to become so much background noise. Figures of this sort do not just look down on what is small and petty and ephemeral—they despise such things and take no account of them (ὑπερφρονήσαντες). In short, to reach for greatness requires that we disregard the vulgar obsession with precision and reliability and the cramped perspective to which such an obsession confines itself.[42]

The supremacy of flawed sublimity is the subject of *Subl.* 33–36, and Longinus's analysis shows clear points of contact with Pliny's letter.[43] It is crucial to recognize, however, that the assessments of Pliny and Longinus do not apply only to literary or rhetorical analysis. Indeed, earlier in his treatise Longinus cites a maxim that illustrates the basic moral orientation and has nothing to do with language: "To slip from greatness is still a noble error" (μεγάλων ἀπολισθαίνειν ὅμως εὐγενὲς ἁμάρτημα, 3.3).[44] Again, Longinus's vocabulary is markedly aristocratic, implying that only those who are socially elevated can enjoy a precipitous fall. Throughout the Roman tradition, the sublime manifests itself as risky adventure and uncontrollable excess, the freedom of the nobility to do and say as they please, and a fall from great heights becomes an admirable inevitability—not a fall *from* sublimity, but the fall *as* sublimity. From this perspective, the fall becomes evidence of one's socio-political superiority, which is easily reconfigured as moral superiority, philosophical superiority, aesthetic superiority, etc. A significant consequence of this paradigm is that the adventure of a sublime subject does not need to end in "success." A subject's stated or ostensible goal can become irrelevant if the dual logic of the sublime configures failure as evidence of greatness, of a will to sublimity.[45]

In a philosophical context, the fall indicates that a subject has made a factual mistake, settled on an incorrect explanation, stumbled into self-contradiction, or otherwise erred. Paradoxically, it signals the author's or speaker's simultaneous approval and condemnation. A good example comes

[42] At *Subl.* 7.1, it is a sign of *megalopsuchia* to scorn things that the common crowd values; cf. Dem. *On Style* 53, where *akribeia* is defined as narrow-mindedness (μικροπρεπὲς γὰρ ἡ ἀκρίβεια). For the contrast between *akribeia* and sublimity, see de Angelis (2014, 95–101).

[43] See Quadlbauer (1958, 108–109), Russell (1964, xli–xlii), and Sherwin-White (1966, 508). For the Younger Pliny and the sublime more generally, see Armisen-Marchetti (1990).

[44] Mazzucchi (2010, 143) compares Herod. 8.100.1, Dion. Hal. *Pomp.* 2.15, Plut. *Crass.* 37.2 (567a–b).

[45] I have discussed this position at Glauthier (2022, 96–100; 2023, 24–25). For additional perspectives on the imagery of falling and the sublime, see Leigh (2006, 236–239) on Statius, Hardie (2009, 212–217) on Lucretius and Horace, Porter (2016, 198–199) on Plato and Longinus, Chomse (2018, 408–409) on Martial, and Chomse (2019, 164–165, 178–181) on Rome itself. Quint (2004) analyzes the falls of Icarus and Phaethon in Milton via Lucretius and Horace.

from Lucretius's doxography of early natural philosophers (1.635–920). When Lucretius turns to Empedocles, he celebrates his predecessor's homeland and poem (1.716–733). The Sicilian landscape teems with energy and power—the agitated sea, gaping Charybdis, Aetna threatening to explode. The description of awe-inspiring natural phenomena (*magna modis multis miranda*, 1.726) culminates in Empedocles himself, who is the most marvelous (*mirandum*) of the island's many features. At this point, the ominous sounds and angry rumblings of the volcano give way to the prophetic voice and otherworldly proclamations of the self-proclaimed god (θεὸς ἄμβροτος, οὐκέτι θνητός / πωλεῦμαι, D4.4–5 *EGP*, B112.4–5 D-K), who scarcely seems human (*ut uix humana uideatur stirpe creatus*, 1.733) and threatens, like a Giant, to assault the heavens with his elemental theory.[46] Empedocles exerted a major influence on Roman conceptions of sublimity, and Lucretius's narrative aims to re-create the experience of reading Empedocles's poem by equating it with the natural world in a state of devastating exuberance.[47]

In the present context, however, what matters comes next. After celebrating Empedocles as a force of nature, Lucretius returns to his doxography. Although Empedocles and other pluralists made many divine discoveries and proclaimed the truth with more sanctity and reliability than the Pythia, "nevertheless, when it comes to the first beginnings of things, they made a downward plunge, and despite their greatness, they fell heavily in that matter with a great fall" (*principiis tamen in rerum fecere ruinas / et grauiter magni magno cecidere ibi casu*, 1.740–741).[48] The inability of Empedocles and his companions to apprehend the fundamental reality and constitutive nature of atoms and void is a profound failure that Lucretius figures as a monumental collapse, a fall from a great height. Given the rapturous praise of Empedocles in the preceding lines, we should read this fall from the perspective of the sublime: to tumble from the elevation of divine knowledge and now to lie low on the earth, like rubble, enacts the dual logic of the sublime. But instead of moving from below to above, from oppression to

[46] Thus Hardie (1986, 211–212; 2009, 88–90). The Gigantomachy imagery links Lucretius's Empedocles with his Epicurus, who successfully triumphs over a "Gigantic" *religio*; see Hardie (1986, 209–211), Gale (2000, 120–122), Volk (2001, 107–108), and Chaudhuri (2014, 58–60). For "intellectual Gigantomachy" more broadly, see Volk (2001, 100–106).

[47] For Empedocles and the sublime at Rome, see again Hardie (2009, 88–90, 99–109, 116–125, 136–152, 172, 197–98, 214–217). For the Greek tradition that identifies Empedocles's writings as sublime, see Porter (2016, 416–418).

[48] For *ruina* = "headlong fall, downward plunge," see *OLD* 2. For Lucretius's use of *ruina*, see Hardie (2009, 217 n.134).

exaltation, the narrative dramatizes the sublime's logic in the opposite direction. The epic echoes of Lucretius's language reinforce the point. The image of the philosophers "great in their great fall" looks back to several Homeric passages where a hero "lies great in his greatness" (*Il.* 16.776, 18.26–27, *Od.* 24.40). The philosophers are fallen heroes, whether they triumph or not, and whether they are "correct" in their theories or not.

Phaethon

As the following chapters will demonstrate, the fall plays a central role in the history of the scientific sublime in the Roman world. A key figure in this history is Phaethon, whose chariot ride repeatedly figures the risks and dangers of the study of nature.[49] The first half of Phaethon's journey is well suited to the logic of the sublime, seemingly a ready-made template for the flight of the mind, the ascent, and the view from above: the sublime subject, claiming his birthright, elevates himself to the level of the divine and finds himself able to gaze at the universe from a new perspective. When everything goes awry, the logic of the sublime begins to work in reverse: instead of moving from below to above, the narrative enacts the sublime's logic in the opposite direction. This movement bears witness to Phaethon's greatness of soul and simultaneously problematizes his intellectual achievements. Two concrete examples set the stage for later readings.

At *De rerum natura* 5.392–406, Lucretius tells Phaethon's story, introducing it as mere rumor (*ut fama est*, 5.395) and explicitly rejecting it as a poetic falsehood (5.405–406). It is precisely in his fall, however, that Phaethon's status as a sublime subject shines through (5.399–401):

> at pater omnipotens ira tum percitus acri
> magnanimum Phaethonta repenti fulminis ictu
> deturbauit equis in terram.

But then the all-powerful father, aroused with fierce anger, dashed high-minded Phaethon from his horses to earth with a sudden stroke of thunder.

[49] For Longinus's thoughts on Euripides's Phaethon/*Phaethon*, see *Subl.* 15.4. Schiesaro (2014, 86 n.48) observes that "the journey of the Sun's horses was called a 'sublime iter' already in Ennius" (fr. 169 Jocelyn, 57 Goldberg-Manuwald). Icarus is another important paradigm for the fall (e.g., Hor. *C.* 4.2.1–4), although he has fewer associations with the study of nature; see, however, La Penna (2001) and Lucian, *Astr.* 14–15, 19.

The epithet *magnanimus* indicates the state of elevated or expanded consciousness that is a marker of sublime subjectivity (cf. La Penna 2001, 540–541). Since thunderbolts are a regular way of describing the force of the sublime (Day 2013, 107–111) and since being thunderstruck is a common metaphor for the sublime experience, the sudden burst of thunder that strikes *magnanimum Phaethonta* underscores the nature of his mental state. Phaethon's lack of control suggests something similar. The hero enters the text at the mercy of the Sun's powerful horses, which, like a raging river, are in the process of sweeping him away (*Phaethonta rapax uis solis equorum... raptauit*, 5.397–398). Phaethon is spatially elevated and physically overwhelmed.

To apprehend Phaethon's role as a philosopher, we should start by noting the similarities between his journey and that of Epicurus. Both are totalizing in scope: Epicurus travels through "all of the infinite" (*omne immensum*, 1.74), Phaethon travels through "all of heaven and all the lands" (*aethere... toto terrasque per omnis*, 5.398). The contrasts, however, are revealing. While thunderbolts did not keep Epicurus in check (*nec fulmina nec... compressit caelum*, 1.68–69), Phaethon succumbs to a sudden bolt of thunder (*repenti fulminis ictu*). While Epicurus brings back a lesson about nature, the limits of what can and cannot come to be, Phaethon destroys the natural world, and his death only lends credence to erroneous beliefs about the gods. If Epicurus defeats religious superstition, Phaethon inadvertently confirms humankind's worst suspicions about the *pater omnipotens*. I suggest that Phaethon's story here represents the difficulty inherent in challenging traditional ideas about the gods, particularly their relationship to celestial and meteorological phenomena, which can trigger philosophical backsliding. Despite his best efforts, Phaethon yields to the power of mythology and *religio*, thereby dramatizing the seemingly unavoidable retreat into superstitious thinking that overtakes all of us from time to time. This nexus of concerns is very much at the heart of *De rerum natura* 5–6, which makes Phaethon's fall a cautionary tale for the reader who has made it this far in Lucretius's poem. Ancient readers will have had no difficulty reading Lucretius's Phaethon in this way; as we have seen, a chariot ride can figure the sublime investigation of nature already in the fifth century BCE.[50]

[50] Although the text is problematic, Lucretius himself appears to embark on an intellectual chariot ride at 6.47, making him a counterpart of sorts to Phaethon.

In Ovid's *Metamorphoses*, Phaethon's fall is once again sublime (1.747–2.332).[51] Logical and thematic markers of sublimity permeate the narrative. For instance, Phaethon's ambition is framed as a desire for transcending the limits not just of his own mortality but of what is permitted even for the gods ("Your fate is mortal; what you desire is not mortal. You are unaware that you are pursuing something more than that which the gods are permitted to have," *sors tua mortalis; non est mortale quod optas. / plus etiam quam quod superis contingere fas sit / nescius adfectas*, 2.56–58). When he crosses that impossible boundary and gazes at the multitudinous wonders of heaven (*in uario passim miracula caelo*, 2.193), Phaethon predictably experiences a combination of terror and bewilderment (*mentis inops gelida formidine*, 2.200; cf. *stupet* at 2.191) that encodes the sublime. To underscore the point, as the chariot veers off course and the flames begin to spread, the Moon looks on in wonder (*admiratur*, 2.209). This brings us to the fall. In his headlong plunge (*uoluitur in praeceps*, 2.320), Phaethon resembles a shooting star (2.321–322), an ambiguous phenomenon that blurs the line between the celestial and sublunary worlds, and when a group of minor divinities buries him, they affix the following epitaph to his tomb (2.327–328):

> HIC • SITVS • EST • PHAETHON • CVRRVS • AVRIGA • PATERNI
> QVEM • SI • NON • TENVIT • MAGNIS • TAMEN • EXCIDIT • AVSIS

Here lies Phaethon, driver of this father's chariot. Even if he did not control it, nevertheless he fell amidst deeds of great daring.

To fall in a deed of great daring is no reason for disappointment, for it is itself proof of one's greatness of soul. The basic sentiment recalls Longinus's dictum that to slip from greatness is still a noble error (*Subl.* 3.3).[52] Significantly, Alessandro Schiesaro has demonstrated numerous connections between Ovid's narrative and *De rerum natura*.[53] This allows him to argue that "Ovid's commentary on Phaethon's hubristic—and sublime—attempt at reaching a divine level of knowledge should be read as a specific critique of the

[51] See Delarue (2006), Barchiesi (2009), Schiesaro (2014, 86–87), Hannay (2016), Lagière (2017, 79–83), Garani (2020a, 110–111), and Hardie (2022a, 80–82; 2022b §§15–29); cf. Garani (2023).

[52] Thus Russell (1964, 72).

[53] Schiesaro (2014). For instance, Lucretius's *magnanimum Phaethonta* (5.400) becomes Ovid's *magna loquentem* (1.751) and is then repeated in the nominative (*magnanimus Phaethon*, 2.111). Hannay (2016, 60–71) builds on Schiesaro's approach.

Lucretian sublime" (Schiesaro 2014, 87, emphasis original). On my reading of the fall, the upshot is that Lucretius's atomic vision of reality is riddled with errors, and yet to inhabit its mental space, to read *De rerum natura*, is to embark on a thrilling adventure that takes one beyond limits of the possible and sets fire to everything that you thought you knew about the world; one-to-one correspondence between theory and material reality matters less than the mental rollercoaster ride. At the same time, Phaethon figures Ovid himself. Lleweyln Morgan and Alessandro Barchiesi have argued that Phaethon's adventurous spirit and ostentatious disregard for tradition correspond to elements of Ovid's own persona.[54] Through Phaethon's fall, Ovid and Lucretius merge, the *Metamorphoses* and *De rerum natura* collide, mythological aetiology and philosophical physics become indistinguishable—and who knows whether any of the narratives are accurate. In the *Amores*, Ovid explicitly labels Lucretius sublime and declares that his poem will perish only when the world meets its own end (*carmina sublimis tunc sunt peritura Lucreti, / exitio terras cum dabit una dies*, 1.15.23–24); one suspects that Ovid felt similarly about himself and the *Metamorphoses*.[55]

In addition, we should note that in the *Metamorphoses* Phaethon's chariot ride is implicitly full of Augustan symbolism; at the very least, the son who longs to inherit his father's power drives through a skyscape that Barchiesi (2009, 170–188) has argued recalls the architecture of contemporary Rome. Without attempting to develop a political reading of Ovid's story, we can simply highlight once again the discursive multiplicity of the sublime, the fact that multiple modes of sublimity interact with one another—the political, the architectural, the philosophical, the epic, etc. When we encounter Phaethon in subsequent chapters, he comes loaded with rich and varied associations.

The Leap

To leap headlong into a dangerous situation can also indicate sublime subjectivity. Hence, Virgil's Hercules leaps into Cacus's abyssal lair precisely where the monster's volcanic fury rages most intensely (*Aen.* 8.256–258).[56]

[54] Morgan (2003, 75–82) and Barchiesi (2009, 163–170). In the *Tristia*, Ovid explicitly likens himself to Phaethon as the victim of Jupiter-Augustus (1.1.69–82, 3.4.29–30, 4.3.65–66).

[55] On *Am.* 1.15, see Conte (1994, 32–33), Schiesaro (2014, 78–79), and Hardie (2022b, §9).

[56] Cf. Hardie (1986, 116).

Ancient writers can, explicitly or implicitly, dare the reader to make an intellectual leap—to demonstrate high-mindedness and expansiveness of thought by following a disorienting argument, accepting a counterintuitive premise or conclusion, and having the courage to view the *mundus* from the vantage point that such a change in perspective constructs. Lucretius does this again and again, most notoriously at the end of the poem, the abrupt termination of which challenges the reader to cultivate *ataraxia* amidst unfathomable suffering.[57] A more concrete example occurs earlier in the same book, at the end of the explanation of earthquakes (6.603–607). Lucretius describes the seemingly unavoidable fear, triggered by a seismic jolt, that the earth may be snatched away from below our feet and carried away into the abyss (*pedibus raptim tellus subtracta feratur / in barathrum*), and that the sum of things may be destroyed from the foundation, leveling everything into a confused pile of cosmic rubble (*fiat mundi confusa ruina*). To dread such a scenario is to fail to have internalized Lucretius's relentless logic and overarching lessons; such a failure reveals a narrowness of vision that construes a fall as inglorious and final. By contrast, to frame such a scenario as an opportunity to leap, metaphorically, into the chasm that the text visualizes amounts to a demonstration of the high-mindedness and expansiveness of thought that come from a thorough assimilation of Lucretius's Epicurean doctrine. Such a mindset transforms a helpless fall into a bold leap and reconfigures a panicked glimpse of the end of the world as the exhilarating survey of atomic truth dramatized at the opening of Lucretius's third book—there, too, the earth withdraws, and we get a panoramic survey of the void that stretches out beneath us (*omnia dispiciantur, / sub pedibus quaecumque infra per inane geruntur*, 3.26–27). Lucretius's challenge to the reader activates a long-standing tradition that extends all the way back to Parmenides, who similarly confronts and plunges headlong into a confounding abyss (χάσμ' ἀχανές, D4.17 *EGP*, B1.17 D-K), taking the reader with him.

A skeptic can easily dismiss the sublimity of the fall or ridicule anyone who leaps into a deep hole. Frequently, such criticism takes the form of the anecdote about Thales, first attested in Plato's *Theaetetus*: while engaged in astronomical observation, Thales fell into a well, and a passerby mocked him for being so preoccupied with what was up above that he did not see

[57] Gale (1994, 227–228) reads Lucretius's plague as a warning of the misery of the life of the non-Epicurean and as a protreptic to continue (re)reading Lucretius's own text to fortify oneself.

what was in front of him and at his feet.[58] By the Roman period, the dimensions of the incident have increased and the consequences have become more dire: according to Diogenes Laertius, Thales was studying the stars when he inadvertently stepped over the edge of a cliff and fell to his death (ἐς τὸ κρημνῶδες ἐκβὰς καταπίπτει).[59]

The anecdote's humorous potential is most effectively exploited when the subject actively leaps. A clever example is Callimachus's epigram about Cleombrotus of Ambracia: Cleombrotus said goodbye to the sun and leapt from a high wall into Hades, not because he had a good reason to commit suicide, but because he had read Plato's "book about the soul" (εἴπας,"Ἥλιε χαῖρε, Κλεόμβροτος ὡμβρακιώτης / ἥλατ' ἀφ' ὑψηλοῦ τείχεος εἰς Ἀΐδην, / ἄξιον οὐδὲν ἰδὼν θανάτου κακόν, ἀλλὰ Πλάτωνος / ἓν τὸ περὶ ψυχῆς γράμμ' ἀναλεξάμενος, *Ep.* 23 Pfeiffer, 53 Gow-Page). The book about the soul is *Phaedrus*, which contains a stirring account of the soul's postmortem chariot ride to the ridge of heaven and its tantalizing glimpse of what lies beyond (*Phdr.* 246a–248b). Inspired by the dialogue's sublime vision, Cleombrotus is eager to make the journey in the quickest way possible, but he has not noticed that most of Plato's souls do not reach the top; in fact, many end up breaking their wings in an undignified stampede (*Phdr.* 248a–b). While Cleombrotus's leap from a high wall (ἥλατ' ἀφ' ὑψηλοῦ τείχεος) suggests that he fancies himself a sublime subject, Callimachus's verdict is surely that he is a raving lunatic; Plato's Socrates, after all, tells the myth to explain the mechanics of philosophical madness (*Phdr.* 249c–250c).

At the end of the *Ars Poetica*, Horace expands the Callimachean paradigm (*Ars* 453–476). Horace imagines a diseased poet falling into a pit or well "while, head in the clouds, he belches up verses and wanders around" (*dum sublimis uersus ructatur et errat*, 457); the clause, complete with *sublimis*, cues up the same deflationary use of the sublime that we have seen in Aristophanes and Callimachus.[60] If such a figure were to call for help, writes Horace, no one would come to his aid. However, if someone were so minded, Horace would ask: "How do you know whether he threw himself

[58] Pl. *Theaet.* 174a (P12 *EGP*, A9 D-K,). The story may be based on an Aesopic fable (40 Perry); cf. Diog. Laert. 1.34, Hipp. *Ref.* 1.1.

[59] Diog. Laert. 2.4. The story is embedded in an apocryphal letter from Anaximenes to Pythagoras (Anaxim. R11 *EGP*).

[60] Brink (1971, 423–424) compares *sublimis* to μετέωρος and cites Plato's anecdote about Thales. *Sublimis* here might be accusative plural with *uersus*; the ambiguity contributes to the humor. Hardie (2009, 180–228) is excellent on Horace's deflationary use of Lucretius's aspirations to sublimity. Hardie (2022a, 99–105) explores parodic and comedic versions of the ascent topos.

down there on purpose and doesn't want to be saved?" (*qui scis an prudens huc se deiecerit atque / seruari nolit*, 462–463). In other words, perhaps we have another Cleombrotus on our hands. Stylistically, the poet's fall indicates a failure to achieve sublimity, a lapse into bathos (Hardie 2009, 212–217). But Horace is also offering a broader critique of claims to sublime subjectivity. The thought experiment leads seamlessly into reflections on the death of Empedocles: eager to be considered a god, Empedocles leapt (*insiluit*) into the fires of Aetna (463–466).[61] The narrative comments not only on style but also on Empedocles's bombastic claims to high-mindedness and elevation of thought, which Horace reduces to arrogance and a kind of pathetic madness (cf. *certe furit* at 472). Poets who follow Empedocles's model cease to be human; if you pull them out, they do not regain their mortal status (*nec si retractus erit iam / fiet homo*, 468–469). But rather than attain the majesty of a god, these figures are reduced to the level of mere animals, a bear or a blood-sucking leech (472–476). Hardie (2009, 200) has argued that Horace's portrait points to Empedocles's most exuberant Roman reader, Lucretius. At one level, then, Horace undoes the sublimity with which Lucretius endows Empedocles, and rewrites the heroic fall that Lucretius had attributed to him (*principiis tamen in rerum fecere ruinas / et grauiter magni magno cecidere ibi casu*, 1.740–741) as the pitiful fate of a sick and contemptible mind. At another level, though, Horace laughs at Lucretius himself, who, we may conclude, has more than a little in common with Callimachus's Cleombrotus. From Horace's perspective, at least at the end of the *Ars poetica*, the will to sublimity resembles the delusional ravings of a madman.

1.6. The Scientific Sublime: Theoretical Perspectives from Plato to Hermogenes

As we have seen, the logic of the sublime can frame the systematic investigation of nature already in the fifth century BCE. In the Roman world, writers and readers had access to numerous works in which the scientific sublime played a prominent role, including those of Parmenides, Empedocles, and Aristophanes, to name only those fifth-century figures whom we have already considered. In addition, a discourse on the scientific sublime helped

[61] For the legend, see P29 *EGP* (the relevant portions of A1 D-K); cf. A2, 16 D-K.

establish and disseminate its logic and imagery, and not always positively. Indeed, a profound ambivalence toward the sublime often marks theoretical discussions. Familiarity with the basic contours of this tradition is necessary for making sense of the complex ways in which Roman literary culture thought and argued in and through the scientific sublime.[62]

Plato and Aristotle

Our earliest extant evidence for the theorization of the relationship between philosophy and the sublime occurs in Plato. We have seen that Diotima's speech in *Symposium* frames the upward journey of philosophy through the dual logic of the sublime: the irresistible force of eros pulls the student upward, from beautiful bodies to beautiful customs and eventually to the form of the beautiful itself. *Phaedrus*'s myth of the soul tells a similar story, and Callimachus's epigram suggests that ancient readers recognized its engagement with the sublime.

Theaetetus takes a different approach. Here, the dialogue's eponymous character has been reduced to aporia (*Theaet.* 155c–d):

THEAETETUS: By the gods, Socrates, I wonder like mad (ὑπερφυῶς ὡς θαυμάζω) at what these things might be, and sometimes, when I look at them, I feel truly dizzy (ἐνίοτε ὡς ἀληθῶς βλέπων εἰς αὐτὰ σκοτοδινιῶ).

SOCRATES: Yes, my friend, it seems that Theodorus did not make a bad guess about your nature. For this experience, wonder, is very much characteristic of a philosopher (μάλα γὰρ φιλοσόφου τοῦτο τὸ πάθος, τὸ θαυμάζειν). There is no other starting point for philosophy than this, and the man who said that Iris was the offspring of Thaumas seems not to have been a bad genealogist.

Theaetetus's vertiginous awe encodes the sublime, and Socrates makes an openness to this experience the defining characteristic of the philosopher.[63] Throughout the Platonic corpus, interlocutor after interlocutor experiences wonder, stupefaction, enchantment, mental paralysis, and other forms of

[62] For the history of the sublime as a critical concept in antiquity, Wehrli (1946), Quadlbauer (1958), and Shuger (1984) remain helpful. For the fifth century, see again Porter (2016, 319–350, 415–444).

[63] For wonder in Plato, see most recently Lightfoot (2021, esp. 1–5, 17–31, 200–205).

intellectual breakdown or alienation when they engage with Socrates and allow him to guide their thoughts. Such moments of sublime disorientation register a confrontation with a new reality, a change in perspective that reveals that we do not know what we think we know. In the present passage, Socrates's genealogy indicates the transcendent nature of the enterprise: the offspring of wonder (Thaumas) is the rainbow (Iris), which connects earth with heaven and allows for communication between the gods and men.[64]

Later in the dialogue, Socrates states that the goal of philosophy is to try to escape from earth to the dwelling place of the gods: "The escape is an assimilation to god, as far as possible" (φυγὴ δὲ ὁμοίωσις θεῷ κατὰ τὸ δυνατόν, 176b). The assimilation to god (*homoiōsis theōi*) is an effort to embody moral ideals, but symbolically it suggests an elevation of consciousness and expansion of mind that reinvents the philosopher as the sublime subject.[65] Under the right circumstances, wonder or awe sets this transformation in motion, launching the philosopher on an upward journey, along the rainbow of inquiry, toward virtue, truth, and being. Theaetetus's vertigo does not guarantee that he will make the ascent; it is merely the foundation without which philosophy cannot get off the ground.[66]

While Plato is very much aware of the allure of the scientific sublime, he firmly rejects the methodology and basic assumptions of fifth-century natural philosophy. In *Phaedo*, Plato's Socrates reveals his youthful passion for the inquiry into nature (96a–b):

> νέος ὢν θαυμαστῶς ὡς ἐπεθύμησα ταύτης τῆς σοφίας ἣν δὴ καλοῦσι περὶ φύσεως ἱστορίαν· ὑπερήφανος γάρ μοι ἐδόκει εἶναι, εἰδέναι τὰς αἰτίας ἑκάστου, διὰ τί γίγνεται ἕκαστον καὶ διὰ τί ἀπόλλυται καὶ διὰ τί ἔστι. καὶ πολλάκις ἐμαυτὸν ἄνω κάτω μετέβαλλον σκοπῶν πρῶτον τὰ τοιάδε.
>
> When I was young, I had a wondrous longing for the kind of knowledge that people call the inquiry into nature. For it seemed to me to be magnificent, to know the causes of each thing, on account of what it comes to be

[64] The genealogy comes from Hes. *Th.* 780, where Iris appears to be identified as the messenger of the gods.

[65] For *homoiōsis theōi* in Plato, later Platonism, and Stoicism, see Reydams-Schils (2017). For antecedents to Plato's formulation, see Roloff (1970). The concept became a significant ethical goal for a wide range of intellectual traditions; see, conveniently, Erler (2002, 159–160), Heath (2013, 179), and Edmonds (2019, 331–333).

[66] Socrates's genealogy hints at the sublimity of the rainbow itself. Aetius 3.5.7–9 (Mansfeld and Runia 2023, 272) cites rainbow theories for Anaximenes, Anaxagoras, and Metrodorus, any of whom might have framed the phenomenon through the scientific sublime. For the history, see Sen. *NQ* 1.3–8. For the sublimity of the rainbow, see Cic. *Nat. D.* 3.51 and Ch. 2.2 on Man. 1.713.

> and on account of what it perishes and on account of what it exists, and often I used to throw myself up and down when I first looked at the following kinds of things.

Socrates's wondrous longing encodes the affective intensity of the sublime, which ὑπερήφανος names.[67] The inquiry is not a linear progression from uncertainty to truth but rather a recursive journey that moves from high to low and back again, over and over. Eventually, Socrates sours on the experience and adopts a fundamentally different methodological approach, thereby launching his "second voyage" into the examination of causation (τὸν δεύτερον πλοῦν ἐπὶ τὴν τῆς αἰτίας ζήτησιν, 99d). According to an ancient tradition, "the phrase δεύτερος πλοῦς refers to the use of oars in the absence of a fair wind, suggesting... a slower and more laborious, but more reliable, method of getting to the same destination";[68] in particular, it signals the use of the theory of forms and Socrates's familiar style of dialogue. For Plato's Socrates and for Plato himself, the inquiry into nature ends in disappointment and the thrill of the scientific sublime amounts to a dangerous distraction—better to row than to sail.

In *Metaphysics* 1.2, Aristotle famously reflects on the origins of philosophy and the extent to which knowledge of causes and first principles exceeds the limits of the human. In a sentence regularly read with *Theaetus* 155c–d, Aristotle declares: "On account of wonder (διὰ γὰρ τὸ θαυμάζειν), human beings both now begin and first began to study philosophy. In the beginning, they wondered at easily manageable difficulties, but they gradually advanced and raised questions about greater things, such as the phenomena of the moon and matters related to the sun and the stars and the genesis of the universe" (982b). The sublimity of nature sets philosophy in motion, and the greater the subject, the greater the intellectual pursuit. To be more specific, Aristotle suggests that such knowledge might be too great for human beings—it might be divine. The reasons are twofold: "God seems to everyone to be among the causes and a kind of first principle, and god alone or most especially can possess this kind of knowledge" (983a). To study first principles and causes is to be driven by wonder toward the divine; this is the scientific sublime. In a consequential twist, however,

[67] The adjective ὑπερήφανος can mean "magnificent, splendid" or "arrogant, proud," which mirrors the range of *megalopsuchos*. For *huper-* words that name the sublime, see Porter (2016, 181).
[68] Rowe (1993, 238, citing Eust. *In Od.* 1453.20–22 Stallbaum).

Aristotle configures the elimination of wonder as the goal of philosophical inquiry. At the outset, he equates wonder with ignorance and suggests that people engage in philosophy to flee ignorance (διὰ τὸ φεύγειν τὴν ἄγνοιαν ἐφιλοσόφησαν, 982b). At the end of the discussion, he returns to this idea (983a):

> The acquisition of this knowledge, however, must in a way end up in a state that is the opposite of our original investigations. For everyone begins, as we have said, from wondering whether this is how things are, such as in the case of automata or the solstices or the incommensurability of the diagonal.... But we must end in the opposite and proverbially better state, as actually happens in these matters when people understand them. For a geometer would wonder at nothing as much as if the diagonal should be commensurable.

Knowledge of first principles and causes replaces wonder, and the sublime confrontation with nature becomes a curious relic of a naïve past.

For a positive theorization of the sublimity of studying the physical world, we must turn elsewhere. Early in the first book of *Parts of Animals* (1.5.644b–645a), Aristotle distinguishes between the philosophical study of eternal and imperishable works of nature (the heavens) and those that are subject to generation and decay (plants and animals). To study the former is difficult since the heavenly bodies exist at a great distance, but because of their excellence our imperfect knowledge of them gives more pleasure than everything that exists down here on earth (ἥδιον ἢ τὰ παρ' ἡμῖν ἅπαντα). From Aristotle's perspective, the sublimity of celestial study needs no further elaboration. The study of terrestrial things, however, requires an argument. This branch of philosophy, claims Aristotle, has a certain advantage because we live among earthly things and our natures are more closely related to theirs, all of which means we can gain much more complete and reliable knowledge of them. At this point, Aristotle urges the reader not to recoil, like a child, from studying seemingly ignoble creatures that do not charm the senses; for "in the study [of animals] nature the demiurge provides inconceivable pleasures for those who are able to establish links of causation and are philosophically inclined by nature" (κατὰ τὴν θεωρίαν ὅμως ἡ δημιουργήσασα φύσις ἀμηχάνους ἡδονὰς παρέχει τοῖς δυναμένοις τὰς αἰτίας γνωρίζειν καὶ φύσει φιλοσόφοις). The very act of studying animal life and identifying causes activates the affective intensity of the sublime. The emphasis on causation is important. Plato's Socrates is thrilled by the analysis of nature's causes but finds the methods of the natural philosophers unsystematic and their results unsatisfying. Aristotle, by contrast, stresses that nature is not haphazard and that the end for the sake of which nature's works have been constructed or come into being has its place among the beautiful. To become aware of that end is to glimpse nature's purpose and an object's beauty, whether up in heaven or down on earth. It is a movement that takes us out of or beyond ourselves and reveals a hidden reality—it is a recognition that "in every part of nature there is something marvelous" (ἐν πᾶσι γὰρ τοῖς φυσικοῖς ἔνεστί τι θαυμαστόν). Such is the basic viewpoint of the scientific sublime, and here, at least, Aristotle wholeheartedly embraces it.[69]

Stoics, Epicureans, and the *nil admirari* Tradition

In the Hellenistic and Roman worlds, it becomes possible to frame the goal of philosophical study or even the good life more generally as the eradication of wonder. The *nil admirari* principle derives its name from the opening lines of Horace, *Epist.* 1.6: "To wonder at nothing—that is perhaps the one and only thing, Numicius, that can make and keep you happy" (*nil admirari prope res est una, Numici, / solaque, quae possit facere et seruare beatum*).[70] Ancient sources ascribe the basic idea to multiple figures, particularly Pythagoras and Democritus, but the influence of Aristotle's position in *Metaph.* 1.2 is unmistakable. In *On Listening to Lectures*, Plutarch criticizes those who misunderstand Pythagoras's dictum that what he had gained from philosophy was the ability to wonder at nothing (τὸ μηδὲν θαυμάζειν): "While philosophical thinking, through knowledge and an inquiry into the cause in each case, removes wonder and astonishment that originate from perplexity and ignorance, it does not destroy our serenity, moderation, and humanity" (13.44b–c). As with Aristotle, the philosophical investigation of causes removes wonder that is rooted in ignorance. In addition, this investigation, according to Plutarch, leaves us serene, calm, and humane; "to wonder at nothing" is shorthand for achieving a kind of moral and epistemological perfection.[71]

[69] For more on nature as demiurge, see Ch. 5.3, 5.5 ("*Natura artifex*").

[70] For background to the *nil admirari* tradition, see McCarter (2015, 107–115) and Lightfoot (2021, 200–208).

[71] The passage from *On Listening to Lectures* does not appear in the Pythagoras sections of D-K or *EGP*. The relevant testimonia and fragments of Democritus are ATOM. D230, 295 (A169, B216 D-K; cf. A168 D-K).

Whether or not Pythagoras or Democritus formulated a similar principle, it spreads from Aristotle into the Hellenistic world and is frequently invoked in contexts where celestial and/or meteorological phenomena are sources of fear. In *Epist.* 1.6, Horace's first examples of things at which one should not wonder are precisely these: "The sun and the stars and the seasons that move in fixed times—there are those who can view these things filled with no terror" (*hunc solem et stellas et decedentia certis / tempora momentis sunt qui formidine nulla / inbuti spectent*, 1.6.3–5). But, implies Horace, such spectacles fill most people with anxiety and dread, as Lucretius well knows (5.1205–1217). The antidote is causal analysis, combined with a kind of exposure therapy. So Strabo, for instance, explicitly states that his explanation of the causes of certain geological changes and his accumulation of parallel examples will lead to freedom from wonder (πρὸς δὲ τὴν ἀθαυμαστίαν) and put a stop to astonishment (παύσει τὴν ἔκπληξιν) since that which is unfamiliar terrifies the senses (1.3.16 [57]). Again, experiences of wonder and astonishment are directly linked with ignorance and fear, and the philosophical study of nature will resolve the subject's discomfort.[72]

The same logic appears in the Stoics and Epicureans. According to Diogenes Laertius, the Stoics declared that the sage wonders at nothing that appears marvelous (τὸν σοφὸν οὐδὲν θαυμάζειν τῶν δοκούντων παραδόξων), such as entrances to the underworld, tides, hot springs, and eruptions of fire (7.123, *SVF* 3.642). The basic premise is that wondering at nature's marvels is a form of psychic disturbance.[73] In his *Letter to Pythocles*, Epicurus states that the only reason to study meteorological phenomena is to acquire freedom from disturbance and a secure conviction about reality (ἀταραξίαν καὶ πίστιν βέβαιον, Diog. Laert. 10.85). In the *Letter to Herodotus*, however, too much study leads to adverse effects "when the astonishment that is based on the prior consideration of these phenomena cannot find a resolution in accordance with the orderly management concerning the most important issues" (ὅταν τὸ θάμβος ἐκ τῆς τούτων προσκατανοήσεως μὴ δύνηται τὴν λύσιν λαμβάνειν κατὰ τὴν περὶ τῶν κυριωτάτων οἰκονομίαν, Diog. Laert. 10.79). Wonder, disturbing and terrifying, impedes *ataraxia*

[72] Cf. the ethical goal of being without wonder or astonishment (τὸ ἀθαύμαστον καὶ ἀνέκπληκτον) at M. Aur. *Med.* 1.15.

[73] For more on this passage, see Ch. 5.5 ("Aetna and the Paradoxographical Tradition"). Cf. *SVF* 1.239, where Zeno is said to urge the cultivation of a fearless and unawed disposition of soul (ἀδεῆ καὶ ἀθαύμαστον... τὴν διάθεσιν τῆς ψυχῆς).

and requires resolution. This is precisely the goal of the Epicurean method of multiple explanations. Due to the difficulty of observing celestial and meteorological phenomena, absolute certainty about the cause of this or that phenomenon remains impossible. But it is also completely unnecessary, since a list of multiple explanations that agree with appearances here on earth is enough to calm a mind that is anxious about divine interference: in an infinite universe, each physical account must be true somewhere at some time, and one of them must be true now. The student, then, should wonder at nothing.[74]

As these examples illustrate, *nil admirari* differs significantly from the basic assumptions of the scientific sublime, but the two traditions are not incompatible. The apparent difficulty is that wonder is often taken to be rooted in ignorance and is generally treated as synonymous with negative emotions like fear, terror, and dread. When Lucretius (or a later imitator) exhorts the reader not to wonder at a particular idea or phenomenon (*nec mirum, nimirum, non est mirandum, mirari mitte*, etc.), he warns against an experience of nature that is grounded in misconceptions or muddled logic and that terrorizes the unsuspecting student with concerns about the divine; such an experience contracts the mind (*animus formidine diuum / contrahitur*, 5.1218–1219; cf. *compressit* at 1.69).[75] To dispel this fear and, by implication, to expand the mind is the formal rationale that motivates the Epicurean examination of celestial and meteorological phenomena. Such admonitions, however, do not undermine the centrality of the sensuous shudder. To be seized by *diuina uoluptas* and *horror* is to be alive to the reality of nature, to see it clearly in a way that ordinary human beings cannot; it has nothing to do with the terror that Burke and Kant locate at the heart of the sublime. The experience is proof of one's greatness of soul, and it is an expansion, rather than a contraction, thereof.[76]

In the Epicurean tradition, the exuberance of the scientific sublime is not confined to Lucretius. In one of the *Vatican Sayings*, the study of nature is a never-ending pursuit that allows the student to pass beyond their own mortality (*Gnom. Vat.* 10):[77]

[74] For the nature and role of multiple explanations in Epicureanism and the origins of the methodology in Aristotle and Theophrastus, see Bakker (2016, 8–75). He underscores the ethical motivation early on (8).
[75] For Lucretius's "rhetoric of *mirum*," see Conte (1994, 20–21).
[76] Cf. Hardie (1986, 171) and Porter (2016, 452).
[77] I cite the text from Bailey (1926, 106).

> μέμνησο ὅτι θνητὸς ὢν τῇ φύσει καὶ λαβὼν χρόνον ὡρισμένον ἀνέβης τοῖς περὶ φύσεως διαλογισμοῖς ἐπὶ τὴν ἀπειρίαν καὶ τὸν αἰῶνα καὶ κατεῖδες "τά τ' ἐόντα τά τ' ἐσσόμενα πρό τ' ἐόντα."
>
> Remember that, although you are mortal in nature and have received a limited time [of life], you have ascended through your discourses on nature to the infinite and to eternal time and you have looked down upon "what is, what will be, and what was before."

The imagery is familiar (the flight of the mind, the ascent, the view from above). What is striking is that the text explicitly frames the inquiry into nature as a movement toward infinity, the transgression or even dissolution of all spatial and temporal boundaries. The knowledge that one gains is divine—it is the knowledge of Homer's Calchas (*Il.* 1.70) and Hesiod's Muses (*Th.* 38)—and it assimilates the student to (a) god.[78] Again, this is a form of *megalopsuchia* or *magnitudo animi*, even of Platonic *homoiōsis theōi*, but the author configures the experience dynamically, as an impossible adventure, radically annihilating the minutiae of everyday life. Whether the saying derives from Epicurus or his pupil Metrodorus matters little. Other Epicurean texts make such assimilation a goal, and the basic idea was appealing, not least of all to Lucretius.[79]

From a Stoic perspective, the study of nature—namely, physics—is the study of the cosmos and everything in it. Significantly, Zeno and Chrysippus equated the cosmos with god, and when one combines the two propositions, physics itself becomes a kind of theology or study of the divine.[80] Within such a worldview, the investigation of any aspect of the natural world, however small, inconsequential, or impermanent, has the potential to inspire inconceivable pleasures, to use Aristotle's phrase, and to connect the student with a reality that transcends the here and now. At the same time, contemplation of the heavens tends to be construed as a particularly

[78] On the use of the epic phrase, see Rutherford (1989, 160–161) and Glauthier (2020, 753). For parallel expressions, see Pease (1979 [1920]) on Cic. *Diu.* 1.63. Edmonds (2019, 193 n.17) adds Hippoc. *Progn.* 1.1. to Pease's list of parallels. Cf. Sen. *Dial.* 10.15.5.

[79] The attribution to Metrodorus is based on Clem. Alex. *Strom.* 5.138.2; see Bailey (1926, 376). For Epicurean formulations of transcendence that do not involve natural inquiry, see *Ep. Men.* (Diog. Laert. 10) 135; Diog. Oen. fr. 63 William (52 Chilton), coll. iii–iv; and Warren (2000). Most (2012, 250) characterizes the Lucretian sublime as a way for humans to achieve greatness and "thereby become themselves gods of a kind."

[80] For the definition of physics, see *SVF* 2.35 (L-S 26A). For Zeno and Chrysippus, see *SVF* 2.1022 (L-S 43A); cf. *SVF* 1.102 (L-S 46B).

ennobling enterprise that promises to expand the mind and to bring the spectator into touch with god.[81]

The Stoic position, however, is not unique. In the Hellenistic and Roman periods, the urge to equate the universe with god and/or to treat nature as divine is widespread. It is an idea that can be traced back to Plato, extends into the furthest reaches of late antiquity, and figures prominently in the debates that are the focus of the present study.[82]

That the study of nature has ethical value is a similarly common conviction. Plato promotes the idea that contemplation of the regularity and order of the heavens can produce regularity and order in the spectator, and a variety of philosophical traditions construe cosmic spectatorship in general as the ideal mode of human life.[83] At *Fin.* 4.11–12, Cicero's character Marcus complains that Epicurus studies the natural world only to eliminate the fear of death, and vibrantly celebrates the way in which contemplation of the night sky generates greatness of soul and even a sense of justice. Marcus then exclaims: "In the same study of nature, there exists a certain insatiable pleasure that comes from beginning to learn about the world, so that in this pursuit alone, when our necessary business has been completed, we are able to live honorably and nobly, free from obligations" (*inest in eadem explicatione naturae insatiabilis quaedam e cognoscendis rebus uoluptas, in qua una, confectis rebus necessariis, uacui negotiis honeste ac liberaliter possimus uiuere, Fin.* 4.12). The study of nature allows one to live the life of the nobility, and it is breathtaking. Marcus encodes the sublimity of this modus vivendi as *insatiabilis quaedam uoluptas,* which is a correction of Lucretius's *quaedam diuina uoluptas…atque horror.* For Cicero, the scientific sublime

[81] Good examples are Sen. *Dial.* 8.5.3–8, *NQ* 1 praef. 11–17. Stoicism can use the rhetoric of sublimity in ethical contexts without reference to physics: Zeno (*SVF* 1.216) describes the sage as elevated (ὑψηλός) because he shares in the elevation (ὕψος) that belongs to the noble and wise, and multiple sources claim that the Stoics defined *megalopsuchia* as a quality that makes one "superior to" or "above" (ὑπεράνω) events that naturally befall everyone (*SVF* 3.264–265, 269–270; cf. 274–275). Such passages configure sublime subjectivity as an ethical goal.

[82] The classic survey of the evidence for "le dieu cosmique" is the second volume of Festugière (1944–1954); for a more succinct account, see Pépin (1986). A similar idea appears already in a fragment of Aeschylus that equates Zeus with aether, earth, and heaven and asserts that "Zeus is all things and whatever is beyond that" (fr. 70 *TrGF*). Within this worldview, to study nature and to disseminate one's knowledge thereof can constitute a form of piety/worship; the conceit appears in authors as diverse as Manilius (1.1–24) and Galen (*UP* 3.10 at 3.237.11–17 Kühn).

[83] Pl. *Ti.* 47b–c, 90c–d. According to fragments of Aristotle's *Protrepticus*, Pythagoras and Anaxagoras promoted observation of the heavens as the virtual *telos* of human existence (frr. 18–19 Düring). The same position is attributed to Anaxagoras at Arist. *Eth. Eud.* 1.5.1216a (P35 *EGP*, A30.18–21 D-K) and Diog. Laert. 2.10 (P40 *EGP*, A1 D-K). For cosmic spectatorship, see Pease (1955–1958) on Cic. *Nat. D.* 2.37.

consists in endless and ennobling pleasure, a far less confounding mode of life than that which Lucretius presents at the opening of his third book. For some in the senatorial elite, the wreckage of the Roman Republic may make *horror* an unattractive component of the idealized life of the mind, and *magnitudo animi* can be constructed as precisely the virtue that allows you to remove yourself from the cares and concerns of those who are situated below you.[84]

Cicero's celebration of *explicatio naturae* participates in a debate about the value and goals of studying nature. As this text suggests and the following chapters demonstrate, it is a debate in which the sublime plays a central role.

Longinus and the Rhetorical Tradition

These traditions meet with combustible force in Longinus. The context is the discussion of flawed sublimity (see Ch. 1.5, "Acrobats, Heroes, and Flawed Sublimity"). Here, Longinus wonders about the mindset of the best writers of the past, those godlike figures who reached for greatness and scorned accuracy. They understood the unique calling of the human mind (*Subl.* 35.2–3):

> ἡ φύσις οὐ ταπεινὸν ἡμᾶς ζῷον οὐδ᾽ ἀγεννὲς ἔκρινε τὸν ἄνθρωπον, ἀλλ᾽ ὡς εἰς μεγάλην τινὰ πανήγυριν εἰς τὸν βίον καὶ εἰς τὸν σύμπαντα κόσμον ἐπάγουσα, θεατάς τινας τῶν ἄθλων αὐτῆς ἐσομένους καὶ φιλοτιμοτάτους ἀγωνιστάς, εὐθὺς ἄμαχον ἔρωτα ἐνέφυσεν ἡμῶν ταῖς ψυχαῖς παντὸς ἀεὶ τοῦ μεγάλου καὶ ὡς πρὸς ἡμᾶς δαιμονιωτέρου. διόπερ τῇ θεωρίας καὶ διανοίας τῆς ἀνθρωπίνης ἐπιβολῇ οὐδ᾽ ὁ σύμπας κόσμος ἀρκεῖ, ἀλλὰ καὶ τοὺς τοῦ περιέχοντος πολλάκις ὅρους ἐκβαίνουσιν αἱ ἐπίνοιαι, καὶ εἴ τις περιβλέψαιτο ἐν κύκλῳ τὸν βίον, ὅσῳ πλέον ἔχει τὸ περιττὸν ἐν πᾶσι καὶ μέγα καὶ καλόν, ταχέως εἴσεται πρὸς ἃ γεγόναμεν.
>
> [They saw that] nature did not deem us as humans to be a lowly or ignoble creature; in bringing us into life and into the entirety of the cosmos as though into a great festival, intending us to be spectators as well as highly

[84] McConnell (2017a, 53–61) is a good overview of *magnitudo animi* in Cicero's corpus. For a broader assessment of Cicero's estimation of the value of physics, see Reydams-Schils (2016).

ambitious competitors in the contests, she planted in our souls from the outset an irresistible passion for everything great and, from our perspective, more than human. This is why not even the entire cosmos is enough for the reach of human contemplation and thought, but our intuitions often go beyond the boundaries even of the whole world, and if one looks around at life and sees how that which is exceptional, great, and beautiful stands out in all domains, one will soon recognize the purpose of our existence. (Trans. Halliwell 2022, modified)

The universe itself is a sublime spectacle in which we are both audience members and active participants, constantly observing and indefatigably asking questions. Indeed, we are driven at an instinctual level to investigate the world and to reach for what lies beyond it, like Lucretius's Epicurus. For Longinus, this is our sublime vocation, and the present passage is a summons to cultivate the will to sublimity.[85]

As he continues, Longinus hymns the force that propels us toward nature's greatest sights, some of which play important roles in this study (the stars, the Nile, mount Aetna). Critics have long recognized that Longinus himself strives to produce sublimity, and the aspirational rhetoric of *Subl.* 35 is a case in point.[86] We should note, however, that Longinus continues to think in terms of social hierarchies and to equate moral, intellectual, and aesthetic greatness. The investigation of the cosmos is an opportunity to display and to enhance one's excellence before the eyes of an adoring crowd. While Longinus focuses on the intoxicating freedom that he finds at the top of the social system, elsewhere he employs language that brings the darker realities of that system to the fore: the sublime works through mastery and irresistible force (δυναστείαν καὶ βίαν ἄμαχον, *Subl.* 1.4), and it not only persuades but also enslaves its audience (οὐ πείθει τὸν ἀκροατὴν μόνον ἀλλὰ καὶ δουλοῦται, *Subl.* 15.9). Enslavement is a metaphor for the overwhelming power of sublimity, but Longinus does not object to the historical reality

[85] For the connection with Lucretius, see Porter (2007, 174; 2016, 453–454). Longinus's festival imagery is related to the supposed saying of Pythagoras (or Diogenes the Cynic) that life resembles the Olympic games (or a festival); see Porter (2020, 230 with nn. 11 and 13) and Halliwell (2022, 373).

[86] See Porter (2016, 63 n.10), who cites Boileau, Pope, and Gibbon. First-person plural expressions of wonder and awe saturate *Subl.* 35.4–5 (θαυμάζομεν, ἐκπληττόμεθα, ἀξιοθαυμαστότερον νομίζομεν), vividly encoding the sublime and thereby facilitating such judgments. This is another example of de Bolla's (1989, 12) "natural tendency" for the discourse on sublimity to generate sublime experience.

from which the metaphor is drawn.[87] Moreover, the relationship between slavery, freedom, and the sublime is the focus of *Subl.* 44, which purports to record a conversation in which Longinus conspicuously fails to criticize historical slavery or political repression and again uses enslavement metaphorically (*Subl.* 44.9–10). While Longinus's treatise can, at times, sweep us off our feet, we need to stay alert to its underlying assumptions and draw attention to the implications of its logic.

Subl. 35 contains a particularly vivid distillation of the aspirations of the scientific sublime, but similar assumptions about the ennobling power of the study of nature permeate the Roman rhetorical tradition.[88] This makes the scientific sublime foundational to elite education, which helps explain how such a wide range of authors and readers came to be familiar with that mode of sublimity. Longinus differs from the authors of other handbooks in his enthusiastic register, his own will to sublimity, but the similarities demonstrate how familiar the outlook must have been; just a few examples will suffice. At *De partitione oratoria* 56, Cicero claims that things can be grand (*magna*) by nature or by experience. The first classification includes "things in the heavens, divine things, things whose causes are obscure, things on the earth and in the sky that provoke wonder" (*ut caelestia, ut diuina, ut ea quorum obscurae causae, ut in terris mundoque admirabilia quae sunt*). Celestial and meteorological objects and phenomena, including a consideration of their causes, are axiomatically sublime. More complex is *Orator* 15, where Cicero attributes Pericles's preeminence as a speaker to his time spent as "a pupil of Anaxagoras, the natural philosopher" (*Anaxagorae physici... auditor*). The exposition is introduced as a summary of a passage from Plato's *Phaedrus*:[89] "From [Anaxagoras], [Socrates] thinks that [Pericles] not only learned many things that were splendid and magnificent (*praeclara quaedam et magnifica*), but also became rich and fertile and knowledgeable of the kinds of speech through which the individual parts of

[87] Halliwell (2022, 251) writes that the "positive application of such a metaphor is necessarily, *for educated Greeks*, piquant and paradoxical" (emphasis mine).

[88] Innes (1979, esp. 165–166) remains essential. For the influence of the paradoxographical tradition on *Subl.* 35, which complicates its relationship to the scientific sublime, see Ch. 5.5 ("Aetna and the Paradoxographical Tradition").

[89] Pl. *Phdr.* 269e–270a. The language of the sublime permeates the passage, with ironic intent: great speech requires "chatter and *meteōrologia* concerning nature"; from here arises "high-mindedness" (ὑψηλόνουν); Pericles fell in with Anaxagoras, became "filled with *meteōrologia*," and advanced toward the nature of "mind and mindlessness" (ἐπὶ φύσιν νοῦ τε καὶ ἀνοίας). The final expression plays on Anaxagoras's theory of *nous* and hints at Pericles becoming an inspired, out-of-his-mind speaker, like a poet; for the levels of irony, see Yunis (2011, 209–210). Pericles is a sublime speaker already at Ar. *Ach.* 530–531, where he hurls rhetorical thunderbolts from Olympus.

the soul are driven this way and that, which is the most important part of eloquence." Familiarity with *praeclara* and *magnifica* points to the scientific sublime, as Cicero subsequently makes clear: "What can I say about the study of nature, knowledge of which supplies the orator with a great resource?" (*quid dicam de natura rerum, cuius cognitio magnam oratori suppeditat copiam*, 16). Knowledge of the mechanics of nature, in the manner of the Pre-Socratics (Anaxagoras) and Lucretius (*de natura rerum*), infuses speech with the overwhelming abundance and inexhaustible excess of sublimity, and therefore it has something quite valuable to offer to the would-be statesman.[90]

Greek treatises tend to be more explicit. In Demetrius's *On Style*, which is difficult to date, certain topics simply possess elevation or greatness (ἔστι δὲ καὶ ἐν πράγμασι τὸ μεγαλοπρεπές), such as a famous land or sea battle "or when the discourse concerns the heavens or the earth" (ἢ περὶ οὐρανοῦ ἢ περὶ γῆς λόγος, 75). While it remains possible to mishandle the topic, the inclusion of such material causes the listener to suppose that the speaker aims for sublimity.

Hermogenes's *On Types of Style* (late second or early third century CE) is more precise in its discussion of solemnity (σεμνότης). Solemnity is one of the types that can produce grandeur (μέγεθος), majesty (ὄγκος λόγου), and dignity (ἀξίωμα)—in other words, the sublime.[91] The first order of solemn thoughts (ἔννοιαι σεμναί) are those concerning the gods qua gods. "The second order of solemn thoughts are those that discuss truly divine things, such as if one should examine the nature of the seasons and how and through what causes they come to be, and the circular motion and nature of the universe, or if one should inquire into the movements of the earth and the sea and how they come to be, or the movements of thunderbolts or things of this sort."[92] Like Socrates in *Phaedo* and Aristotle in *Parts of Animals*, Hermogenes emphasizes causation. Like Aristotle and Longinus, he highlights the active investigation of the natural world. To ask questions about the mechanics of the heavens, meteorological phenomena, or the universe itself is an attempt to understand the divine and, through thought,

[90] On Cicero's appraisal of physics here, see Reydams-Schils (2016, 100–101). *Brut.* 44 contains similar remarks about Pericles. For Anaxagoras and the sublime, see Porter (2016, 421–431).

[91] Hermog. 1.5 (pp. 241–242 Rabe). Longinus uses *semnotēs* to mean "sublimity" at *Subl.* 12.3, and Porter (2016, 181) includes it on a list of terms that are "names, synonyms, and equivalents" for *hupsos*; cf. Innes (1979, 165) and Fantham (1992a, 19).

[92] Hermog. 1.6 (pp. 243–244 Rabe); trans. based on Wooten (1987).

to approach it. For these authors, such modes of speech produce sublimity for the same reasons, or for some combination of the same reasons, that Plato, Aristotle, the Stoics, and the Epicureans frame the study of nature as sublime. Although the philosophical and ethical dimensions mostly go without emphasis, they implicitly undergird each lesson. Collectively, Cicero, Demetrius, Hermogenes, and Longinus constitute evidence of a rich and varied Roman discourse about the scientific sublime that grounds itself in language but remains aware of and sometimes explicitly shifts focus to the phenomenon's broader implications about nature and humanity's place in the cosmos.

The values that structure the rhetorical tradition come to life in the Latin poetry of the late Republic and early Empire. The *recusatio* tradition, for one, repeatedly sets the implicit grandeur of poetry that explores the *mundus* and its causes in opposition to elegy and other less exalted genres.[93] At the same time, imperial epic often contains narratives that can be read as philosophical allegory, that exploit the reader's philosophical knowledge of nature, and/or that explicitly discuss the mechanics and causes of nature.[94] In part, this impulse stems from the mode of Homeric exegesis that identifies in the *Iliad* and *Odyssey* deep philosophical knowledge, including the systematic knowledge of physics.[95] It also derives energy from the powerful tradition of hexameter poetry that did in fact discuss the natural world according to philosophical principles, from the poems of Parmenides and Empedocles to those of Aratus and Lucretius. In a kind of feedback loop, mythological epic, which sits atop the hierarchy of genres, imbues the study of nature with its own quintessential magnificence, while the study of nature amplifies and diversifies the greatness of heroic *epos*. Over time, this interrelationship becomes normative, and its centrality to the educational curriculum guarantees that the magnificence of both epic and the study of nature percolate through the ranks of imperial authors and audiences. Roman students of poetry, rhetoric, and philosophy will have had ample opportunity to become familiar with the logic and imagery of the scientific sublime. When a text engages in this tradition, to formulate its own vision

93 See Virg. *G.* 2.475–494; Prop. 2.34.27–28, 51–54, 3.5.23–48; *Ciris* 1–53, esp. 5–8, 29–41. For discussion, note Innes (1979) and Volk (2005b, 161–163).

94 Ovid's *Met.* is a prime example, and Schrijvers (2005, 31–34) is a succinct entry point. For a detailed case study in epic and natural inquiry, see Ch. 4.

95 The classic studies of this tradition are Buffière (1956) and Lamberton (1986). Exemplary texts include [Plut.] *De Homero* and Heraclit. *Quaestiones Homericae*.

and/or to respond to an alternative vision, the Roman reader will have had the necessary tools to identify the relevant intellectual history and interpretive constructs; we should not hesitate to use the scientific sublime to reframe the history of Latin literature.

In the following chapters, we will consider some of the pressing questions that are addressed through the scientific sublime, from the late Augustan Principate into the Flavian period. These questions include: What areas of the natural world deserve our attention? Where does the scientific sublime lead? What values or ideologies does it promote or conceal? What kind of world or what kind of history does it bring into being? With what intellectual or literary traditions does it create dialogue? With what other modes of sublimity is it in competition? Should we give in to its unrelenting power? Is it even possible to achieve that state of transcendence that it so alluringly promises? At the end of the day, is it all a lie? Most of these questions, in one form or another, have already surfaced in the preliminary analyses of this chapter. Through the rest of this study, we will see how such questions link authors, texts, and genres, shaping the evolution of Latin literature and Roman attitudes toward the study of nature in the early Empire. To see how the conversation continues to unfold, we turn to Manilius's *Astronomica*.

The Scientific Sublime in Imperial Rome: Manilius, Seneca, Lucan, and the Aetna. Patrick Glauthier, Oxford University Press. © Oxford University Press 2025. DOI: 10.1093/9780197787588.003.0002

2
Manilius and the Aesthetics of the Infinite

2.1. Introduction

The dazzling lights of the night sky constitute a sublime spectacle already in the Homeric poems, and the connection between stargazing and sublimity was to continue unabated throughout antiquity.[1] From an ancient perspective, by simply writing about the stars Manilius almost unavoidably engages with the sublime. But sublimity is central to the vision of celestial inquiry that Manilius espouses, and its most distinctive features have not received sufficient attention. Manilius configures the study of the stars as an exhilarating exercise that confronts readers with an experience of the infinite and thereby puts them in touch with the divine. The ecstatic accumulation of astronomical and astrological minutiae, the relentless calculation of unimaginably large and impossibly small magnitudes, the performance of a never-ending series of mathematical tasks—all of this creates an impression, now empowering, now overwhelming and disorienting, of the infinite vastness of the *mundus*, and to experience the infinite is to experience god. This basic approach to the cosmos constitutes Manilius's aesthetics of the infinite.[2] The flipside of transcendence, however, is transgression, and Manilius repeatedly represents celestial inquiry as a dangerous undertaking that violates boundaries, runs great risks, and ends in spectacular falls. Mythological paradigms for the astrologer, both as composer and student of astrological poetry, include the norm-smashing daredevils Phaethon, Salmoneus, and Bellerophon. Such figures underscore the thrill that comes from breaking with tradition, going where one implicitly should not go, and seeing the universe in its entirety. If the numbers do not always add up, that's a small price to pay. Manilius grants himself considerable leeway in

[1] For Homer, see *Il.* 8.555–559 with Glauthier (2023, 18–19).
[2] I borrow "aesthetics of the infinite" from Nicolson (1959).

his calculations in the fervent belief that the emotional rollercoaster ride cannot be distinguished from the experience of divinity. At the same time, Manilius's program is thoroughly implicated in contemporary Augustan culture, which maximizes the potential of astrological iconography and symbolism.[3] Manilius maps the cosmic sphere and the territory of the Roman Empire onto one another, which equates celestial inquiry with imperial expansion; both processes are infinite and unstoppable. From this perspective, Manilius represents a particular strain of cultural triumphalism, one that imagines itself as an integral part of the cosmic design and that simultaneously delights in the fiery erasure of borders.

A significant portion of my discussion focuses on Manilius's mathematics. While most critics are familiar with A.E. Housman's (1903–1930, 2.xiii) quip that the poet's "eminent aptitude for doing sums in verse... is the brightest facet of his genius," few readers have tried to interpret Manilius's numbers. The major exception is Duncan Kennedy (2011), who observes a connection between transcendence and mathematics throughout the poem. The goals and scope of Kennedy's argument differ from mine. In particular, Kennedy (2011, 175) is concerned to establish Manilius as a mathematical realist who subscribes to the idea that "mathematical entities... exist independently of the human mind."

Conversely, James Porter offers the only sustained account of the sublime in Manilius but does not consider mathematics.[4] Still, Porter advances the study of the sublime in the *Astronomica* in crucial ways. Above all, he demonstrates that Manilius exploits some of the modes of sublimity that Lucretius so powerfully develops in *De rerum natura*. Without adopting Lucretius's philosophical commitments, Manilius hijacks the language and imagery of Lucretius's poem, allowing his stars, for instance, to shine in the vast expanses of heaven like atoms hurtling through the void.[5] On Porter's view, this maneuver fills the *Astronomica* with the same sense of vital danger and the same awareness of the frightening discord of nature that

[3] Volk (2009, 137–161) persuasively argues that books 1–4 are Augustan in date and allows the possibility that book 5 may be Tiberian. Fratantuono (2012), by contrast, reads the poem as "a reflection on the transition from Augustus to Tiberius" (314). For astrology in Augustan literature and culture, see Volk (127–173) and Green (2014).

[4] Porter (2016, 483–508). For a possible connection between the proem of Man. 2 and Long. *Subl.* 13.3–4, see Ellis (1899) and Volk (2010). Based on this connection and other parallels, Goold (1961, 171) asserts that Manilius had read *Subl.* and that he and Longinus were acquaintances.

[5] Lühr (1969) amply demonstrates Manilius's familiarity with Lucretius. In Glauthier (2022), I argue that Manilius is well attuned to Aratus's use of the sublime; Lucretius is not the only influence.

animate Lucretius's poem. In a similar vein, Dunstan Lowe (2014) emphasizes the disorder of Manilius's earth and its ability to impinge upon the orderliness of the heavens, and Bill Gladhill (2016, 97–118) draws attention to the enmity and disharmony that define the relationships between Manilius's stars. The emphasis on Manilius's cosmos as a *discordia concors* ("a harmonious discord," 1.142), with its vibrant echoes of Empedoclean Strife and Love (cf. the juxtaposition of *concordia discors* with *Empedocles* at Hor. *Epist.* 1.12.19–20), constitutes a welcome corrective to earlier scholarship that tended to treat the *Astronomica* as a Stoic hymn to cosmic order.[6] This approach, however, can go too far in the opposite direction—the poet insists, indefatigably and at crucial junctures, that behind chaos stands *ratio*. Careful attention to Manilius's use of mathematics will help balance these competing critical impulses. The calculations can appear chaotic, but they simultaneously reveal or perhaps conjure into being the essential order of the *mundus*. To struggle through Manilius's mathematics is to experience *discordia concors*, to come to see how *discordia* is in fact *concors*; it is an experience structured according to the dual logic of the sublime.

Before we leap into the text, a few words about Manilius's cosmos and brand of astrology are in order.[7] Manilius's cosmos is rational, material, and fundamentally interconnected. He describes it as a living animal and a machine, characterizes it as divine or filled with god, and sometimes equates it with god himself.[8] In addition, the poem's lessons and pronouncements are grounded in absolute determinism; everything is fated and inescapable, everything unfolds through an unbroken chain of cause and effect. Within this cosmos, the stars are primarily the agents that set in motion and so control that unbroken chain of cause and effect, although Manilius sometimes describes them as the instruments through which god or cosmic reason operates.

Against this background, Manilius discusses some of the principles of genethlialogical astrology, the branch of astrology that analyzes the

[6] Housman (1903–1930, i.11–12) connects Man. 1.142 with Empedocles, and Porter (2016) suggestively connects the two at several points (pp. 484, 495, 500). For Stoic influences, see Lühr (1969), Reeh (1973), Lapidge (1979, 1393–1397), Salemme (2000, 10–21, 27–56), Volk (2009, 226–234), and Ramelli (2014). MacGregor (2005) and Habinek (2011) discuss material in the poem that is potentially incompatible with Stoicism.

[7] For Manilius's worldview, see Volk (2009, 58–67, 216–226). The most important passages for understanding his conceptions of fate and astrology are 2.60–135, 4.1–121, 866–935.

[8] Manilius promotes the pantheistic "cosmic religion or religious cosmology" (Volk 2009, 251) that originated in the Hellenistic period and spread throughout the Roman world; see Ch. 1.6 ("Stoics, Epicureans, and the *nil admirari* Tradition").

positions of the heavenly bodies at the precise moment of an individual's birth as seen from the exact terrestrial location of that birth.[9] The process is known as establishing a birth chart, nativity, or horoscope, and the individual is referred to as the native. The underlying idea is that certain celestial configurations determine the native's fate, in all its details, at every level, and the more complementary and/or overlapping principles the astrologer can use to interpret those configurations, the more completely the native's fate can be mapped. The poem, however, does not furnish enough information to allow a reader unfamiliar with genethialogical astrology to establish a birth chart or to tie all the disparate lessons together into one meaningful whole; we get tantalizing fragments, not a complete system. The incomplete nature of the exposition is due, at least in part, to the fact that Manilius is more interested in the deeply felt experience of studying the stars than in the mundane process of establishing a nativity. It is the study of the stars that puts us in touch with god, that reveals and thereby activates our own participation in the cosmic divinity.[10] The experience of the sublime is crucial to the ethos of the poem, so much so that it subsumes the practical concerns of the practicing astrologer. The poem is an exercise in *megalopsuchia* and *megalophrosunē*, in cultivating a will to sublimity, and it has no time for such humdrum operations.

2.2. The Milky Way

The first book of the *Astronomica* contains a lengthy discussion of celestial circles (1.539–804).[11] After an introduction, Manilius discusses the five parallel circles, the two colures, the meridian, the horizon, the zodiac, and the Milky Way, a passage that occupies over one hundred lines (1.684–804). Here, Manilius describes the Milky Way's path through the northern and

[9] For ancient astrology in general, see Bouché-Leclercq (1899), Barton (1994), and Beck (2007). Hübner (1984, 135–213) locates Manilius within ancient astrological traditions. At 2.453–465 and 4.701–710, Manilius briefly discusses *melothesia*, a system that gives the signs of the zodiac control or influence over specific body parts. These two passages are the only hints in the poem of astrological medicine (iatromathematics). At 3.142–144 and 149–155, Manilius briefly assumes the viewpoint of catarchic astrology, which uses the position of the stars to determine the opportune time to undertake an enterprise.

[10] Cf. Volk (2009, 181): "Manilius's poem . . . is not really about how to do astrology, but about the idea of astrology itself," "Manilius creates an alluring vision of astrology as a divine science."

[11] The circles are theoretical lines that divide the celestial sphere; see Evans and Berggren (2006, 149–160). Note that my discussion here builds off Glauthier (2022, 89–96).

southern constellations (1.684–700), celebrates its appearance through a series of similes (1.701–717), and reviews seven explanations of its origins and nature (1.718–804). The account of the Milky Way is an outstanding point of entry into Manilius's aesthetics of the infinite since it dramatizes the dual logic of the sublime and prominently encodes the sublimity of contemplating the infinite stretches of the night sky.

After tracing the Milky Way's path through the heavens, the poet declares: "You will not have to seek it out—it meets your very eyes of its own accord and is its very own teacher and compels observation" (*nec quaerendus erit: uisus incurrit in ipsos / sponte sua seque ipse docet cogitque notari*, 1.701–702). The Milky Way possesses an unexpected agency through which it actively seeks out its own observers and teaches them a lesson, forcing itself upon them. This realization leaves the poet stupefied, and he seeks in vain for an appropriate image through which to convey the circle's radiant appearance against the darkness of the night—it resembles a well-worn path that cuts through a green field, the foamy wake that a ship leaves behind in the sea, the bright rainbow that Iris traces through the clouds (1.714–717):

> sic super incumbit signato culmine limes
> candidus et resupina facit mortalibus ora,
> dum noua per caecam mirantur lumina noctem
> inquiruntque sacras humano pectore causas.

Thus the shining path lies overhead, marking heaven's high point, and turns the faces of mortals upwards, while they wonder at the strange light through the dark night and seek, with human mind, explanations of the divine.

The Milky Way towers up above, in the midst of an attack (*incumbit*), and physically prostrates those on earth, reminding them of their mortality (*resupina facit mortalibus ora*). This is not a metaphor. The Milky Way's unexpected power makes visible the normally hidden mechanisms of fate, the material force through which the stars control everything that lies below them.[12] But the Milky Way's overwhelming might simultaneously elevates

[12] Manilius does not discuss this material force explicitly, but other sources do. At *Tetrabiblos* 1.2.1–3, Ptolemy writes of the power (δύναμις) that emanates from the ether and is dispersed throughout the region surrounding the earth and declares that of the heavenly bodies the moon transmits the most powerful effluence (ἀπόρροια) to all that lies below.

the subjects' gaze and expands their minds as they start to think philosophically about the heavens (*inquirunt...causas*). The idea brilliantly reimagines the familiar contrast between animals that live their lives facedown and human beings who raise their eyes toward the heavens and so contemplate higher things.[13] This process constitutes a striving for transcendence, an endeavor through the limited resources of the human mind to understand the forces of divinity that, both textually and in the real world, surround and constrain us (*sacras humano pectore causas*).

To drive the point home, Manilius specifies that the entire experience is one of wonder (*mirantur*), which encodes the sublimity of both the Milky Way and the intellectual project that it sets in motion.[14] In Ch. 1.6 ("Plato and Aristotle"), we encountered the idea that wonder can lead to philosophy and the investigation of nature's causes. For Plato's Socrates in *Theaetetus* (155c–d), the experience of wonder (τὸ θαυμάζειν) is the hallmark of the philosopher, and an openness to vertiginous awe is the precondition without which philosophy cannot get off the ground. Socrates symbolizes the process with a genealogy: wonder (Thaumas) is the parent of the rainbow (Iris), which connects earth to heaven. Manilius has this passage in mind when he compares the Milky Way to a rainbow (*utque suos arcus per nubila circinat Iris*, 1.713) and then pivots to the proposition that wonder (*mirantur*) leads to natural philosophy (*inquirunt...causas*). The protreptic function of the scientific sublime is inscribed within a long-standing tradition that enriches the symbolic resonances of the present passage. To gaze at these strange lights through the darkness of heaven is to become aware of and to move upward toward the divinity that orders the cosmos; it is the study of astrology, the process of writing or reading the *Astronomica*.[15]

But Manilius is not content simply to encode the sublimity of celestial inquiry; he stages its dual logic, too, and at great length. After declaring that the Milky Way impels human spectators to search for explanations of the divine (1.714–717), the poet launches into open-ended speculation about

[13] The conceit appears already in Plato (*Rep.* 9.586a), a passage that Longinus reckons sublime (*Subl.* 13.1). For raising one's eyes toward heaven in Manilius and his contemporaries, see Rossetti (2017, 80–91) and note Man. 2.99–104.

[14] The careful reader perhaps hears the adjective *sublimis* in *super... limes*, placed at the opening and end of line 714.

[15] Hübner (1984, 250 n.376) and Schindler (2000, 233) note the connection with *Theaetetus*. At 225, Schindler well observes that Manilius "versucht also auf sprachlicher Ebene dieselbe Wirkung hervorzurufen, wie sie die «reale» Milchstraße auf den Betrachter hat." Note that Kant (2000, 140, from *Critique* §26) considers the Milky Way in his discussion of the mathematically sublime.

the phenomenon's *causae*: perhaps it is a crack in the celestial sphere (1.718–722), or the seam where the sphere's two halves join (1.723–728), or the scorched path where the sun used to move in former times (1.729–734), or the scorched path that Phaethon created on his wayward chariot ride (1.735–749), or a stain on the heavens left behind by milk spilled from Juno's breast (1.750–54), or the concentrated light of a dense patchwork of small stars (1.755–757), or the location of certain virtuous souls that have been deemed worthy of heaven after death (1.758–804).[16] The massive scale of the list explodes the format of Epicurean or Lucretian multiple explanations, and that sense of boundless and uncontrollable excess is precisely the point.[17]

This becomes clear when we consider the structure and details of the list of potential *causae*. The list begins as follows (1.718–722):

> num se diductis conetur soluere moles
> seminibus,[18] raraque labent compagine rimae
> admittantque nouum laxato tegmine lumen;
> quid sibi non timeant, magni cum uulnera caeli
> conspiciant feriatque oculos iniuria mundi?

Perhaps the whole structure's constituent elements have been broken apart and it is attempting to undo itself, and with the loosening of the joints cracks are giving way and letting in new light now that the roof has been opened up. What result would people not fear for themselves when they behold the wounds of great heaven and when the injury done to the *mundus* strikes their eyes?

The language is Lucretian,[19] but Manilius is also thinking of Virgil's description of sinking ships in *Aeneid* 1: "With the loosening of the side joints,

[16] For the individual theories and Manilius's possible sources, see Salemme (2000, 13–19) and Musso (2012, 91–135). The virtuous soul theory draws on Cicero's Dream of Scipio, among others; see Baldini Moscadi (1981, 52–55), Landolfi (1990), and Abry (2011). Among ancient theorizers, only Democritus (ATOM. D98 *EGP*, A91 D-K) appears to have identified the Milky Way's light with that of regular stars; note, however, Anaxagoras D4.10 *EGP* (A42.10 D-K; cf. A1.9).

[17] For Epicureanism and multiple explanations, see Ch. 1.6 ("Stoics, Epicureans, and the *nil admirari* Tradition"). From the perspective of Latin poetry, the methodology is esp. connected with Lucretius; see Hardie (2009, 231–263, on Man. at 245).

[18] Goold (1998), following Scaliger and Housman, unnecessarily emends *seminibus* to *segminibus*.

[19] Lucretius regularly describes the atoms as *semina*, refers to heaven's *tegmen* (e.g., 1.988, 2.662, 5.1016), and likens the collapse of the universe to that of a building (5.95–96). Similarly, the narrative of cosmic collapse at Lucr. 1.1102–1113 uses the imagery of architectural collapse (*ruo*) and rubble (*ruina*).

all [the ships] take on hostile water and gape at the seams" (*laxis laterum compagibus omnes / accipiunt inimicum imbrem rimisque fatiscunt*, 1.122–123). Manilius's cosmos is a foundering boat, and the flood that pours in through the cracks figures the Milky Way. If we read Virgil's fleet as political allegory,[20] the implication is that the universal ship of state is breaking up and that Rome cannot last. Either way, it's only a matter of time before we are all submerged or simply swept out into the cosmic sea. The terror of the earthly spectators (*timeant*) encodes the ominous sublimity of the image, which pushes the reader's mind upward and outward, beyond the limits of the *mundus*.[21]

As Manilius continues, a subtle but reassuring progression soothes the anxious reader. The first two hypotheses (crack in the heavens, celestial seam) suggest the perishability of the whole cosmos, while the following three (former path of the sun, Phaethon, Juno's breastmilk) admit only the possibility of superficial alterations to the appearance of the *mundus*, moving from the disastrous to the relatively insignificant. By contrast, the final two hypotheses (small stars, home of virtuous souls) simply declare that the Milky Way constitutes an ordinary part of the universe that has always existed and always will. At the end, rather than the sinking of the universal ship of state, we behold the dwelling place of blessed souls (*suumque habitantia caelum*, 1.760), from the heroes of the generation before the Trojan War to Julius Caesar and the recently (or soon to be) deified Augustus.[22] At this point, Manilius must correct himself, for these two spirits, along with Jupiter and Quirinus, dwell "higher up" (*altius aetherii quam candet circulus orbis*, 1.802): "That is the home of the gods, this the home of those who, resembling the gods in their own virtue, touch the nearest peak" (*illa deis sedes: haec illis, proxima diuum / qui uirtute sua similes fastigia tangunt*, 1.803–804). Caesar and Augustus, as catasterized divinities, live on the sphere of the fixed stars, just beyond the Milky Way.[23] The list closes on a note of serenity that emphasizes intellectual and political triumph, with poet and

[20] Luc. 1.498–503 reads the Virgilian passage allegorically; see the relevant notes in Roche (2009). For a satiric take, see Pers. 3.58–59.

[21] Manilius perhaps hints at the Stoic distinction between the whole (i.e., the cosmos) and the all (i.e., the cosmos plus the infinite void that exists outside of it); see Sext. Emp. *Math.* 9.332 (*SVF* 2.524).

[22] The tenses of the verbs are problematic and have implications for dating; see the discussion at Volk (2009, 141–145).

[23] For connections between this image and Augustan ideology, see Montanari Caldini (1981, 80–94, Landolfi (1990, 97), Flores (1991), and Salemme (2000, 64–74).

reader confidently surveying the outer limits of the Roman *mundus* and what it means to assimilate oneself to the divine, a form of that *homoiōsis theōi* that figures so prominently in Plato's *Theaetetus*. The movement from political implosion to imperial supremacy dramatizes the sublimity of celestial inquiry and equates Manilius's cosmic gaze with that of the Roman state. Through the verses of the *Astronomica*, the reader can learn to behold the *mundus* with all the authority and power of a Caesar.

At the same time, the extravagance of the entire narrative stakes out a position in an ongoing debate about the study of the stars. The most important literary context in which to read Manilius's Milky Way is the Aratean tradition. The immediate model for 1.714–717 is Aratus, *Phaen.* 469–479.[24] At the beginning of his account of celestial circles, Aratus imagines a second person singular "you" filled with wonder as you gaze at a night sky that is split all around by a broad circle (περὶ φρένας ἵκετο θαῦμα / σκεψαμένῳ πάντῃ κεκεασμένον εὐρέϊ κύκλῳ / οὐρανόν, 473–475). I have argued elsewhere (Glauthier 2022, 84–87) that the image encodes the sublimity of celestial study. Unlike Manilius, however, Aratus does not narrate the Milky Way's path through the constellations and does not consider its causes. Instead of losing himself in the vastness of the Milky Way, Aratus treats it as entirely practical—as a visible circle in the heavens, it provides readers with an analogy that allows them to visualize the actual celestial circles, which are important for keeping track of the progression of the seasons and predicting the weather. The wonder that the Milky Way inspires is subordinated to its utilitarian value.

A similar attitude prevails in the Latin Aratean tradition. In Cicero's *Aratea*, the Milky Way shines with excessive brightness and illuminates the cavernous expanses of the night (*nimio fulgens candore* at fr. 33.249, *late caeli lustrare cauernas* at fr. 33.252). But Cicero describes no wonder when he imagines a second person singular "you" looking up at the heavens (fr. 33.245–252). Moreover, Cicero corrects Aratus by pointing out that the Milky Way is not a perfect circle that traces a continuous path around the celestial sphere (fr. 33.250 vs. *Phaen.* 474–475). Cicero's concern for factual accuracy lends the passage a tone very different from that of both Aratus and Manilius.

The corresponding passage in Germanicus's *Aratea* is short and to the point (*Aratea* 455–458).[25] This Milky Way has a unique color, shines in the

[24] See Glauthier (2022, 89–92); to the bibliography cited there, add Wilson (2020, 144–145). For Manilius and the Aratean tradition, see Liuzzi (1988), Gee (2013, 116–122), and Glauthier (2017). For the worldview of the *Phaen.*, see Hunter (2008).

[25] The relative dates of Germanicus and Manilius remain uncertain. For the many similarities between the two poems, see Abry (1993, 179–183).

shadows (*mediis uia lucet in umbris*, 457), and is not smaller than the other circles, and that's all. The text eliminates Aratus's second person singular verbs and pronouns, which elides the opportunity for wonder, and mutes the markers of sublimity. To be sure, Germanicus's skyscape is full of energy, vibrant commotion, and blinding lights, all of which register as sublime, and several passages, such as the magnificent account of a winter storm (*Aratea* 290–305), vividly display nature's terrifying might. Germanicus can even encode the scientific sublime. At the start of the poem, thanks to the emperor's establishment of peace, "now there is time to raise bold faces toward the heavens and to come to know the stars and the various movements of the cosmos" (*nunc uacat audacis ad caelum tollere uultus / sideraque et mundi uarios cognoscere motus, Aratea* 11–12). The elevation of bold faces activates the same trope that Manilius so splendidly reconfigures at 1.714–717. Germanicus's verb, however, is significant: *uacat*, "there is time." In similar situations, Manilius, Seneca, Lucan, and others use a stronger impersonal verb that underscores the affective intensity and confounding nature of the experience of philosophical study and/or celestial inquiry: *iuuat*, "it is thrilling."[26] Although Germanicus's poem reverberates with markers of the sublime, he does not consistently imbue the study of the stars with the same affective intensity that, as we have begun to see, Manilius brings to bear. The contrast between these two visions of the Milky Way makes this clear, and it is all the more striking when we consider that Germanicus himself was a prominent member of the imperial household. By writing about the heavens, Germanicus asserts intellectual mastery over the stars, bringing them within the emperor's sphere of influence. And yet, from Manilius's point of view, the outlook of the *Aratea*, symbolized through the exposition of the Milky Way, is altogether practical and lacking in sublime elevation.[27]

From the perspective of late Augustan culture, the sublimity of celestial inquiry and the symbolic value of the Milky Way are contested. Even this

[26] For *iuuat*, see Ch. 4.2 ("Nero-Phaethon"). In his proem, Germanicus looks back to Lucretius, who prays for peace so that he and Memmius may study the nature of things (1.38–43). Lucretius's poem unfolds in a world of civil war, which fills the text with tumultuous energy. Germanicus, by contrast, situates his *Aratea* in calmer political times.

[27] Apropos of the political implications of writing about the stars in Augustan Rome, note the confluence of terms referring to astronomy, military power, and domination at Ov. *Fast.* 1.295–310 (e.g., *sub ducibus* [i.e., earlier writers on astronomical topics] *caelum metabimur illis, / ponemusque suos ad uaga signa dies*, 309–310). The activity is flagged as sublime, not least of all because the first humans to study the heavens possessed *sublimia pectora* (1.301). I return to celestial study as celestial conquest in Ch. 2.6.

brief consideration of the fragments of the *Aratea* of Cicero and Germanicus shows that the Latin Aratean tradition is not as invested in the sublimity of studying the heavens as Manilius, and when we read Manilius's Milky Way against those of Cicero and Germanicus, the *Astronomica* can be seen to strike back against this relatively narrow mindset and its cramped vision of celestial inquiry.[28] The narrative contains a surfeit of information, from the description of its path through the stars, to the list of similes that try to capture its majesty, to his enumeration of *causae* that might explain its nature and origin, to the names of individual souls that may or may not account for its radiance. This stream of data does not lead inexorably to any surefire conclusion, for Manilius does not actually endorse any particular hypothesis. By refusing to pass judgment on its true nature, he complicates a familiar phenomenon. What's more, since Manilius is not interested in keeping track of time and the Milky Way serves no function in Greco-Roman astrology, Manilius's narrative appears to go nowhere. Despite all his knowledge and all his theories, Manilius's Milky Way remains fundamentally non-utilitarian and unknowable; all we can do is stare at it in rapt wonder, a symbol of imperial conquest and control.

The Milky Way narrative encapsulates Manilius's aesthetics of the infinite. By developing one perspective after another on the same topic, the poet creates the impression of the infinite multiplicity and never-ending expanse of the *mundus*. This is Manilius's basic approach to the cosmos throughout the *Astronomica*. Just as the accumulation of detail defamiliarizes the Milky Way and allows the reader to transcend the boundaries of earthly life in search of the divine, which is simultaneously Roman and imperial, the poem's ceaseless elaboration of astrological minutiae defamiliarizes the constellations and invests them with an aura of infinite greatness, magnitude, and power, all of which flows to readers, elevating their gaze and expanding their consciousness. The result is all-encompassing but not logically consistent, totalizing at a distance but fractured up close and consisting of incompatible parts. The experience of the sublime, represented by the experience of the Milky Way, matters more than the intellectual values that we tend to think of as necessary for the functioning of philosophical discourse.

[28] In Glauthier (2017), I develop a similar argument at greater length, focusing on Man. 1 but not from the perspective of the sublime.

2.3. Phaethon and the Celestial Daredevils

The Milky Way exposition includes a representation of this very approach to celestial inquiry, Phaethon's chariot ride.[29] While drawing on the Phaethon narratives of Lucretius and Ovid (see Ch. 1.5, "Phaethon"), Manilius assimilates himself to the exuberant hero and configures his own brand of celestial inquiry through that most conspicuous exemplar of the fall.

One of Manilius's favorite tropes for studying the stars is the celestial journey.[30] At the opening of the poem, Manilius announces that the *mundus* itself encourages students to examine it up close ("now the universe shows favor to those who probe it up close," *iam propiusque fauet mundus scrutantibus ipsum*, 1.11). He celebrates the thrill that comes from traveling through and studying the heavens ("it is thrilling [*iuuat*] to go through the air itself and to live roaming about in the boundless sky and to get to know the constellations and opposing movements of the planets," 1.13–15).[31] At the opening of his second book Manilius advertises his independence from tradition and proudly proclaims that he soars into the heavens in a solitary chariot ("in a solitary chariot, we fly into the heavens," *soloque uolamus / in caelum curru*, 2.58–59). In such passages, the celestial journey represents the sublimity of celestial study by raising the poet aloft and bringing him into physical proximity with the divine, which is the subject of his song.

Manilius's account of Phaethon's fiery joyride picks up on this tradition (1.735–739, 743, 740–741):

> fama etiam antiquis ad nos descendit ab annis
> Phaethontem patrio curru per signa uolantem,
> dum noua miratur propius spectacula mundi
> et puer in caelo ludit curruque superbus
> luxuriat nitido, cupit et maiora parente,
> deflexum solito cursu, curuisque quadrigis
> monstratas liquisse uias orbemque recentem
> imposuisse polo.

[29] Note that my discussion here builds off Glauthier (2022, 96–100).

[30] See Volk (2001, 86–92; 2002, 225–234; 2003) and Landolfi (2003, 11–28). Kuhn-Treichel (2020) approaches the theme from the perspective of cognitive studies.

[31] For significance of *iuuat* here, repeated at 1.17, see Ch. 4.2 ("Nero-Phaethon").

There is also a story that comes down to us from long ago, that Phaethon, flying through the constellations in his father's chariot, while he wondered at the strange spectacles of heaven from a new proximity and played childishly in the sky and proudly reveled in the shining chariot and longed to surpass his father, veered from the customary path and, with his swerving team, abandoned the way he had been shown and imposed a new circle upon the heavens.

Phaethon's celestial journey (*curru per signa uolantem*) resembles Manilius's own celestial journey in book 2 (*soloque uolamus / in caelum curru*, 2.58–59). In the process of encoding the sublimity of Phaethon's experience (*dum noua miratur propius spectacula mundi*), the text echoes the sublimity of staring at the Milky Way (*dum noua per caecam mirantur lumina noctem*, 1.716) and Manilius's own claim, from book 1, that the *mundus* itself encourages close examination (*iam propiusque fauet mundus scrutantibus ipsum*, 1.11).[32] Phaethon figures the experience of studying the heavens through the text of the *Astronomica*.

From this perspective, certain differences between Manilius's account and those of Lucretius and Ovid acquire significance.[33] In particular, Lucretius and Ovid make the horses drag Phaethon off course, which leads to disaster (Lucr. 5.397–398, Ov. *Met.* 2.161–168, 201–209; cf. Diod. Sic. 5.23.2). But Manilius's Phaethon leaves the beaten track of his own accord (*Phaethontem... deflexum*[34] *solito cursu... monstratas liquisse uias*), which makes him alone responsible for the creation of something entirely new, the Milky Way (*orbemque recentem / imposuisse polo*). With this in mind, the customary path that Phaethon abandons metamorphoses from the path of the sun into that of song, a metapoetic commonplace with which Manilius is familiar.[35] In this context, the idea that Phaethon undertakes his journey in a spirit of intergenerational rivalry (*cupit et maiora parente*) hints at the sense of *aemulatio* that fuels Greek and Roman literary production. Manilius's Phaethon, then, is a poetic and programmatically innovative investigator of the night sky whose literary adventures have a lasting impact on the universe—he is very much at home in the late Augustan world, where Phaethon is a potent symbol.

[32] Schrijvers (1983, 145) connects the *dum* clause at 1.737 with 1.11.

[33] For an extended analysis with a different emphasis, see Kyriakidis (2018).

[34] *OLD* s.v. *deflecto* 2a classifies *deflexum* here as middle. The *Thesaurus Linguae Latinae* (5.1.357.66) takes it as reflexive.

[35] For the Pindaric/Callimachean path of song, see Asper (1997, 21–107. For Manilius's use of the trope, see 2.49–52 and Volk (2002, 212–213; 2009, 202–203).

If Phaethon's flight encodes the sublime—if he experiences it as sublime (*miratur...spectacula*, 1.737)—his cataclysmic crash-and-burn fulfills the same function. According to the logic of the fall, sublimity demands a willingness to run risks that often results in spectacular failure, which is regularly construed as a fall. That fall, however, is itself sublime, evidence of greatness of thought and expansiveness of mind, a spectacle that overwhelms and elevates. Conspicuously, Manilius insists that we should not complain about the devastation that Phaethon unleashed: "Why do we lament that flames raged over the entire world and that the earth became a funeral pyre and burned through all the cities?" (*quid querimur flammas totum saeuisse per orbem / terrarumque rogum cunctas arsisse per urbes*, 1.744–745). If there is no cause for sorrow, perhaps there is reason to celebrate. At the end of the narrative, the poet emphasizes the enduring legacy of the fall itself: "Neighboring stars blazed with new flames and now, too, bear the mark of a past fall" (*uicina nouis flagrarunt sidera flammis / nunc quoque praeteriti faciem referentia casus*, 1.748–49). Phaethon, the explorer of the night sky, has changed the world, and subsequent generations see the evidence of his *magnitudo animi* whenever they gaze at the stars.[36]

Manilius embraces this ethos. His Phaethon-style daring consists in his willingness to quickly shift intellectual modes, to explain the Milky Way now through the principles of natural philosophy, now with the help of mythological narrative, now by invoking social and political history—after emphatically declaring that Phaethon's fall has left a permanent mark on the heavens, Manilius breathlessly pivots to a conflicting explanation ("I must not conceal a well-known story," *nec mihi celanda est uulgata fama*, 1.750), and will offer up two more besides. This is no accident. Throughout the poem, Manilius makes room for a multiplicity of voices that cannot be reconciled, and his enthusiastic eclecticism bears interpretation through sublimity.[37] As we have seen (Ch. 1.5, "Acrobats, Heroes, and Flawed Sublimity"), Longinus scorns moderation, soundness, and that which is impeccable; perfect precision runs the risk of triviality, whereas true greatness contains items that must be overlooked (*Subl.* 33.1–2). Manilius represents

[36] The phrase *nunc quoque* (1.749) occurs regularly in aetiological narratives. For such aetiological tags, see Myers (1994, 66–67) and Glauthier (2020, 757 n.59).

[37] For other perspectives on Manilius's willingness to espouse positions and deploy modes of explanation that conflict, see esp. Lühr (1969, 96–110), Volk (2002, 196–245, esp. 207–208, 222, 233–234, 242, 245), Lehoux (2011), Volk (2009, 12–13, 111–115, 169–170, 251–258, 263–265; 2011), and Fanti (2019, 65, 70–71). Mann (2011) and Ramelli (2014) rely on Stoicism to resolve at least some apparent tensions in the poem.

a similar ideal at the level of content or basic methodological orientation. To put it simply, Manilius shows an inspired disregard for the principle of moderation and revels in the infinite accumulation of theories and ideas. To communicate successfully to humankind the majesty of the stars, one must traverse the entirety of heaven, cross all boundaries, smash all norms. The lessons might not always be consistent, the reader might not learn how to cast a nativity, but the experience will be unforgettable, and the night sky will never look the same. Phaethon, then, gives us a better purchase on Porter's (2016, 501) argument that "the coincidence of opposites [in the *Astronomica*] is meant to jar, to shock, and to display something of the world's immense grandeur" as well as "its unresolved complexity." Phaethon anchors this attitude firmly in Manilius's own self-representation, his aesthetics of the infinite.

Phaethon is just one example of the kind of bold daredevil with whom Manilius associates his particular brand of inquiry. The narratives of Salmoneus and Bellerophon (5.91–101) provide two additional models. For Manilius, Salmoneus is a charioteer who attempts to re create the sound of Jupiter's thunder by driving his four-horse team across a bronze bridge. In other sources, Salmoneus's thunder-play forms part of an explicit desire to imitate the king of the gods and/or to be honored as a god.[38] For Virgil, such a desire is a symptom of mental derangement (*diuumque sibi poscebat honorem, / demens, Aen.* 6.589–590); Salmoneus is an example of sublime subjectivity, *magnitudo animi*, gone awry. Manilius does not make this element of the story explicit, but rather states more abstractly that Salmoneus's crime was to have imitated heaven on earth (*caelum imitatus in orbe*, 5.91). Such divine role playing lowers the gods to ground level, a transgression that Manilius deplores in other authors (2.25–27) but celebrates when he himself is the agent (1.1–4). Perhaps, then, it should not be surprising if Manilius does not condemn Salmoneus outright. As a charioteer with heavenly associations and a well-known aspiration for *homoiōsis theōi*, Salmoneus resembles Manilius, and Salmoneus's transgression and ultimate fate are described in terms that resonate with other passages in the *Astronomica*.

Manilius states that Salmoneus "imagined that he had represented the sound of the heavens" (*expressisse sonum mundi sibi uisus*, 5.93). The literal reference is to Salmoneus's thunder-play, but *sonum mundi* suggests the

[38] Virg. *Aen.* 6.585–594, Apollod. *Bibl.* 1.9.7; cf. Hes. fr. 30 Merkelbach-West, Soph. fr. 10c *TrGF* (cf. frr. 537–539 *TrGF*).

music of the spheres.[39] At the beginning of the poem, Manilius contrasts his own singing, which obeys the fixed rules of meter, with the deafening roar of the *mundus* that literally surrounds the poet ("all around the poet who sings in accordance with a fixed law, the *mundus*, with its boundless sphere, resounds," *certa cum lege canentem / mundus et immenso uatem circumstrepit orbe*, 1.22–23). The sound of the *mundus* is the music of the spheres,[40] and the experience is both disorienting and ennobling, since few are those whose ears are attuned to such music.[41] Salmoneus, then, imagined that he had reproduced the music of the spheres, and who is to say that he did not?[42]

Salmoneus's death is suggestive as well: "While he was fabricating thunderbolts, he felt [real ones], and following after the fires that he himself had flung, he learned in death of Jupiter" (*dum fulmina fingit / sensit, et immissos ignes super ipse secutus / morte Iouem didicit*, 5.94–96). Salmoneus, like Phaethon, is thunderstruck (*fulmina...sensit*), a fate that literalizes a widespread trope for the experience of the sublime.[43] Moreover, he seems to become a thunderbolt himself, falling to earth in the wake of his own counterfeit flames. In other words, he becomes an agent of stupefaction, a sublime subject. Most noteworthy, though, is the revelation that Salmoneus "learned of Jupiter" (*Iouem didicit*). As we will see (Ch. 2.4), Manilius frames the goal of celestial inquiry as just this. In book 2, the subject of his song is god (*canam...deum*, 2.60–61), and in book 4, what the student seeks to know is god (*quod quaeris, deus est*, 4.390). On a first reading, Manilius's account of Salmoneus sounds ironic, but we do not have to take it that way. Thanks to his crime and through his fall, Salmoneus achieves a kind of wondrous transcendence.

Following hard upon Salmoneus is Bellerophon, who "put a road in heaven while flying through the constellations" (*imposuisse uiam mundo per signa uolantem*, 5.98). The phrase *per signa uolantem* assimilates Bellerophon to Manilius and Phaethon, in their capacities as explorers of the night sky, and *imposuisse uiam mundo* looks back to Phaethon's dazzling disaster

[39] Cic. *Rep.* 6.19 calls the music of the spheres *mundi...sonitus*, which seems a closer parallel than Virg. *Aen.* 6.586, cited by Housman (1903–1930, 5.12), where Salmoneus's thunder is *sonitus...Olympi*.

[40] So Schrijvers (1983, 148–150), Wilson (1985, 293–294), and Volk (2002, 235–236; 2009, 214–215).

[41] For the deafness of humans to the music of the spheres, see Arist. *Cael.* 2.9.290b and Cic. *Rep.* 6.19.

[42] Note that at Cic. *Rep.* 6.18, musicians are said to be capable of imitating the music of the spheres and thereby gaining reentry to heaven; cf. Zetzel (1995, 241).

[43] For thunderbolts and the sublime, see again Day (2013, 107–111).

(*orbemque recentem / imposuisse polo*, 1.740–741). Of course, Bellerophon soared through the sky on the back of Pegasus, but Manilius makes no mention of the winged horse or any of the hero's exploits. This silence allows Bellerophon to resemble Manilius and Phaethon more closely. Significantly, Bellerophon is guilty of the same transgression as Salmoneus, the confusion of earth and heaven ("the sky was his playground," *cui caelum campus fuerat*, 5.99), which leads to thoughts of his fall. In various sources, Bellerophon attempts to ride Pegasus through the heavens and/or up to Olympus, which invariably ends in disaster when Pegasus throws his master.[44] In Hyginus's telling, Bellerophon looks more than a little like Ovid's Phaethon: "While he was striving to fly to heaven and was no longer far off, he looked down at the earth, was struck with fear, and fell, and there he is said to have died" (*cum ad caelum contenderet euolare neque longe iam abesset, despiciens ad terram, timore permotus, decidit ibique perisse dicitur*, *Astron.* 2.18). Manilius's final image is tantalizingly imprecise: "Earth and sea were below his feet, his journey allowed no footprints" (*terraeque fretumque / sub pedibus, non ulla tulit uestigia cursus*, 5.99–100). Manilius writes Bellerophon's fateful fall out of the picture, leaving him to sail through the skies indefinitely. While Bellerophon might have tumbled all the way to earth or into the sea, some authorities identified him with the constellation Heniochus, the Charioteer, whose simultaneous rising with Aries is the subject of discussion at this point in the text (5.67–101).[45] To become a celestial body, a god, is an appropriately sublime flourish with which to cap Bellerophon's journey—Bellerophon falls upward, as it were, and those on earth can gaze at the evidence of his assimilation to the divine.

Phaethon, Salmoneus, and Bellerophon serve as models for celestial inquiry, not despite their transgressions and failures, but because of them.[46] While these individuals operate on an epic scale—they are attested as early as Homer and Hesiod—Manilius also invests a much humbler figure with the same symbolic resonances: the tightrope walker. The influence of Engonasin rising simultaneously with Pisces has the potential to turn the native to *artes* (5.650). Such an individual will devote himself to

[44] Pindar has Pegasus throw Bellerophon but is silent regarding his death (*Isth.* 7.44–47, *Ol.* 13.91–92). In Euripides's *Bellerophon*, the hero survives the fall long enough to deliver a speech (frr. 306–311 *TrGF*). Homer does not mention this adventure, leaving Bellerophon to wander the earth hated by the gods, shunning the paths of men (*Il.* 6.200–202).

[45] Goold (1997, 308) cites Schol. vet. in Arat. *Phaen.* 161.

[46] Elsewhere, Manilius likens celestial inquiry to Gigantomachy, which is a parallel phenomenon; see Volk (2001, 106–114).

"precipitous" pursuits (*in praerupta dabit studium*, 5.651) and will sell his natural abilities in exchange for danger (*uendetque periclo / ingenium*, 5.651–652). In particular (5.652–655):

> tenuis ausus sine limite gressus
> certa per extentos ponet uestigia funes
> et caeli meditatus iter uestigia perdet
> paene sua et pendens populum suspendet ab ipso.

Daring narrow steps without a path, he will place sure footsteps along extended ropes. Practicing a path to heaven, he will lose his footing—almost!—and, suspended, keep the crowd hanging in suspense upon himself.

The bold tightrope walker, hovering in midair, holds the audience in suspense (*pendens populum suspendet ab ipso*), simultaneously immobilizing them with anxious anticipation and directing their minds upward and outward; the image succinctly encodes the sublime subjectivity of the acrobat and the sublimity of the spectacle itself.[47] The constant expectation of a fall from great heights, the excruciating oscillation between equilibrium and a fatal loss of balance, makes the tightrope walker a potent symbol of sublimity, representing the sublime orator in the Younger Pliny (*Ep.* 9.26.3) and the sublime poet in Horace (*Epist.* 2.1.208–213).[48]

In the *Astronomica*, the image works on multiple levels. Throughout the poem, stars hang (*pendere*) in the night sky, which makes the tightrope walker a heavenly body and the crowd a collection of stargazers.[49] The audience's reaction to the show recalls the wonder of primitive humans who were stupefied and kept in suspense, both grief-stricken and joyous, every morning when the sun rose anew (*stupefacta* [sc. *uita*] *nouo pendebat lumine mundi / tum uelut amisso maerens, tum laeta renato*, 1.68–69). Since the tightrope walker practices a journey to heaven (*caeli meditatus iter*),

[47] Hübner (2010, 2.382) reads the acrobat's near fall as a deliberate maneuver; the funambulist controls the audience.

[48] For the Younger Pliny, see Ch. 1.5 ("Acrobats, Heroes, and Flawed Sublimity"). Horace's acrobatic poet moves *per extentum funem* (*Epist.* 2.1.210), which Manilius's *per extentos…funes* (5.653) echoes. When Horace envisions a performance by such a poet, he describes an experience of affective intensity that operates according to the dual logic of the sublime (*meum qui pectus inaniter angit, / irritat, mulcet, falsis terroribus implet…magus ut*, *Epist.* 2.1.211–213). For echoes of Gorgias's *Helen* here, see Brink (1982, 231–236).

[49] Thus the stars hang in the sky (*pendentia sidera caelo*, 1.529) and the planets hang between the celestial sphere and the earth (*terram caelumque inter uolitantia pendent*, 1.806).

he is also a student of the night sky, a Phaethon, a Salmoneus, a Bellerophon—a Manilius. In book 4, Manilius describes himself as "hanging in midair" (*pendentem*) on a journey to the stars ("[the steps] are able to lead the poet, hanging in midair, along a curved path to the stars," *qui* [sc. *gradus*] *ducere flexo / tramite pendentem ualeant ad sidera uatem*, 4.120–121).[50] In book 2, a chariot ride to heaven (2.136–140) leads to a self-portrait where Manilius plays the performer whose audience is heaven itself: "I will sing things for the heavens to know, while the stars look on in wonder and the *mundus* exults in the song of its own poet" (*caelo noscenda canam, mirantibus astris / et gaudente sui mundo per carmina uatis*, 2.141–142).[51] Through the sublime acrobat, then, Manilius unexpectedly assimilates himself to someone who performs tricks in the street and literally sells his *ingenium*. Of course, this is how and where astrologers plied their trade in the Roman world.[52] From this perspective, Manilius invests the day-to-day life of the practicing astrologer with the thrill of the sublime, the electric shock that comes from the pursuit of the divine.[53]

In his tireless search for self-representation, Manilius careens wildly from the stylistic heights of weighty mythological exempla to the lowly level of popular entertainment and everyday tricks. In his quest for celestial knowledge, he darts from one end of the cosmos to the other, collecting and relaying a seemingly infinite amount of data before suspending judgment. The scope of his movements and his total disregard for all sense of balance and propriety simply take one's breath away.

2.4. Mathematical Poetry and the Revelation of God

Phaethon, Salmoneus, Bellerophon, and the tightrope walker symbolize Manilius's daredevil approach to celestial inquiry. But Manilius is explicit about what the study of the stars looks like: it is mathematical and takes

[50] For these lines, which editors since Bentley have bracketed, see Volk (2004).

[51] The image suggests Orpheus singing to the animals. More specifically, the combination *mirantibus...gaudente* recalls *gaudet...miratur* in the reaction to Silenus's song at Virg. *Ecl.* 6.29–30, where Silenus himself is contrasted with Orpheus; see Ch. 1.1. I have discussed Manilius's performance for the heavens in a different context at Glauthier (2011, 199–200).

[52] Elite literature repeatedly represents astrologers as selling their predictions from stalls in the Circus; see Volk (2009, 129 n.4).

[53] For a complementary reading of the tightrope walker, see Hübner (1984, 194–196). He notes that Scaliger connects this scene with the effect of the Milky Way at 1.715.

place in verse. The poem's elaborate and expansive use of numbers and calculations runs counter to traditional aesthetic norms, which makes it a risky proposition. It's also transgressive for another reason. For Manilius, mathematical poetry about the heavens lays bare information that surpasses human understanding and yet can be mastered by the student. Programmatic statements suggest that Manilius's "eminent aptitude for doing sums in verse" stages the dual logic of the sublime and cultivates in the reader a *magnitudo animi* that not only reveals god but elevates the student to the level of the divine. Through the sublime, Manilius transforms his eminent aptitude from a tedious liability into an astounding asset.

At the beginning of book 3, which is the most mathematically dense in the poem, Manilius lists mythological and historical themes that he will not treat in verse.[54] "It is commonplace," he declares, "to write poetry about pretty things and to compose an uncomplicated work" (*speciosis condere rebus / carmina uulgatum est, opus et componere simplex*, 3.29–30). Manilius rejects prettiness and simplicity in favor of an aesthetic program grounded in the sublimity of mathematical calculations.[55] He makes his position clear in the following boast (3.31–34):

> at mihi per numeros ignotaque nomina rerum
> temporaque et uarios casus momentaque mundi
> signorumque uices partesque in partibus ipsis
> luctandum est. quae nosse nimis, quid, dicere quantum est!

But I must struggle through numbers and the unknown names of things and the seasons and the various fortunes and movements of the heavens and the successions of the signs and degrees in degrees themselves. Why, how great a task it is to speak of what it is too much to know!

Manilius must "struggle" (*mihi... luctandum est*) through a series of difficult mathematical topics (*numeros*), extending from the calculation of infinitely large sums (*momentaque mundi*) to the infinitesimal division of degrees of the ecliptic (*partesque in partibus ipsis*). The language likens the poet to one

[54] For the proem, see Volk (2002, 214–215, 241–242; 2009, 208–210) and Landolfi (2003, 61–76).

[55] A distinction between beauty and sublimity is fundamental to Burke (e.g., "The Sublime and the Beautiful Compared," 1998, 157–158) and Kant (e.g., "Analytic of the Beautiful" vs. "Analytic of the Sublime," 2000, 89–127 vs. 128–159). In ancient contexts, a similar distinction sometimes occurs (Porter 2016, 402–403, 550–551), but beauty and sublimity can and do overlap (Porter 2016, 33–34, 566–567).

of the planets, which "struggle" through the sky against the daily revolution of the celestial sphere ("heavenly bodies struggling against the firmament that turns in the opposite direction," *aduerso luctantia sidera mundo*, 1.259, 670).[56] The metaphor characterizes the arduousness of Manilius's task as cosmic in scope and eternal in duration. Significantly, the topics themselves and the poet's treatment of them are transcendent, surpassing human understanding (*quae nosse nimis*). Through mathematics, Manilius will convey infinite, impossible knowledge.

As he continues, he stresses that readers must expend great effort to learn the lesson and that they should not expect "sweet songs" (*impendas animum; nec dulcia carmina quaeras*, 3.38). Several critics have compared this entire passage to two well-known passages in *De rerum natura*, one where Lucretius reflects on the difficulty of presenting Greek material in Latin verse (1.136–139),[57] and another where he characterizes his poetic medium as the honeyed cup that allows the reader to swallow his bitter teachings (1.936–950; cf. 4.11–25).[58] One possible conclusion is that Manilius distinguishes himself from Lucretius by insisting that the difficulty inherent in treating astrology surpasses the difficulty involved in expounding Epicurean physics to such an extent that Manilius is forced to give up any hope of rendering his poem sweet. I propose a different approach. As the poet of *numeri* and *partes in partibus*, Manilius sets himself in opposition to an entire tradition of Latin poetry that presents itself as *dulcia carmina*. Lucretius's decision to smear his verses with the sweet honey of the Muses (*quasi musaeo dulci contingere melle*, 1.947; cf. 4.22) demonstrates that *dulce* is an appropriate epithet for his own brand of hexameter poetry. Manilius, by contrast, disdains sweetness. His tone is not apologetic or defensive but defiant. There will be nothing sweet or pretty or simple in Manilius's mathematical song; it will be complicated and difficult, and therefore truly sublime.

The mastery of numbers elevates the poet of the stars and turns him into a sublime subject. In book 4, Manilius catalogs the ways in which each of the zodiacal signs controls the character traits and professional lives of those born under its influence. Here, we learn that Gemini produces poets. Manilius's initial description makes such individuals resemble the bucolic

[56] For planetary motion as "struggle," see Volk (2009, 45).
[57] Lühr (1969, 38) and Volk (2002, 242).
[58] Effe (1977, 113–14), Neuburg (1993, 258), Schindler (2000, 250), Volk (2002, 242), Landolfi (2003, 63–64), and Green (2014, 29).

shepherd and the elegiac lover (4.152–57), familiar literary clichés. Then Manilius adds a twist (4.158–160):

> inueniunt et in astra uias numerisque modisque
> consummant orbem postque ipsos sidera linquunt:
> natura ingenio minor est perque omnia seruit.

They also discover paths to the stars and, with numbers and measurements, add up the world and then leave the stars behind them. Nature is weaker than their intellect and is made their slave in all things.

These poets undertake the same journey through the heavens as Manilius himself, and they do so through their mathematical operations. In a poetic context, *numerus* and *modus* usually refer to rhythm and meter. In book 1, after he declares that it is a thrill to travel through the sky, Manilius announces that it is even more deeply thrilling "to render into meter (*numerum*), with Phoebus regulating the rhythm (*modulante*)," what he knows about the stars (*in numerum Phoebo modulante referre*, 1.19; cf. 2.766–767). Poets born under Gemini are particularly well suited to their task because the *numeri* and *modi* with which they carry out their calculations cannot be distinguished from the *numeri* and *modi* with which they compose their verses (cf. Kennedy 2011, 174–175, 179). This intimate connection between poetic form and the mathematical nature of the *mundus* makes such poets particularly potent. By "adding up" the contents of heaven or by "calculating" various astrological values, they "perfect" the world around them (*consummant orbem*). At one level, this means that their "cosmic sum" reveals the mathematical regularity of the *mundus*.[59] But the natural world is also inferior to their minds (*natura ingenio minor est*), which means that they create something better or higher than nature, and that in turn reduces nature to the status of a slave (*perque omnia seruit*). This is not the soft pursuit of bucolic or elegiac poetry (*mollius e Geminis studium est*, 4.152)—it is an exercise in cosmic overcoming, in rising above the limits of nature and displacing the stars as the most powerful figures in the universe. The power that such poets wield through mathematics has the potential for shock and awe, the ability to control and to dominate; it is the very essence of

[59] I take "cosmic sum" from Hübner's (1984, 191) "kosmische Summe." Trying to capture the nuances of *consummo*, Kennedy (2011, 174) writes that Manilius's mathematical poets "give a full account of the heavens."

Longinus's vision of the sublime (see Ch. 1.6, "Longinus and the Rhetorical Tradition," on *Subl.* 1.4, 15.9).

Manilius's portrait of Gemini contains traditional elements. Ancient sources regularly identify the twin figures of Gemini as Apollo and Hermes, the god of song and the *primus inuentor* of astrology, respectively, and frequently ascribe poetic and/or astrological interests to those born under the influence of this sign.[60] More than that, though, Manilius's sketch reflects astrology's close connections with mathematics and the simple fact that in Latin the practicing astrologer was a *mathematicus.* Manilius implicitly claims this role for himself, and we should see in his account of Gemini another defiant declaration of the sublimity of his mathematical project.[61]

But in what sense does Manilius claim to displace nature and the stars? The key lies in properly framing the subject of the *Astronomica.* Near the end of the account of decans (4.294–407), the student bursts into the text complaining about Manilius's instruction. The student begins by stating that Manilius bids him endure an extensive and subtle task (*multum… tenuemque… laborem*, 4.387), extensive because of the celestial dimensions involved, subtle because of the minute sums to be calculated.[62] The student then protests that Manilius has plunged him back into mental darkness just when he thought he was beginning to see the light thanks to an easy calculation (*cernere cum facili lucem ratione uiderer*, 4.389). No calculation is ever easy; the mathematics of the poem has overwhelmed the pupil.[63] To this outburst the poet offers an august rejoinder (4.390–392):

> quod quaeris, deus est: conaris scandere caelum
> fataque fatali genitus cognoscere lege
> et transire tuum pectus mundoque potiri.

What you are investigating is god. You endeavor to ascend to heaven and to come to know fate, even though you are born subject to the law of fate, and to pass beyond your own body and to master the universe.

[60] See Feraboli, Flores, and Scarcia (1996–2001, 2.329–331). Although Virgo regularly carries similar associations, Manilius's sketch makes those born under her influence natural philosophers (4.189–196).

[61] For the relevance of the vignette to Manilius's own project, see Hübner (1984, 191), Abry (2002, 71–72), and Volk (2002, 220; 2009, 108 n.109).

[62] The student has internalized Manilius's own rhetoric (e.g., the dodecatemories are "slight in appearance, great in importance," *tenuem uisu rem, pondere magnam*, 2.693).

[63] Green (2014, 41) well comments that lines 4.387–389 "reveal first-hand the student's own reflection on the entire learning experience since the beginning of the poem." The experience, I would add, is specifically that of mathematical *ratio.*

Manilius operates from a philosophical perspective that equates god, nature, fate, reason, and even the universe as a whole.[64] Here, the subject of the poem is god (cf. 2.60–62). Manilius expands this conceit with parallel thoughts: to ascend to heaven is to transcend our own materiality, and to learn the rules of fate is to gain control over the world in which we live. Such thoughts activate the aesthetics of the infinite, enlarging the dimensions of the human to the scale of god, denying any imposition of limits on human potential. A few lines later, Manilius frames the matter slightly differently: "Man must expend himself so that god may be able to dwell in him" (*impendendus homo est, deus esse ut possit in ipso*, 4.407).[65] These kinds of assertions erase the distinction between the divine and the human, between that which is up above and that which is down below. The mind-boggling play with proportion and magnitude—the human expands to the dimensions of the *mundus*, the *mundus* shrinks to the size of the human—takes the reader through the looking glass and represents the process of celestial inquiry as the pursuit of impossible knowledge and the actualization of fantastical transformations. The implicit goal is *magnitudo animi* and assimilation to god, and it will happen through a mathematical poetry whose calculations are neither easy nor sweet.

The following section of the poem continues in the same vein. Treatment of the *partes damnandae* (4.408–501) requires Manilius to list by name the individual degrees of each sign that exert a negative influence due to cold, heat, or an improper amount of moisture (4.411–414). Eventually, he names a total of 102 degrees; the text of the *Astronomica* is defined by seemingly endless repetition and mathematical excess. But the introduction to the topic is revealing. It is not enough to understand how the decans work, since all the signs rise with a mixture of powers and varied texture (4.414–415). The basic principle is that nothing in heaven or on earth is *aequale*, "equal, uniform, balanced, consistent" (*est aequale nihil*, 4.416), which requires Manilius to go through number after number after number (*tot numeros totiens...referre, / tot partes iterare...tot dicere summas*, 4.431–432). The repetition and mathematical excess of the cosmos are reflected in the repetition and mathematical excess of the *Astronomica* itself. Manilius's slogan

[64] Helpful discussions include MacGregor (2004) and Volk (2009, 217–226). The position has Stoic associations but is not confined to Stoicism; see Ch. 1.6 ("Stoics, Epicureans, and the *nil admirari* Tradition"). The modes of transcendence outlined here and elsewhere in the poem have striking parallels in Hermetic texts; see Salemme (2000, 21–26) and Volk (2009, 234–239).

[65] A similar idea appears at 2.107–108. For the thought that man must not simply "expend himself" but "weigh himself out in payment" (*impendendus*), see Glauthier (2011, 200–201).

est aequale nihil, which he repeats later in the context of eternity (*immenso nihil est aequale sub aeuo*, 4.821), reflects the same basic value as Longinus's preference for flawed sublimity over τὸ σύμμετρον, "equality, proportion, moderation" (see Ch. 1.5, "Acrobats, Heroes, and Flawed Sublimity"). It is no surprise, then, that the numbers of the *partes damnandae* allow Manilius to reveal the divine (*ostendisse deum*, 4.439).

There is, however, another way in which the student of the stars approaches god. The system of dodecatemories divides each 30° sign into 12 units of 2.5°. The doctrine is essential because it is one of several systems that collectively explain why individuals born under the same sign can have different fates, a question that ancient skeptics of astrology routinely raised.[66] In this case, when the practitioner considers the location of a given planet among the signs, he must consider both in what sign and in what *dodecatemorium* of that sign it is located, since both will affect the influence of the planet. The infinite variety of human life is reflected in the infinite divisions of the heavens. For Manilius, this all constitutes part of a grand cosmic plan (2.701–706):

quae mundi conditor ille
attribuit totidem numero fulgentibus astris,
ut sociata forent alterna sidera sorte,
et similis sibi mundus, et omnia in omnibus astra,
quorum mixturis regeret concordia corpus
et tutela foret communi mutua causa.

The great founder of heaven distributed [the dodecatemories] to the shining signs, equal in number, so that the constellations might be united by alternating allotment and heaven might resemble itself and all the signs [might be present] in all [the other signs], so that through their mixings harmony might rule the body [of the zodiac] and there might be mutual guardianship based in common cause.

The passage is tied together by the language of community building (*conditor*) and political alliance (*sociata, concordia, tutela, communi causa*), which overlaps with Stoic terminology that denotes cosmic sympathy and interconnectedness (*mixturis, con-cordia, com-muni*).[67] Geometrical

[66] For criticism of astrology in antiquity, see Long (1982).

[67] For the Stoic use of συν-/*con-* compounds, see Lapidge (1989, 1383–1384, 1388); he cites several examples from Manilius at 1393–1397. For Manilius's political language in general, see Gladhill (2016, 97–118).

knowledge, the ability to divide and distribute the signs of the zodiac, allows one to perceive the order that lies behind the apparent chaos of everyday life. This is more than just theory or speculation. Cosmic order is real, the dodecatemories are actual objects that exist in the heavens, created by the *mundi conditor*. By calculating and mapping out the dodecatemories, the practitioner reproduces the original creative activity of god, and the goal is precisely the same, to produce order. The astrologer, in a sense, becomes god. This is the ultimate act of transcendence, the perfection of the world, the enslavement of nature. It is what Phaethon, Salmoneus, and Bellerophon all attempt in their own ways.

There are two important contexts for making sense of Manilius's claims about the power of his mathematical song and its transgressive nature. For Manilius, the study of the stars is simultaneously religious and philosophical.[68] More specifically, the enthusiastic attempt to explain all things, including god, through mathematics and the treatment of mathematical expertise as a means of transcendence have associations with Pythagorean and Platonic traditions that were in vogue in certain circles in the late Republic and early Empire—one thinks, for instance, of Nigidius Figulus and the Sextians.[69] Such movements existed on the margins of Greco-Roman religious and intellectual life, and Manilius is happy to locate himself in a similarly liminal cultural space. Indeed, Manilius's claims about the extraordinary potency of his art potentially characterize him as an occult figure who practices magic rather than as a mainstream philosopher who studies nature or certain methods of divination.[70] The author of the so-called Mithras Liturgy, for instance, strikes a similar pose. At the beginning of the spell, the magician requests temporary immortality "so that I alone may travel through heaven as an investigator and explore all things" (ὅπως ἐγὼ μόνος αἰτητὴς οὐρανὸν βαίνω καὶ κατοπτεύω πάντα, *PGM* IV.484–485); subsequently, he draws in divine breath, becomes a star, traverses the cosmos, and converses with the great

[68] MacGregor (2004, 149) rightly states that Manilius's math is "more than" Housman's knack for doing sums in verse: "This is theology." Volk (2009, 211–215) discusses Manilius's "sacralization" of poetry.

[69] For Pythagoreanism at Rome, including Nigidius Figulus and Q. Sextius, see Flinterman (2014). For Platonism/Pythagoreanism in Manilius, see MacGregor (2005, 59–65), Habinek (2007, 234–235), and Volk (2009, 239–246).

[70] For Manilius's positive representations of magic and the occult, see Volk (2009, 246–251). For astrology as magic, including a discussion of Manilius, see Edmonds (2019, 236–268). Hor. *Carm.* 1.11.1-3 nicely demonstrates how one might disparage and thereby marginalize astrology by emphasizing its use of mathematics (*Babylonios… numeros*).

god himself.[71] In the masterful hands of the magician, the flight of the mind and the view from above are literalized, and the result resembles Manilius's own self-representation. Astrology does not have to present itself in this way; the opening chapters of Ptolemy's *Tetrabiblos*, for one, read quite differently. Manilius's mathematical dexterity contributes to a self-portrait that positions him above and beyond the ordinary and the mundane.

The second context for thinking about Manilius's depiction of mathematical poetry is that of generic decorum. The Homeric epics sanction the limited deployment of numbers and the bare intimation of simple calculations. The numbers involved tend to be formulaic,[72] and the point is often to convey largeness or greatness rather than a specific sum. The Homeric catalog of ships (*Il.* 2.494–759) is the prime example of numbers and implied calculations that work in this way.[73] The catalog contains the names of the leaders of each Achaean contingent that sailed to Troy and the number of ships in that contingent. Paradoxically, it is the citation of specific figures that conveys the general sense of largeness: "of these, fifty ships went forth" (2.509), "with these, thirty hollow ships were marshalled" (2.516), "forty black ships followed along with these" (2.524), etc. Theoretically, one could crunch the numbers, taking 120 and 50 to be maximum and minimum figures for each ship's crew (2.510, 719) and then multiplying by the total number of ships (1,186)—this, in fact, is what Thucydides does (1.10.3–5).[74] But that is the approach of historians and philologists, not members of a live audience listening to an oral performance. The numbers vouchsafe the reliability of the narrative while simultaneously gesturing toward the uncountable greatness of a sum that remains forever hidden and inaccessible.

Manilius, by contrast, performs actual calculations, over and over again, and his numbers are by no means formulaic. If the limited use of mathematics can stage the sublime in hexameter poetry, Manilius dramatically sets fire to this principle. To a certain extent, we can read this maneuver against the Hellenistic interest in mathematical poetry. High-powered mathematical figures such as Archimedes, in his *Cattle Problem*, and Eratosthenes,

[71] I cite the text from Preisendanz and Henrichs (1973, 88). The papyrus that contains the well-known spell (*PGM* IV) dates to the fourth century CE.

[72] Rubincam (2003, 462) defines formulaic numbers as "the numbers 3, 5, 7, 9, 10, 12, and their respective multiples in the series of 10s, 100s, and 1,000s."

[73] See Leventhal (2015, 214–216). Ford (1992, 72–79) connects Homer's Muses, invoked immediately before the catalog proper (*Il.* 2.484–493), with Kant's mathematically sublime.

[74] Thucydides rounds the total number of ships up to 1,200 but does not perform the calculation. For discussion, see Kim (2010, 40–42).

in his verses about doubling the cube, composed poems that asked readers to perform wildly complicated mathematical tasks, conjuring up, in the words of Reviel Netz, "a sense of dazzlement, of the carnivalesque."[75] As we will see, Manilius indulges in some arduous calculations, and his self-presentation as a mathematically dexterous daredevil implicitly channels the ludic element that Netz identifies as central to this tradition of Hellenistic mathematics. But even with Archimedes and Eratosthenes in the background, Manilius remains Phaethon-like in his ostentatious disregard for the aesthetic norms of Latin hexameter poetry; Lucretius and Virgil have no time for decans and dodecatemories.[76]

The passages and contexts that we have been considering point in the same direction: Manilius expects the mathematical operations of the *Astronomica* to register as sublime. By reading the poem and performing its calculations, we become defiant daredevils, astrological magicians, and poetic outsiders, darting boldly across the literary sky.

2.5. Calculating the Horoscope

Manilius repeatedly advertises the sublimity of his mathematical poem, promising the reader an experience of transcendence and the knowledge of god. How does this work? The contemplation of unimaginably large and impossibly small magnitudes and the performance of a never-ending series of mathematical tasks, including the accumulation and evaluation of alternative methods for calculating the same sum, give the reader an intimation of infinity. The result is an experience of stupefaction that is also an experience of the divine. The most extended and dramatic example of Manilius's modus operandi is the calculation of the horoscope (3.203–509), a passage that has not received much sympathetic attention.

The horoscope is defined as the degree of the ecliptic that is rising above the horizon at the moment of birth. Without accurately and precisely calculating this value, the astrologer cannot perform numerous basic tasks.

[75] Netz (2009, 58). Archimedes's *Cattle Problem* sets out conditions for calculating the number of Helios's cattle; per Netz 34, the simplest solution involves a number that contains 206,545 digits. For further discussion, see Benson (2014) and Leventhal (2015). For Eratosthenes's poem, see Netz 160–166 and Leventhal (2017).

[76] Hor. *Ars* 325–332 furnishes a limited model for the simultaneous rejection of and indulgence in mathematical poetry. For an opposition between Manilius's aesthetic program in book 4 and Horace's *Ars*, see Landolfi (2003, 70–71).

Accordingly, Manilius sets the stakes quite high: "Unless this is seen and grasped with an exact calculation, the foundations of the system collapse and the sequence does not harmonize" (*quod nisi subtili uisum ratione tenetur, / fundamenta ruunt artis nec consonat ordo*, 3.206–207). In theory, a nighttime birth does not present an impossible challenge. Sextus Empiricus describes a scenario in which a man, who sits beside the woman in labor, uses a gong to communicate the exact moment of birth to a Chaldaean astrologer who sits on the roof gazing at the stars (*Math.* 5.68). Sextus may intend the scene to appear comical, but it vividly suggests the lengths to which one might go to ascertain the horoscope with precision.[77]

In *De diuinatione*, Cicero's character Marcus criticizes the Chaldaeans for trying to make such determinations with their eyes, the most fallible of senses (2.91). Instead, he argues, they ought to rely on the *ratio mathematicorum*. This makes sense, since numerous factors make observation unreliable: meteorological phenomena and topography can obscure or block the view of the horizon, and if the birth occurs during the day, the ecliptic is not visible at all. Luckily, if one knows the time of the birth and the position of the sun in the ecliptic at that time, it is possible to calculate the horoscope. Manilius discusses several methods of calculation, the first of which, the so-called "common calculation" (*uulgata ratio*), he immediately invalidates. The remaining methods do not produce consistent results, and the final method yields values that are identical to those of the repudiated *uulgata ratio*. In a moment of anguish and despair, Housman (1903–1930, 3.xxi) laments:

> Alas, alas! This alternative method of yours, my poor Marcus, is none other than the vulgar method which in 218–24 you said you knew, and which in 225–46 you exposed as false. The wolf, to whom in his proper shape you denied admittance, has come back disguised as your mother the goose, and her gosling has opened the door to him.

Based on this apparently sad state of affairs, Pierre Brind'Amour (1983) concludes that the entire passage that sets forth the final *ratio* is likely spurious. Critics have also puzzled over certain superfluous details. For instance, in addition to calculating the rising times of the signs in both hours and stades,

[77] For rooftop astrologers at the highest level of Augustan culture, see Suet. *Aug.* 94.12.

when only one set of numbers would have sufficed, Manilius provides their setting times, numbers that are entirely irrelevant to the computation of the horoscope.[78] At a more fundamental level, a practicing astrologer in Manilius's day simply would not have relied on his own calculations to determine the horoscope—he would have consulted astronomical tables—which renders the entire exercise pointless, at least from a practical perspective.[79] It would be a mistake, then, to assume that Manilius aims merely for internal consistency or straightforward exposition. Reading Manilius's treatment of the horoscope through the aesthetics of the infinite allows us to formulate rather different conclusions.

This is the most mathematically complex passage of the *Astronomica.* It will be helpful first to summarize Manilius's calculations:

(1) The *uulgata ratio* (3.218–246) assumes that all the signs rise at the same rate, which Manilius rejects as false. As he points out, the obliquity of the ecliptic causes the signs to rise at different speeds, the length of day and the rising times themselves vary depending on the latitude of the observer, and the length of day itself varies depending on the season.

(2) Manilius resolves these issues through two sets of calculations, the first of which has multiple parts. Manilius establishes the length of a standard hour at the solstices and equinoxes for the latitude of Alexandria and, using this information, calculates the rising and setting times for the signs in both stades and hours (3.247–300).[80] But these figures are only valid at one latitude, for the length of day and night differ as one moves between the equator and the North Pole (3.301–384). Consequently, Manilius provides two formulae, one that allows the reader to calculate the rising and setting times of the signs in both stades and hours at any latitude (3.385–442), and another that facilitates the determination of the rate at which daylight hours increase from midwinter to the summer solstice at any latitude (3.443–482).

[78] A stade is "half a degree of the celestial equator, that portion of the celestial equator which takes two minutes to rise" (Goold 1997, 184).

[79] See Volk (2009, 91). Baccani (1992, 77–79) discusses the papyrological evidence for how practicing astrologers dealt with the horoscope.

[80] Manilius's values correspond to the latitude of Rhodes, not Alexandria; see Housman (1903–1930, 3.xiii). For the treatment of rising times in ancient astrological texts, see Abry (1998).

(3) Manilius adduces one last set of calculations (3.483–509), the first for a daytime birth and the second for a nighttime one, that also allow the reader to determine the horoscope. It is this last pair of calculations that are the subject of Housman's lament.

Manilius develops his exposition in such a roundabout manner and includes so much mathematical detail, much of which is either irrelevant or unnecessary, that we cannot treat it simply or even primarily as instructional. What, then, is the point?

Manilius frames the entire exercise as the contemplation of both impossibly large and infinitesimally small magnitudes (3.211–215):

> sed, quanta effectu, res est tam plena laboris
> cursibus aeternis mundum per signa uolantem,
> ut totum lustret curuatis arcubus orbem,
> exprimere et uultus eius componere certos
> ac tantae molis minimum deprendere punctum.

It is as laborious as it is consequential to represent heaven flying in its eternal course along with the signs, as it circles through its entire orbit with its curved arcs, and to fashion its precise features and to apprehend in so great a mass the smallest point.

The passage moves from the incomprehensibly and eternally great to the unimaginably and unbelievably small, from the heavenly sphere, which spins ceaselessly around the earth, to the *minimum punctum* of that sphere. Manilius is looking for a needle in a cosmic haystack, and it is precisely the juxtaposition of magnitudes that initially stuns the reader.

As we continue, Manilius becomes a kind of inspired tour guide, directing our attention and controlling our movements as we pivot from one stupefying calculation to the next. But the poet reveals his mastery of these unwieldy sums only gradually, which has the effect of slowly establishing his control over the student; it is not about persuasion but domination. The strategy is readily grasped through the moods and tenses of the verbs. The *uulgata ratio* and the calculations for the latitude of Alexandria use present tenses of the indicative (e.g., the *uulgata ratio* "assigns two hours to the signs as they rise and arranges equal constellations in matching spaces," *binas tribuit signis surgentibus horas / et paribus spatiis aequalia digerit astra*, 3.219–220). At the end of the calculations for Alexandria, Manilius also writes in the indicative (3.295–300):

> haec ubi constiterint uigilanti condita mente
> iam facile est tibi, quod quandoque horoscopet astrum,
> noscere, cum liceat certis surgentia signa
> ducere temporibus propriasque ascribere in horas,
> partibus ut ratio signo ducatur ab illo,
> in quo Phoebus erit, quarum mihi reddita summa est.

When these things stand fixed and are settled in your wakeful mind, then it is easy for you to know what sign is horoscoping and when, since it is possible to compute the signs that are rising at sure times and to assign them to their proper hours so that from that sign in which Phoebus will be the calculation may be made in the divisions of which I have rendered the sum.

Initially, the poet encourages "you," the student, but by the end all the weight falls squarely on Manilius: it will be easy for you only because "I," the poet (*mihi*), have performed the arduous task of tallying the hours and stades through which the signs rise and set. Manilius is an intermediary between the student and the mathematics of the *mundus*, and hence he is god-like in his control not just of the stars but also of his mortal audience.

When he pronounces the first universal formula, Manilius gives orders in the subjunctive. This is true for his mathematical instruction (e.g., "let him calculate," *deducat*, 3.396). It is also true for the poet's address to the student: "Let him pick up the path that I have laid down, and let each one follow it for himself and advance through his own tracks, and let him owe his art to me" (*a me sumat iter positum, sibi quisque sequatur / perque suos tendat gressus, mihi debeat artem*, 3.393–394). Manilius's calculations lay the groundwork upon which the student will draw in the future. Manilius himself stands at both the beginning and the end of this progression, encircling the student (*me... sibi quisque... mihi*) with his mathematical authority. Finally, in his last set of calculations, which are also universally valid, Manilius writes primarily in future indicatives ("you will heap up this number on itself multiplying it ten times," *hunc numerum reuocabis in ipsum / multiplicans decies*, 3.486–487). Instead of an exhortation, which leaves room for the student to fail to comply, Manilius writes with all the confidence of one who knows and hence controls the future. The sheer weight of his accumulated calculations crushes all other points of view.

Manilius's mastery of the mathematics of the horoscope also establishes him as a gatekeeper of astrological discourse itself. If one does not calculate the horoscope correctly, astrology as a discipline collapses (*fundamenta ruunt artis*, 3.207). As a result, Manilius begins his exposition by announcing his awareness of the *uulgata ratio* (*nec me uulgatae rationis praeterit ordo*, 3.218) and immediately signaling its inaccuracy (*sed iacet obliquo signorum circulus orbe*, 3.225). In other contexts where Manilius develops multiple methods of calculation, the existence of a second *ratio* is simply taken for granted. In the case of the dodecatemories, the second *ratio* just exists (*nec genus est unum, ratio nec prodita simplex*, 2.722), and it is implied that if anyone devised this *ratio*, it was *natura* ("nature placed the truth in many methods/measurements/musical modes," *pluribus inque modis uerum natura locauit*, 2.723). In book 3, however, the reader understands that there are other astrologers and that this method of calculation is widespread (*uulgata*) among them. Given the evidence of Sextus Empiricus and Cicero, it is not hard to imagine the kind of situation that Manilius has in mind. Manilius's laborious calculations are evidence of his professional competence and serve to exclude other practitioners, whether real or symbolic, from the arena of astrological practice.

When viewed from this perspective, Manilius's rejection of the *uulgata ratio* takes on added significance. The phrase looks back to the proem of the book, where, as we have seen, Manilius defiantly declares that "it is commonplace (*uulgatum est*) to write poetry about pretty things and to compose an uncomplicated work" (3.29–30). The *uulgata ratio* assumes two hours for the rising of every sign—it takes no account of the obliquity of the ecliptic or changes in daylight hours depending on the time of year or latitude, it does not give the reader values in stades, it does not provide setting times, even if those numbers are neither here nor there. By comparison with Manilius's own calculations, the *uulgata ratio* is common, simple, and pretty, and his rejection of it symbolizes his repudiation of *dulcia carmina* (3.38). Manilius's treatment of the horoscope is the crystallization of his planetary struggle through mathematical subjects in search of knowledge that surpasses understanding, requires great intellectual expenditure, and refuses to be sweet. All of this means that Housman's dramatic rebuke will not have mattered to Manilius in the least. The final *ratio* may produce figures identical to those of the *uulgata ratio*, but the procedure itself is more complicated, involves more steps, requires more numbers, and is not so well known. "The following path," writes Manilius, "will also be able to lead [you] to the star that is being born" (*illa etiam poterit nascens uia ducere ad astrum*, 3.483). Two ways to heaven will always be better than one. The point is not logical consistency but the impression that Manilius conveys of control, the experience of a never-ending series of calculations, an intimation of the infinite.

Manilius's elaboration of calculations of the horoscope is overwhelming, but it is also uplifting, and a key passage encodes its sublimity. A basic premise of the *Astronomica* is that readers observe the stars from a Mediterranean latitude. This means that the signs overhead revolve at an angle to the observer's horizon: every night six signs rise and six set, and those signs change over the course of the year as the sun circles through them (cf. 1.255–262). When Manilius explains how to determine the rate at which the daylight hours increase from midwinter to the summer solstice (3.301–384), he forces readers out of that Mediterranean comfort zone, driving them first to the equator and then northward, until they eventually reach the North Pole; the result is a precursor of sorts to the Romantic and Gothic traditions of the arctic sublime.[81]

As Manilius directs the reader's footsteps, he emphasizes the different celestial spectacles that define each locale. At the equator, the signs rise and set perpendicular to the horizon, and night and day are always of equal length. As the reader moves northward, the ecliptic begins to tilt, the ratio of day to night shifts, and the curvature of the earth gradually removes from view one southern constellation after another. The journey culminates at the North Pole (3.356–361):

> si uero natura sinat sub uertice caeli,
> quem gelidus rigidis fulcit compagibus axis,
> aeternas super ire niues orbemque rigentem
> prona Lycaoniae spectantem membra puellae,
> stantis erit caeli species, laterumque meatu
> turbinis in morem recta uertigine curret.

[81] For the intellectual background, see Abry (2006, 154–158). For the arctic sublime, see Rix (2023).

But if nature should allow [you] to travel beneath the summit of heaven, which the icy pole supports with stiff joints, over eternal snow and a stiff earth that gazes up at the downward-facing limbs of Lycaon's daughter, there will be the spectacle of heaven standing straight, and with the movement of its sides it will rotate like a top in an upright spin.

At the North Pole, the signs neither rise nor set. The celestial sphere rotates around the earth, parallel to the horizon, like a top, which means that it is only ever possible to see the same six zodiacal constellations. Here, day and night each last for six continuous months (3.365–374). The image is utterly unfamiliar, and Manilius encodes the awesomeness of this celestial spectacle (*caeli species*) through the imagined gaze of the earth itself (*orbemque...spectantem*). The language is surprisingly erotic. The earth grows hard (*rigentem*) as it stares at the limbs of the girl who lies facedown (*prona...spectantem membra puellae*). At the literal level, what is seen is the outline of the Great Bear, which Jupiter created when he transferred Lycaon's daughter Callisto to the heavens. But Manilius writes as if Callisto herself were still visible, bringing to the fore the constellation's dark aetiology and perhaps hinting at the erotic passion that is the driving force of philosophy. The change in latitude, affected by Manilius's knowledge of mathematical astronomy, radically alters our perception of space and time and pulls the reader upward into a strange new sky that is charged with the echoes of sexual violence. Amidst the unforgiving ice and snow of an extreme northerly latitude, our minds are opened and expanded, even as we contemplate the violence and brutality of (a) god.[82]

The image represents the experience of Manilius's intellectual program. To read through and to attempt to keep track of Manilius's never-ending sequence of numbers is as disorienting as traveling to the North Pole and looking up, in frozen wonder, to see the heavens spinning around you like a top. As Josèphe-Henriette Abry (2006, 167) writes, "Manilius réussit à transformer la théorie scientifique en un spectacle cosmique fascinant."

[82] For astronomical eros, see 5.569–570, where an implicitly stupefied Perseus (*facie quem non stupefecerat hostis* [i.e., Medusa]) grows hard (*deriguit*) when he sees Andromeda hanging (*pendentem*) from the rock, like a star that "hangs" in the sky (e.g., 1.529, 806). Note, too, that Perseus subsequently plays in the air (*per aethera ludit*, 5.599), a veritable Phaethon (*puer in caelo ludit*, 1.738). Perseus, in other words, is a star gazer who experiences astronomical eros, a form of the scientific sublime. For Manilius's Perseus and Andromeda, see Uden (2011) and Fratantuono (2012). For another combination of eros and the investigation of nature, see Ch. 3.6 ("Hostius Quadra").

The production of a cosmic spectacle—this is the goal in pursuit of which Manilius performs so many calculations and itemizes so many numbers, even when those calculations are unnecessary, methodologically problematic, or simply inaccurate. This constitutes an aesthetic interpretation of the basic fact, observed by many and noted above, that the *Astronomica* does not provide the necessary information that would allow one to construct a birth chart or to formulate even the most rudimentary astrological prediction. Steven Green has read this situation as part of a deliberate strategy both to indulge in astrological discourse, which was wildly popular at the time, and to avoid offending the emperor, who sought to restrict the use of astrological knowledge and symbolism by others.[83] Such a reading is too grounded in the practical realities of everyday life. I propose a different tack: we should treat the poem's lack of clarity, its occasional incoherence, and its sheer excess of unconnected information from the perspective of the aesthetics of the infinite. The stream of data is vertiginous, which matters more than whether it is accurate.[84]

2.6. Infinity and Empire

Manilius's imaginary journey to the North Pole suggests the extent to which the earth lies open, rhetorically, to a Roman citizen of the late Augustan Principate, when Augustus can claim to have conquered the globe.[85] Various actors, however, push this claim further, suggesting that Roman power fills the universe itself. This is a mode of propaganda with which Manilius enthusiastically colludes. By projecting Roman *imperium* outward from the earth onto the celestial sphere, the poet suggests the infinite expanse of empire and turns Rome herself into a celestial spectacle. The rapturous union of the study of nature with imperial power constitutes a distinguishing feature of the *Astronomica*.[86]

[83] Green (2014, 11–61), who remains agnostic about the identity of the emperor.

[84] Many, if not most, of the passages in which Green detects frustration on the part of the student can be read as experiences of the sublime. In contrast, Neuburg (1993, 257–282) reads Manilius's role vis-à-vis his student as that of an earnest, Stoically informed instructor.

[85] Note the title of the first chapter in Nicolet (1991): "The *Res Gestae* of Augustus: Announcing the Conquest of the World."

[86] For positive readings of Manilius's relationship to imperial power, see Bajoni (2004), Abry (2011), Gale (2011), Kennedy (2011), Green (2014, esp. 59–61), and O'Rourke (2019, 38–40). Gladhill (2016, 97–118) establishes connections between politics, empire, and the heavens in the poem.

Manilius's infinite cosmos is pervasively Roman. Allusions to the works of Republican and Augustan poets cannot be missed, the great heroes and events of Roman history appear repeatedly, the emperor receives appropriate attention, and both Rome and her *imperium* are fated (4.773–774).[87] On its own, this state of affairs allows us to conceptualize Manilius's *mundus* as Roman or simply as Rome writ large. Such is the premise that implicitly bookends the list of the possible *causae* of the Milky Way; the list begins with the universal ship of state and concludes with a parade of Roman political heroes literally occupying a broad swath of heaven (Ch. 2.2). But if such passages merely suggest the equation, the end of the poem, as we will see, equates the infinity of the stars with the population of the *res publica*. This maneuver makes the population of the empire infinite and identifies the infinite expanse of the cosmic sphere with the geographical reach of the Roman state. Thanks to celestial inquiry, "we have surveyed nature in its totality, we have captured the universe, and we control it" (*peruidimus omnem* [sc. *naturam*] / *et capto potimur mundo*, 4.883–884). The poem constitutes a monumental exemplar of astronomical knowledge and astrological iconography deployed in the projection and maintenance of imperial power.[88]

The alliance between the study of the stars and Roman *imperium* is on display in a fascinating way at the end of book 5, where we get a lesson about stellar magnitudes. The lowest magnitude contains the largest number of stars, and Manilius suggests that this number is infinite. On a night when the brighter luminaries lie concealed below the horizon, it is possible to discern these stars (5.726–727, 729–733):

> tum conferta licet caeli fulgentia templa
> cernere seminibus minimis totumque micare
> stipatum stellis mundum nec cedere summa
> floribus aut siccae curuum per litus harenae,
> sed, quot eant semper nascentes aequore fluctus,
> quot delapsa cadant foliorum milia siluis,
> amplius hoc ignes numero uolitare per orbem.

Then it is possible to see gleaming the quarters of heaven, packed and shining with the smallest seeds, and the whole firmament, filled with stars, [and to see

[87] For the presence of Rome in the poem, see Volk (2009, 162–173).

[88] For the relationship between political power and astrology in the Roman world, see Barton (1994, 38–51) and Volk (2009, 127–161).

that they] do not yield in their sum to the flowers or to the dry sand along the curved shore, but, as many waves as always are born on the sea and go forth, as many thousands of leaves as drop and fall in the forests, more than this in number are the fires that fly through the heavens.

Manilius combines several proverbial images for the uncountable—the stars in the sky, the sand on the shore, the leaves that fall from the trees.[89] Since the stars are unsurpassed in their *summa* and second to none in their *numerus*, they outstrip all other images of the infinite. Echoes of Lucretius contribute to the impression that the stars cannot be counted (Porter 2016, 503–504). In particular, Manilius lifts *semina minima* from Lucretius, who uses the same phrase to describe the atoms (e.g., 1.487). For Lucretius, the atoms are indeed infinite in number, and the universe extends infinitely in space and time. Manilius applies to his stars and, by implication, to his *mundus* the associations with infinity of Lucretius's Epicurean cosmos. Manilius then compares the stellar magnitudes to the hierarchical divisions of Roman society, equating the brightest magnitude with the *patres* and the next brightest with the equestrians, to be followed by the *populus*, the *uulgus*, and finally the *sine nomine turba*: "Thus a kind of *res publica* exists even in the heavens, the creation of nature, which has founded a city in the sky" (*sic etiam magno quaedam res publica mundo est / quam natura facit, quae caelo condidit urbem*, 5.738–739). The image inscribes Rome's social hierarchies within the fabric of the cosmos, a convenient rationalization for a brutal reality, and thereby collapses the celestial sphere into the Roman state. The contrast between *orbem* (5.733, end of line) and *urbes/urbem* (5.734 and 739, end of line) vividly underscores the equation. Of course, there's a twist, since *orbis* here is short not for *orbis terrarum*, "the world" (*OLD* 12), but for *orbis caeli*, "the vault of heaven" (*OLD* 7c). The universe and the city are mapped onto one another so that when we look up in wonder at the spectacle of the night sky, we find ourselves gazing at celestial Rome.[90]

Manilius's cosmic sphere is now a map of the Roman state. Like the other mapping projects of the Augustan period, both visual and textual, it performs

[89] McCartney (1960) provides a useful catalog of such images. For connections between the present passage and earlier literature, see Schindler (2000, 234–242) and Hübner (2010, 442–444).

[90] The simile is widely discussed; see esp. Landolfi (1991), Schindler (2000, 245–248), Abry (2008), Volk (2009, 109–114), Fratantuono (2012, 308–315), Gee (2013, 116–118), and Kyriakidis (2015, 279–284). For background to the *urbis/orbis* pun, see Hardie (1986, 364–366).

a symbolic function: knowledge of geographical space indicates physical control and political mastery of that space. One of the best-known examples, even if it is not extant, is the map of the *orbis terrarum* that was on display in the Porticus Vipsania.[91] The map was based on the *commentarii* of Agrippa, and Augustus himself was responsible for completing the building project; geographical knowledge and imperial power reinforce and reflect one another, and the interrelationship is displayed for all to see. From this perspective, Manilius's map extends Roman knowledge and hence control of space into the heavens. And what better evidence of Rome's cosmic domination than the fact that mankind now makes divinities and sends them to the stars, allowing heaven itself to grow and prosper under the influence of Augustus (*iam facit ipse* [sc. *homo*] *deos mittitque ad sidera numen, / maius et Augusto crescet sub principe caelum*, 4.934–935)? Naturally, this *imperium* also radiates downward, back to earth, which it blankets in its totality, from the inordinately large (e.g., countries and continents in 4.585–817) to the imperceptibly small (e.g., the minutiae of everyday life in 3.43–159). Manilius's text replicates, on a truly global scale, the symbolic strategy of the so-called horologium of Augustus, which traced the movements of the stars onto the city's topography. There is nowhere, there is nothing, that is not Rome.[92]

Thinking about map making in this way constitutes part of a broader theoretical move. For some time, critics of imperial literature have drawn attention to a conceptual link between the accumulation of knowledge and the accumulation of empire.[93] For instance, the Elder Pliny's impossibly large encyclopedia, which purports to contain all knowledge, not only acts as a symbol of Roman *imperium*, which purports to extend over the whole earth, but also legitimizes and entrenches that *imperium* by monumentalizing Roman intellectual mastery over the disparate peoples, places, and materials that it draws into its textual ambit. This mastery is on full display in Pliny's first book, which lists the facts and authors that the following books contain, starting with the dimensions of the *mundus* and the question whether it is to be considered finite. The impression is one of unlimited scope, coverage, and power. We will return to Pliny and his complicated relationship with the scientific sublime in Ch. 5.5 ("*Natura artifex*") and

[91] See Nicolet (1991, 95–122). Others have connected Manilius with Agrippa's map; see Bajoni (2004, 105–106) and Abry (2011).
[92] For the so-called horologium, see Heslin (2007).
[93] See esp. Nicolet (1991), Flemming (2003), Murphy (2004), the essays in König and Whitmarsh (2007), and Kennedy (2011, 183–184 on Manilius; 2013a, 153–202; 2013b).

the Epilogue ("The Scientific Sublime and the Case of the Elder Pliny"). For now, we can simply observe that Pliny's preface uses numbers to create the same impression of totality (*HN* praef. 17):

> $\overline{\text{XX}}$ rerum dignarum cura (quoniam, ut ait Domitius Piso, thesauros oportet esse, non libros) lectione uoluminum circiter $\overline{\text{II}}$, quorum pauca admodum studiosi attingunt propter secretum materiae, ex exquisitis auctoribus centum inclusimus XXXVI uoluminibus, adiectis rebus plurimis quas aut ignorauerant priores aut postea inuenerat uita.
>
> Twenty thousand items meriting attention (since, as Domi tius Piso says, the task calls for storehouses, not books), thanks to the perusal of about two thousand scrolls (very few of which scholars handle on account of the abstruseness of the subject matter), from one hundred authors whom we have examined—that is what we have included in our thirty-six scrolls, plus very many other items that either had been ignored by earlier generations or had been discovered subsequently by human experience.

The figures convey authority and totality, both Pliny's and Rome's. At the same time, one senses that the end point has been declared arbitrarily. It will always be possible, perhaps it will always be necessary, to extend further the boundaries of the text and the boundaries of empire.

This is a model of accumulation—knowledge and empire can be accumulated indefinitely simply by adding one more fact, by pushing back one more boundary. In contrast to a model of accumulation, Kennedy (2013a, 171–176; 2013b, 64–67) describes a model of compression. At the level of knowledge, a model of compression offers one grand theory that can explain everything and therefore does not have to explain everything. Since atomic theory, for instance, explains everything everywhere without exception, Lucretius does not need to explain everything everywhere without exception, and no one feels the need to add new chapters to his text. At the political level, a model of compression posits universal empire, Virgil's *imperium sine fine* or Augustine's City of God, both of which exist without limits, transcending space and time.

These models are useful in analyzing Manilius's conceptualization of knowledge and power. On one level, Manilius operates according to the compression model. Manilius's astrology is a system that can explain everything everywhere without exception, a transcendent form of knowing that brings those on earth into contact with the divine.[94] This makes astrology the preeminent form of natural inquiry, the *telos* of human intellectual development (1.66–112), in comparison to which all other sciences are to be dismissed (*mitte alias artes*, 2.109). The political structure to which astrology corresponds is equally universal and unlimited. Like astrology, Roman *imperium* is transcendent, because it is to be equated with the heavens, and just as astrology subsumes all other *artes* and obliterates the distinction between man and god, Roman *imperium* absorbs all other political entities and erases the boundaries between *urbs* and *orbis*.[95]

The differences between how Manilius and Virgil construe Rome's infinite expanse are instructive.[96] In the *Aeneid*, Jupiter can easily count the number of years that will elapse from the start of Aeneas's war with the Italians to the birth of Romulus and Remus (1.261–277): Aeneas will control Latium for three summers and three winters, Ascanius will rule for thirty years and move to Alba Longa, and Hector's race will dominate Alba Longa for 300 years after that, until Ilia gives birth to twin sons. But the future of the descendants of Romulus and Remus will be infinite: "For these, I set neither limits to their affairs nor temporal boundaries: I have given them empire without end" (*his ego nec metas rerum nec tempora pono: / imperium sine fine dedi*, 1.278–279).[97] Jupiter makes an important distinction. Rome's prehistory can be calculated, and the numbers involved are formulaic, which means safely and appropriately epic (cf. Rubincam 2003 and Ch. 2.4). Rome itself, however, cannot be so bounded. Jupiter states as fact what Manilius conveys at length. Virgil's poem repeatedly foresees but does not attempt to quantify or comprehend the infinite expanse of space and time allotted to the Roman Empire. Manilius allows readers to experience the sublimity of that very expanse through the tireless performance of mathematical tasks. Manilius, in a sense, offers mathematical proof of Virgil's empire without end.

But the accumulation model is also important for understanding Manilius. Although astrological discourse can explain everything everywhere, it does

[94] Kennedy (2013a, 173–176; 2013b, 66–67) emphasizes that the compression model generally has recourse to the divine, which is conceived of as existing outside time. For the *Astronomica* as a "Weltgedicht," see Lühr (1969, 5–8) and Hübner (1984, 227–242).

[95] For empire as the erasure of boundaries, see Kennedy (2013a, 78–79).

[96] Kennedy (2013a, 43–83) is a stimulating reading of infinity and time in the *Aeneid*.

[97] Virgil's *metas rerum* indicates both "limits to their affairs, fortunes" and "spatial, geographical limits"; see Hardie (1986, 199–200, 364–366) and Nicolet (1991, 29). For *imperium* denoting power "of immense or even infinite extent," see Richardson (2008, 132).

not constitute a grand theory, like atomism. To explain everything everywhere, the astrologer needs more methods of calculation, more geometrical schemes, more divisions and subdivisions of the zodiac. From the outside, astrology operates on the compression model, but once you enter its way of thinking, it requires accumulation, infinite accumulation, as it works its way asymptotically toward universality. Thus, at the start of book 5, Manilius states that another traveler might have stopped his journey after treating the signs of the zodiac (*hic alius finisset iter*, 5.1) and would not have carried his work any further (*non ultra struxisset opus*, 5.4). Manilius is prepared to do what others will not, to violate boundaries, to disregard customary expectations of restraint. Why?[98] The cosmos has witnessed his daring and bids him to continue his expedition (5.8–11):

> me properare etiam mundus iubet omnia circum
> sidera uectatum toto decurrere caelo,
> cum semel aetherios ausus conscendere currus
> summum contigerim sua per fastigia culmen.

The cosmos bids me also to be eager to travel around all the heavenly bodies and to traverse the entire sky, since I have dared once to mount my ethereal chariot and have touched the highest pinnacle [of heaven] along its upward slopes.

Manilius will pursue his bold undertaking forever as he traverses the entire cosmos and surveys all the stars. After listing some of those stars, he circles back to this very point: he must sing of "everything that glides all over the place in the entirety of heaven" (*ceteraque* [sc. *signa*] *in toto passim labentia caelo*, 5.26). Unlike Lucretius's Epicurus, who passes beyond the ramparts of the world and then returns home, there will be no definitive triumph. The implications for Roman *imperium* are straightforward: despite the projection of world domination and all appearances to the contrary, the job is never done. Astrology and empire are open ended and ongoing, continually striving for infinity.[99]

[98] For other approaches, see Effe (1977, 120), Hübner (1984, 245–247), Volk (2002, 210–211), and Kennedy (2011, 170 with n.20).

[99] At 2.765, Manilius similarly claims to fly through the entire cosmos (*mihi per totum uolitanti carmine mundum*). Hübner (1984, 242–245), by contrast, reads the trajectory of the poem as a descent from heaven to earth, structurally parallel to Aratus's *Phaen.* For a critique of this position, see Volk (2002, 232–234). We might also read Manilius's reluctance to return to earth against Plato's allegory of the cave: the enlightened traveler does not wish to return to the dark (note esp. *Rep.* 7.516c–518b).

How, then, to end the poem without putting limits on the infinite scope and absolute power of astrological inquiry and Roman *imperium*? Manilius rounds out the comparison of stellar magnitudes to the Roman social classes with a vision of a different kind of infinity (5.742–745):

> maximus est populus summo qui culmine fertur;
> cui si pro numero uires natura dedisset,
> ipse suas aether flammas sufferre nequiret,
> totus et accenso mundus flagraret Olympo.

The greatest in number is the crowd [of stars] that revolves at the highest peak. If nature had given it strength equal to its number, the ether itself would be unable to endure its own flames, and Olympus would be ignited and the whole cosmos would be on fire.

Nature was wise to limit the power of the *populus*, to stamp out any form of democratic self-government. Astrological discourse, then, provides philosophical support for the Roman social and political organization, and Roman *imperium* tacitly confirms the validity of astrological discourse itself. But what if things had been different? The counterfactual does not present a vision of Stoic *ekpurōsis*, which occurs at regular intervals and constitutes both the end of one cosmic cycle and the beginning of the next. On the contrary, it forces us to imagine a reality completely different from our own, one in which the power of the faintest stars overwhelms and consumes the *mundus*.[100] This cosmic uprising enacts the ultimate erasure of boundaries, but from the perspective of the masses. The possibility of revolution on a universal scale and of the resolution of the *mundus* into an infinite expanse of fire threatens to wipe out both *imperium* and astrological discourse, and Manilius thrills at the thought of such total obliteration. And yet, it is the ultimate expression of power to be able to dabble in such counterfactuals and to deny the apodosis out of hand, to imagine what might have been but was not, what could have come to pass and did not. That exercise of power is the source of the ecstatic rush that animates Manilius's musings about the cosmic conflagration.

This expression of power resonates at the epistemological level as well. As Housman (1903–1930, 5.94) saw, *summo*...*culmine* (5.742) refers to the

[100] The image is often described as *ekpurōsis*, but Lapidge (1979, 357) rightly notes that Manilius nowhere describes this phenomenon. For *ekpurōsis* and the cosmic cycle, see *SVF* 2.596–632 and L-S 46 with D. Furley at *CHHP* 434–439.

Milky Way, which Manilius describes in book 1 as shining overhead and marking heaven's *culmen* (*super incumbit signato culmine limes / candidus*, 1.714–715). With this in mind, we might translate 5.742 as "the greatest in number is the crowd [of stars] that revolves in the Milky Way." Given the unknown nature of the Milky Way in book 1, it is surprising to find out now that it simply consists of dim stars. Of course, this is one of the *causae* that Manilius surveys (1.755–757), but he touches on it only briefly before losing himself in the glories of Roman political history, the history of the elite rather than that of the *maximus populus*. In the present context, the theory heightens the reader's sensation of the infinite—the experience of contemplating the infinity of stars in the Milky Way leads to an image of total cosmic breakdown, the dissolution of all structure, the spread of all-consuming fire. It does not matter if the text of book 5 does not square with the doxography of book 1. Such disregard for consistency is also the privilege of power, the ability to say what one wants when one wants and to make it so. To exercise such unbounded power is the prerogative of god or of one who aspires to become god—a Phaethon who is happy to burn it all down.

This kind of maneuver constitutes a viable position at the beginning of the first century, while Augustus lives on or, at the very least, while his memory remains fresh. But one may already feel uneasy with the idea that nature has made a *republic* in the heavens (*res publica mundo est / quam natura facit*, 5.738–739). Despite protestations to the contrary, the Roman Republic is a thing of the past. Manilius, of course, glosses over this disconcerting gap between rhetoric and reality, all too happy to deploy his calculations in support of the status quo. The corrosive grind of empire, however, will gradually make this kind of alliance between the study of nature and political power less and less tenable. By the time we arrive at the final decade of Nero's reign, as deception and violence engulf the aristocracy, Manilius's supreme confidence in astronomical inquiry and Roman *imperium* look woefully outdated. When Seneca and Lucan turn their attention to the natural world, they register this shift in epistemological and political realities, and the sublimity of the study of nature becomes something radically different.

The Scientific Sublime in Imperial Rome: Manilius, Seneca, Lucan, and the Aetna. Patrick Glauthier, Oxford University Press. © Oxford University Press 2025. DOI: 10.1093/9780197787588.003.0003

3

Seneca's *Natural Questions*

The Promise and Danger of the Sublime

3.1. Introduction

Throughout the *Natural Questions* (*NQ*), Seneca structures the inquiry into nature through the dual logic of the sublime; this is the text's primary aesthetic and moral mode.[1] To study the natural world, in collaboration with others, is a wild journey that promises to strengthen the mind, to lead readers toward wisdom and virtue, and to open their eyes to the greatness of the cosmos, which is identical with god. While certain aspects of this project resemble the ethos of the *Astronomica*, the intervening years have changed the fundamental calculus. In the chaos and paranoia of Neronian Rome, the universe itself has become more erratic, tumultuous, and violent. Under such circumstances, the inquiry into nature is an uncertain enterprise, and reliable answers often remain elusive. Consequently, we may be forced to jettison familiar concepts, reimagine our surroundings, and try to think what had previously been unthinkable. This intellectual journey is full of false starts and dead ends, and so, too, is its ethical counterpart. Seneca views the study of nature as an opportunity to sharpen our intellectual faculties, to begin to see the connections that tie the cosmos together, and to learn to assess more accurately our place within it. In theory, this process leads toward moral perfection. In the reality that the text presents, however, we might briefly attain the serenity of the sage, but the most consistent outcome for which we can realistically hope is a paradoxical state of vital calmness or serene tension, a mode of being in the world that is free from fear and simultaneously alive to the volatile energy of the *mundus*. As we strive for progress, competing sources of sublimity, including Roman *imperium* itself, threaten to derail the project and to plunge the reader into vast seas of moral depravity. Against such degradations of the sublime, Seneca

[1] The treatise dates to the early 60s CE. See Gauly (2004, 19–28) and Hine (2006, 68–72).

hopes that the quickening thrill of natural inquiry will act as a bulwark. The study of nature comes with no guarantee of success, and we often get things wrong, but Seneca eagerly embraces the expansive uncertainty of the cosmos as a symbol of its grandeur and a promise of our own capacity for greatness.

"Seneca and the sublime" is not a new topic.[2] While my focus here is limited, the importance of the sublime just in the *NQ* has received significant attention in the first decades of this century; the work of Fernand Delarue, Gareth Williams, Erik Gunderson, and Myrto Garani stand out.[3] One important theme to have emerged from these studies is the significance for Seneca of the Lucretian sublime.[4] In what follows, I take Lucretius's influence on the treatise for granted; instead of retreading now familiar ground, I attempt to situate Seneca's project in new literary and intellectual contexts.

In particular, I emphasize the spectacle of the fall. Like the *Epistulae*, also addressed to Lucilius and written at roughly the same time, the *NQ* is aimed at and describes the activities of the *proficiens*, the student who is advancing toward wisdom and virtue, not the perfect sage.[5] Although Seneca makes the overcoming of human weakness and limitation the goal of natural inquiry, it is a goal toward which we can only ever progress; for the most part, we all fall short. The spaces through which the treatise moves the reader are primarily atmospheric, some ways off the ground and not yet as high as the heavens, a zone symbolically suited to those who are progressing in virtue and wisdom, and one from which a dramatic fall is almost unavoidable.[6] The tension between philosophical ambition and seemingly

[2] Important general discussions include Guillemin (1954), Michel (1969), Mazzoli (1970, 46–59), Setaioli (1985, 801–811; 2000, 141–155), Armisen-Marchetti (1989, 53–59), Mazzoli (1990; 1991), Setaioli (1991, 144–154; 2000, 242–254), Gunderson (2015), Williams (2016), and Dross (2019). For the tragedies, see Schiesaro (2003, 22–23, 52–54, 127–132), Torre (2003), Littlewood (2004, 121–127), Garelli-François (2007), Torre (2007a), Staley (2010, 42–47), Williams (2017a, 318–325), Littlewood (2017), and Aygon (2019). Often overlooked but very much worth reading are Faggin (1967) and Küppers (1996); both discuss relevant material but name the sublime only in passing.

[3] Delarue (2006), Williams (2012, 113–116, 213–257; 2016, 173–180; 2017a, 312–318), Gunderson (2015, 56–73), Garani (2020a; 2023).

[4] Seneca's use of Lucretius in the *NQ* is well-established. See Mazzoli (1970, 206–209), De Vivo (1992b, 82–89, 91–109), Althoff (2005), and Tutrone (2017).

[5] Williams (2012) is aware of this aspect of the treatise but does not stress its connections with the sublime. Maruotti (2019) suggestively reads the *Epistulae* from this very perspective with an emphasis on sublimity. For Stoic conceptions of moral progress, see P. Donini at *CHHP* 724–736.

[6] For the symbolic value of the atmosphere/*meteōra*, see Gigon (1991, 341–342), Leitão (1998, 159–160), Hine (2006, 68), and Williams (2012, 69, 72–73, 167–168, 335–337). For Seneca's self-characterization as a *proficiens*, see Edwards (1997, 31–34), Mannering (2013, 188, 202), and Wilson (2014, 4).

inevitable setback is itself sublime, and Seneca encodes the sublimity of the experience through Phaethon as well as other more subject-appropriate figures.

From this perspective, Manilius's development of the sublime is every bit as important as Lucretius's, and we can profitably read the *NQ* against or even as a rejoinder to the *Astronomica*. Both Manilius and Seneca see the world as simultaneously discordant and harmonious (*discordia concors* at *Astron.* 1.142, *tota haec mundi concordia ex discordibus constat* at *NQ* 7.27.4), a tension that reflects the inspired energy of Empedoclean Strife and Love.[7] For Seneca, however, the heavens are no longer a source of truth and understanding. The decadence and depravity of Neronian Rome have rendered that which is up above frustratingly inaccessible.[8] Better, then, to turn one's attention to phenomena closer to home, from subterranean rivers to rainbows, winds, and snow. But even these topics can remain inscrutable, as the deceptively theatrical nature of the Neronian age compromises the investigator's ability to formulate reliable answers to important questions. Moreover, it is precisely this corrupted and corrupting landscape that makes the search for moral improvement so urgent. While Manilius is perfectly capable of bemoaning the failings of his contemporaries, he does not construe the pursuit of astronomical knowledge as a pathway to moral improvement, and Seneca will make no such claim on his behalf. Indeed, the modus operandi and the mathematically grounded knowledge of the astronomers are entirely inadequate for the ethical project on which Seneca wishes us to embark. But even if the natural world does not yield reliably to causal analysis, Seneca will charge ahead, collecting information, testing out different theories and modes of analysis, unafraid of intellectual incongruity and the possibility of a crash-and-burn. Progress comes haltingly, and we must be prepared to dust ourselves off and pick ourselves back up again. The similarities with and differences from Manilius will constitute an important leitmotif of this chapter.[9]

[7] Housman (1903–1930, i.11–12) connects Man. 1.142 and Sen. *NQ* 7.27.4 with Empedocles; cf. Mazzoli (1970, 254 n.124) and the additional parallels cited by Vottero (1989, 716). Seneca's familiarity with Manilius is most readily deduced from the tragedies; see, e.g., Boyle (2017, 121, 369, 377, 379, 400).

[8] Cf. *Ep.* 90.42–43, where the sublime spectacle of the night sky, once a source of awe, has been put out of reach thanks to contemporary moral failings.

[9] The influence of Lucretius and his conception of the sublime may account for some of the points of contact between Manilius and Seneca, but to overlook Manilius's role in the development of Roman writing about the natural world is a dereliction of scholarly duty.

Before we begin, a few words must be said about the state of the manuscripts and the original structure of the *NQ*.[10] All modern critical editions print the individual books in the order 1, 2, 3, 4a, 4b, 5, 6, and 7. The scholarly consensus, however, now holds that the original ordering must have been 3, 4a, 4b, 5, 6, 7, 1, and 2. Although none of my arguments depends on the acceptance of this consensus, I assume that Seneca arranged the work in this order. As is customary, I have retained the traditional numbers for each individual book. In addition, the *NQ*, as we have it, is clearly incomplete; at the very least, the end of 4a and the beginning of 4b have been lost.[11] For the most part, Seneca's subject matter is meteorology, in the ancient sense—an inquiry into phenomena that take place between the celestial region and the earth, as well as certain terrestrial phenomena that can be explained through the same physical forces.[12] Aristotle's *Meteorologica* effectively established the parameters of the ancient discipline, and while almost all of Seneca's topics appear in Aristotle's work, a few standard topics are missing from the *NQ*. Might Seneca, for instance, have followed Aristotle and written about the sea? While this remains a possibility, there is no compelling evidence, and ultimately the question does not bear on my own conclusions.

3.2. Progressing toward Transcendence

In the *NQ*, Seneca repeatedly configures the inquiry into nature as an exhilarating adventure that promises to open up a truly "cosmic viewpoint" on the world and our place in it and to elevate the student to the level of the divine.[13] At the same time, it becomes clear that this promise only ever gets

[10] For details and discussion, see Codoñer Merino (1979, 1.xii–xxi), Hine (1981, 4–23), and Gauly (2004, 53–67).

[11] Portions of the end of 4a can be recovered from John the Lydian, *De mens.* 4.107, which Vottero (1989), Hine (1996a), and Parroni (2002) print following the extant portion of 4a; for discussion, see Gross (1989, 174–178). At 318–320, Gross argues from minimal circumstantial evidence that multiple books have been lost.

[12] Graver (2000) and Taub (2003, 141–161) treat Seneca's place in the meteorological tradition. The treatise can also be described as physics (Gigon 1991, 361; Inwood 2000, 24, and 2002, 120; Lehoux 2012, 81) and/or natural philosophy verging on cosmology (Waiblinger 1977, 111–112; Limburg 2007, 80–82).

[13] For Seneca's "cosmic viewpoint," see Williams (2012, esp. 4, 8, 223, 255, 258–259). Numerous studies stress Seneca's combination of ethics and physics; succinct entry points include Scott (1999), Inwood (2000; 2002; 2009), Berno (2003, 15–17), Gauly (2004, 73–85), Williams (2012, 2–5), Wilson (2014, 175–183), Kreuzwieser (2016, 32–37), and Tutrone (2017). Inwood (1995, 69–70) emphasizes the combination of physical and ethical interests in Seneca's immediate intellectual milieu.

fulfilled momentarily, before the baggage of our earthly lives drags us back down to ground level. To experience this tension between elevation and collapse is to experience the sublime.

In the preface to book 3, originally the opening book of the work, Seneca underscores the overwhelming nature the task before him (3 praef. 1–2):

> non praeterit me, Lucili uirorum optime, quam magnarum rerum fundamenta ponam senex, qui mundum circumire constitui et causas secretaque eius eruere atque aliis noscenda prodere. quando tam multa consequar, tam sparsa colligam, tam occulta perspiciam? premit a tergo senectus.
>
> It does not escape me, Lucilius, best of men, how great are the matters whose foundation I am laying in my old age, in that I have decided to travel around the world, to unearth its causes and secrets, and to bring them forth for others to come to know. When will I pursue things so numerous, bring together things so scattered, understand clearly things so hidden? Old age presses at my back.

Seneca is building a mighty structure. Relying on the flight of the mind trope (Ch. 1.1, 1.4), he frames his project as an exploration of the entire world, a truly global undertaking.[14] The awesomeness of the enterprise hints at the sublimity of the journey, for author and reader both, and yet, there's more. Seneca begins his project in his old age, and the prospect of not having sufficient time to complete his investigations fills him with anxiety. The danger that time is running out, that questions will remain unanswered, that the work itself will fail, means "we must press forward all the more" (*tanto magis urgeamus*, 3 praef. 2). Seneca and reader find themselves in the same precarious position.[15]

As we have seen, the prospect of failing in a magnificent task can also invigorate and uplift. Only a short while later, Seneca adopts this very position (3 praef. 3–4):[16]

[14] The tone is well captured by Torre (2007b, 53): Seneca "intenda confrontarsi con la topica immagine del poeta-*vates* spettatore privilegiato del cosmo, viaggiatore del cielo, esploratore degli arcani dell'universo."

[15] The emphasis on old age and the need to hurry echo the preface to Varro's *De re rustica*; see Hine (2006, 48), Limburg (2007, 126–127), and Williams (2012, 31–32).

[16] This passage was first connected with the sublime by Mazzoli (1968; 1970, 48, 258). More recent readings include Williams (2012, 229; 2017a, 313–314) and Garani (2020a, 108–113).

> libet igitur mihi exclamare illum poetae incluti uersum:
>
> tollimus ingentes animos et maxima paruo
> tempore molimur.
>
> hoc dicerem si puer iuuenisque molirer (nullum enim non tam magnis rebus tempus angustum est); nunc uero ad rem seriam, grauem, inmensam postmeridianis horis accessimus. faciamus quod in itinere fieri solet: qui tardius exierunt, uelocitate pensant moram. festinemus, et opus nescio an <in>superabile, magnum certe, sine aetatis excusatione tractemus. crescit animus quotiens coepti magnitudinem aspexit, et cogitat quantum proposito, non quantum sibi supersit.
>
> So, I'd like to shout out the following line from a famous poet: "We raise our mighty spirits and, in a brief span of time, set about the greatest work." This is what I would say if I were setting about my work as a child or youth, for no amount of time is not tight for things that are so great. As it is, however, we have approached a serious, weighty, boundless task in our afternoon hours. Let's do what is usually done on a journey: those who have set out rather late offset their delay with speed. Let's hurry and handle a work that is perhaps impossible, at the very least great, without making an excuse of age. My mind expands whenever it beholds the greatness of the undertaking, and it contemplates how much for its enterprise, not how much for itself, remains.

Setting aside, for the moment, the verse quotation, we again see Seneca drawing attention to the daunting enormity of his journey and emphasizing that time is of the essence. But the very idea of attempting to complete an impossibly large task in a finite amount of time ennobles the subject: "My mind expands whenever it beholds the greatness of the undertaking." Seneca describes the physical expansion of a mind that surges toward *megalopsuchia*. The image encodes the sublimity of natural inquiry, of writing and reading the *NQ*, whether or not time runs out.

In the remainder of the preface, Seneca turns to the idea that what really matters in life are not conventional markers of success, such as political or military achievements, but "to have seen the all with the mind and...to have conquered our faults" (*animo omne uidisse et...uitia domuisse*, 3 praef. 10).[17] The investigation of the natural world is helpful in achieving this end

[17] For Seneca's negotiation of history and historiography in this passage, see Master (2015, 335–343) and Garani (2020a, 113–122). Hine (2006, 43–50) explores Seneca's general dismissal of traditional Roman pursuits and values in the *NQ*.

because it exercises our critical faculties and thereby gives us the intellectual tools with which to see things for what they are and to live accordingly. As we progress, we will have to withdraw from that which is sordid or degrading and "detach the mind itself, which needs to be elevated and great, from the body" (*deinde animum ipsum, quo summo magnoque opus est, seducemus a corpore*, 3 praef. 18). The imagery figures natural inquiry as death, a process whereby the soul is liberated from the materiality of the body and becomes pure.[18] To claim that "my mind expands whenever it beholds the greatness of the undertaking" is to assert that one's mind is working toward becoming *summus* and *magnus*, toward overcoming the imperfections and limitations of our own mortality. To experience this transfiguration of categories, the resolution of the human into the divine, constitutes an animating goal of the *NQ*. In the preface to book 1, Seneca equates the study of nature with theological inquiry.[19] It is a pursuit that allows us to rise above human affairs (1 praef. 5), to come into partnership with god (1 praef. 6) and to begin to know him (1 praef. 13), to transcend our own mortality and to have ourselves enlisted in a better lot (*transilire... mortalitatem suam et in meliorem transcribi sortem*, 1 praef. 17). The study of nature prepares the student for the possibility of such transcendence, even if the journey is recursive rather than linear.[20]

What, then, of the verse quotation, *illum poetae incluti uersum*? Giancarlo Mazzoli points out that at 6.2.9 Seneca quotes a few verses from a famous poem written by a friend named Vagellius (*Vagellius meus in illo incluto carmine*).[21] These two passages are the only instances of *inclutus* in the treatise, and given that Seneca nowhere else applies *inclutus* to a poet or poem, it is reasonable to conclude that he refers to the same individual and to the same text in these two passages. Mazzoli, however, goes further. Based on the citation contexts and the lines themselves, he convincingly argues that verses in question come from a poem about Phaethon and were spoken by Phaethon himself, no doubt at the very start of his wild ride (Vagellius fr. 2 *FLP*); his position has won widespread approval.[22] This makes excellent sense

[18] For the connections between the flight of the mind trope and death, see Rutherford (1989, 157).

[19] Cf. Inwood (2000, 23–25; 2002, 119–121, 149).

[20] Chrysippus (*SVF* 2.42, L-S 26C) regarded physics as the highest branch of philosophy and theology as the culmination of physics. Seneca aligns himself, at the very least, with the latter position.

[21] Mazzoli (1968). On the possible historical identity of Vagellius, see Mazzoli 363–364 and *FLP* 347.

[22] Mazzoli (1968, 364–368), followed by Duret (1988, 141–142), Williams (2012, 227–230), Garani (2020a, 108–112; 2023, 209–210), and others.

on multiple levels. As we saw in Ch. 1.5 ("Phaethon") and 2.3, Phaethon's chariot ride can represent precisely the sort of sublime inquiry into nature that Seneca undertakes in the *NQ*. Moreover, Seneca has already configured his journey as a trip around the globe (*mundum circumire constitui*) during which he will discover profound truths about the natural world (*et causas secretaque eius eruere*); it's a great undertaking for which he has precious little time (*magnarum rerum fundamenta ponam senex*). Phaethon, too, launches on a great adventure that will take him around the earth within the blink of an eye (*maxima paruo / tempore molimur*), and it is precisely Phaethon's youthful boldness that Seneca wishes he could summon when confronted with his own advanced age (*hoc dicerem si puer iuuenisque molirer*). The contrafactual raises a question: does Seneca play the role of Phaethon? The answer must be yes, even though Seneca equivocates, for the daring mindset announced by *crescit animus* corresponds precisely to Phaethon's own declaration *tollimus ingentes animos*, especially if we take the epithet proleptically ("we raise our spirits *and thereby make them great*, we raise our spirits *to greatness*"). By the end of the paragraph, Seneca has worked himself up into a Phaethon-style state of boldness; Phaethon's story stages and implicitly encodes the sublimity of natural inquiry.

This brings us to the fall.[23] The phrase *tollimus ingentes animos* points to a passage in the *Georgics* about breaking spirited horses: "Before the breaking, they will raise their mighty spirits" (*ante domandum / ingentis tollent animos*, 3.206–207). Phaethon raises his spirit, and no doubt those of his horses, but he will be broken soon enough. The fall is integral to Seneca's Phaethon. The verse quotation at 6.2.9 similarly looks to the fall:

> egregie Vagellius meus in illo incluto carmine, "si cadendum est" inquit "mihi, e caelo cecidisse uelim."[24] item licet dicere, "si cadendum est, cadam orbe concusso, non quia fas est optare publicam cladem, sed quia ingens mortis solacium est terram quoque uidere mortalem."
>
> My friend Vagellius puts it outstandingly in that famous poem: "If I must fall," he says, "I should like to have fallen from the heavens." Similarly, one can say: "If I must fall, may I fall with the world having been shattered,

[23] Duret (1988, esp. 147–148) argues that Vagellius, Seneca, and Lucan suppress the fall to make Phaethon a suitable symbol for Nero. The following discussion suggests that no such suppression takes place in the *NQ*. For Phaethon and Nero in Lucan, see Ch. 4.2 ("Nero-Phaethon").

[24] The line is unmetrical and the place of *mihi* varies in the MSS. Hine (1996b, 95) argues that we are likely dealing with "a partially paraphrased quotation" and should therefore refrain from altering the text.

not because it is right to hope for a communal disaster, but because it is a great comfort against death to see that the earth, too, is mortal."

The quotation makes good sense if spoken by Phaethon, presumably in response to his father's warnings about the danger of his proposed adventure (Mazzoli 1968, 366). The consolatory spin echoes the sentiment of Ovid's epitaph: "Here lies Phaethon, driver of this father's chariot. Even if he did not control it, nevertheless he fell amidst deeds of great daring" (*HIC • SITVS • EST • PHAETHON • CVRRVS • AVRIGA • PATERNI / QVEM • SI • NON • TENVIT • MAGNIS • TAMEN • EXCIDIT • AVSIS*, *Met.* 2.327–328). The comfort of spectacular failure resides precisely in the dual experience of the sublime—such failure overwhelms and simultaneously uplifts.

But what is Phaethon's mental state at the moment of the fall? Ovid's character is blinded with fear (*Met.* 2.180–181) and struck out of his mind (*mentis inops*, *Met.* 2.200), a dark mode of sublime *ekstasis*. Seneca provides a clue about his own Phaethon at the end of the book. At 6.32.4, Seneca imagines a natural disaster of cosmic proportions and pictures a subject, free from anxiety (*securus*), standing undaunted over the abyss, where he knows he must fall, and perhaps leaping in of his own accord (*stabit super illam uoraginem intrepidus, et fortasse quo debebit cadere desiliet*, 6.32.4).[25] Reading this figure against Phaethon, we understand that Phaethon will attempt to transform an accidental fall into a willful leap, a moment of paralyzed fear and self-alienation into a burst of sublime agency. Such self-mastery and presence of mind are the ultimate goal of natural inquiry, and Seneca realizes that this attitude may remain out of reach—perhaps the subject will jump rather than fall (*fortasse... desiliet*), perhaps he will transcend the Phaethon paradigm itself.

Two of the *Dialogues* marshal Phaethon to similar ends, both with citations of Ovid's Phaethon narrative from the *Met.*[26] In *De prouidentia*, Seneca creates a miniature drama in which Phaethon twice responds to the warnings of Ovid's sun god. The vignette illustrates that one does not attain

[25] Faggin (1967, 33) reads *NQ* 6.32.4–5 through Kant's dynamically sublime; cf. De Vivo (1992b, 103–104) and Williams (2012, 255–256). The sentiment resembles *Dial.* 1.5.4 (*boni uiri* are not dragged by fortune but follow her and keep pace). For another cosmic *uorago*, see *Ben.* 6.22.1.

[26] Delarue (2006) and De Vivo (2009, 126–137) treat Seneca's use of Ovid here in detail; cf. Lagière (2017, 79–83), Aygon (2019, 193–197), Garani (2020a, 109–113), and Porter (2020, 267–270). For the relevance of Plato's Phaethon (*Ti.* 22b–d) to those of Ovid and Seneca, see Garani (2023).

virtue by avoiding danger (*Dial.* 1.5.10), and Seneca concludes that "it is the mark of one who is low and lazy to pursue the safe path: virtue moves on high" (*humilis et inertis est tuta sectari: per alta uirtus it, Dial.* 1.5.11). For some, this Phaethon represents the Stoic sage, impervious to adversity and hardship.[27] But if virtue moves on high and Phaethon knows that he may fall ("'I approve of the journey,' he said, 'I will go up; it is worth it to take that trip even if I will fall'," *"placet" inquit "uia, escendo; est tanti per ista ire casuro," Dial.* 1.5.11), he has more in common with the *proficiens* than the sage. The high likelihood of Phaethon's collapse does not dissuade him from attempting to make progress. On the contrary, his willingness to embark on a risky adventure bears witness to virtue's fondness for danger (*Dial.* 1.4.4).[28]

In *De uita beata*, Phaethon plays a similar role. Seneca asserts that it is no surprise that those who attempt to scale magnificent heights do not make it to the top: "Look up to those who attempt great things, even if they fall" (*suspice, etiam si decidunt, magna conantis, Dial.* 7.20.2). A quotation of Ovid's epitaph rounds out the argument that the one who attempts to live virtuously "will make a journey toward the gods—indeed, even if he does not hold on to it, 'still he falls in a great undertaking'" (*ad deos iter faciet, ne ille, etiam si non tenuerit, "magnis tamen excidit ausis," Dial.* 7.20.5). The journey of the *proficiens* is marked by setbacks, and for most of us, such spectacular failure is the best for which we can hope. Backsliding is to be expected, and it is imperative that we have the proper outlook.[29]

We should read this approach to Phaethon's fall back into the programmatic quotation of Vagellius at the opening of *NQ* 3. When Seneca worries about his ability to complete his undertaking, to determine every answer with precision and accuracy, to rise above the contaminating hubbub of everyday life, he acknowledges that he may fall, like Phaethon. Conceptualizing this scenario as a failure, though, misrepresents Seneca's aspirational mindset. If Seneca's Phaethon transforms an accidental fall into a willful leap, the

[27] La Penna (2001, 543), Torre (2003, 239, 249–250; 2007a, 66–67), and Lagière (2017, 79–83).

[28] In the *Ep.*, the language of falling can describe the backsliding of the *proficiens*: those advancing toward wisdom "can fall back (*relabi*) into the same condition" (*Ep.* 75.13); "unless he pays close attention, [the man of advancing wisdom] is immediately weighed down and spun back (*grauatur et... reuoluitur*) into the same condition, [but] the wise man cannot fall back" (*sapiens recidere non potest, Ep.* 72.6). For the imagery of ascent in connection with the *proficiens*, see Armisen-Marchetti (1989, 2269–2270).

[29] Note that Wilson (2014, 136–137) detects "a certain sense of anxiety" in Seneca's use of Phaethon in this passage; again, he is not the unflappable sage but rather the all-too-human *proficiens*. For a similar use of mythological narrative, see Delle Donne (2021), who argues that Ulysses figures the *proficiens* in multiple Senecan texts.

investigator of nature reconfigures a haphazard inability to ascertain accurate causes into a daringly provisional exploration of the unknown and the unknowable; the investigation of the natural world is a task without end, certainty can remain elusive, and Seneca hopes that he will be able to advance undeterred, to make meaningful progress, to leap boldly into the abyss. Indeed, Seneca's analyses often leave the search for definitive *causae* unresolved, there is a marked ambivalence about the need to be exhaustive and systematic, and he does not hesitate to adopt conflicting positions and irreconcilable modes of discourse or explanation.[30] In this regard, the analytical fabric of the *NQ*, like that of the *Astronomica*, is shifting, open-ended, and all-embracing.

At the same time, we cannot separate the search for meteorological truth from the pursuit of virtue and the lofty goal of transcending our own mortality (1 praef. 17); as Seneca puts it elsewhere, the promise of philosophy is the assimilation to god (*hoc enim est quod mihi philosophia promittit, ut parem deo faciat*, *Ep.* 48.11).[31] This journey, too, is full of false starts. Virtue is magnificent (*magnifica*, 1 praef. 6) because it expands the mind, prepares it for the study of celestial things, and makes it worthy to come into partnership with god, but it is a quality after which we strive (*adfectamus*, 1 praef. 6), not one that we firmly possess. Whereas Manilius communes with the divine and knows the *mundus*, both mathematically and mystically, Seneca is always trying to make progress, always slipping and hastening to get back up (cf. Leitão 1998, 159; Williams 2012, 336). As soon as he proclaims the that we will transcend our own mortality, he advances even further and articulates a daring proposition: through such study, we will recognize that all things are small once we have measured god (*sciam omnia angusta esse mensus deum*, 1 praef. 17). But no such measurements are forthcoming; the proposition is aspirational. The tension between elevation and collapse, between uplift and deflation, enacts the dual logic of the sublime. Phaethon

[30] See Graver (2000, 52–53, on philosophical grounds, Seneca is "strangely uneven," "sometimes gossipy and undemanding"); Hine (2006, 51, "at the level of terrestrial or meteorological phenomena there is no aspiration to completeness"), (2010, 6–7, on the "disconcerting" effect of self-contradictions); Lehoux (2012, 102–103, on Seneca's multiplicity of voices and subjectivities); Gunderson (2015, 58–59, "Seneca's scientific method admits of multiple orders of explanation" and on the "heterogeneity of Senecan *ratio*"); Tutrone (2017, 775, on Seneca's "philosophically inclusive attitude"). Donini (1979, 219) nicely notes that in the last years of his life Seneca wrote "sotto la tutela della musa dell'ambiguità."

[31] See Setaioli (2007, 363–367) on assimilation to, collaboration with, and surpassing of god in Seneca. Porter (2020, 271), by contrast, concludes that for Seneca the study of nature "exists to teach us humility," rather than to exalt us beyond the limits of the human. My emphasis on precarity and the fall dovetails with Porter's position.

is its avatar, and while he may have seen the all with his mind (*animo omne uidisse*, 3 praef. 10), no one can long maintain that essentially divine perspective (cf. Gunderson 2015, 68). Still, Phaethon's story serves a hortatory function, summoning readers to follow Seneca's lead and to embark on a wild journey that brings them, if only for a moment, closer to truth and to god, closer to that mindset that allows us to leap rather than to fall.

This tension is hardwired into the ancient meteorological tradition precisely because meteorology is concerned with, as Margaret Graver (2000, 45) writes, "just those phenomena that take place above or . . . beneath or beyond the limits of observation." But this tension does not necessarily register as sublime. Aristotle, for one, offers a dispassionate assessment of the limits of meteorology at *Mete.* 1.1.338a–339a that in no way evokes the sublime. But Seneca exploits this tension repeatedly and at key junctures. For instance, as he brings his critical doxography of cometary theory to a close,[32] he makes the following statement (7.29.3):

> haec sunt quae aut alios mouere ad cometas pertinentia aut me: quae an uera sint, di sciunt, quibus ueri scientia est. nos rimari illa et coniectura ire in occulta tantum licet, nec cum fiducia inueniendi nec sine spe.
>
> These are the ideas relating to comets that have inspired others or me. Whether they are true, only the gods know, for they have knowledge of the truth. We are permitted only to explore these things and to enter into the darkness by guesswork, with no confidence of making a discovery but not without hope.

The theories canvased in the preceding chapters rouse and excite (*mouere*) Seneca and his intellectual predecessors and contemporaries. This state of affective intensity signals an approach to the divine—if this theory is correct, what is to separate us from the gods? The process will either elevate us to their level or lead to nowhere. Due to our fundamental ignorance and the basic weakness of our senses and faculties, though, we will never know whether we are making progress. We remain in a state of suspense, forever groping around in the dark, motivated not by certainty but by hope (*nec sine spe*). This assessment applies to the whole of the *NQ* and marks the study of nature as a precarious adventure that leaves the investigator teetering on the brink of god-like triumph and a spectacular fall.

[32] I borrow "critical doxography" from Hine (2006, 56–58), who stresses Seneca's active engagement with the theories he cites. For ancient doxography in general, see J. Mansfeld at *CHHP* 17–19.

Seneca's own definition of meteorology underscores the sublimity of this thrilling balancing act. At the beginning of book 2, Seneca declares that "the whole investigation of the universe is divided into the study of heavenly things (*caelestia*), elevated things (*sublimia*), and earthly things (*terrena*)" (2.1.1). Seneca's topic, for the most part, is *sublimia*, a calque for the Greek *meteōra* or *metarsia* (Hine 1981, 125–126). But *sublimia*, like its Greek equivalents, also means "the sublime" or "things that are sublime"; Seneca is interested as much in the natural phenomena as in our experience of them and their potential to elevate and expand the mind.[33] Throughout the *NQ*, the sublime primarily occupies a middle position between the terrestrial and the celestial (*caelestia, sublimia, terrena*). The experience of sublimity that most interests Seneca and that features most prominently in this treatise unfolds physically, ethically, spiritually below that of the *Astronomica*. The inquiry into *sublimia* is an exhilarating adventure that is marked by intoxicating success and, more often than not, a tumbling back down to earth. It prepares us to learn about higher things and to transcend our own mortality, but we continually return to ground level.

3.3. Working with Others

Phaethon's programmatic role in book 3 sets the tone for the *NQ* as a whole. But while Seneca does imagine himself traversing the heavens at the outset, he does not do so again, and there are good reasons for this. At the political level, the relationship between Phaethon and imperial power appears to have shifted since the late Augustan period. Nero was an enthusiastic charioteer, and multiple texts connect or appear to connect the emperor with Phaethon (see Ch. 4.2, "Nero-Phaethon"). Given Nero's jealous disposition, perhaps the safer course of action is to cede the role of Phaethon to the *princeps*. At the very least, this kind of anxiety might explain why Seneca does not explicitly name Phaethon in the preface to book 3. At the philosophical level, the treatise's primary topic is *sublimia*, not *caelestia*. The emphasis on the sub-celestial world makes Phaethon a less apposite role model for Seneca than for Manilius. From a methodological perspective, however, and this is crucial, Phaethon is a solitary figure who undertakes a

[33] Mazzoli makes the connection multiple times (1970, 49; 1990, 89; 1991, 177–180). Seneca uses *sublimis* in precisely this sense elsewhere (e.g., *cum Attalum audirem... illum sublimem altioremque humano fastigio credidi, Ep.* 108.13).

solitary mission. Seneca, by contrast, conspicuously embeds himself in the history of meteorological inquiry. Throughout the *NQ*, he constantly engages that history, citing a wide range of authors and restlessly testing out and evaluating their theories and positions; a profound sense of dialogue and cooperation with this "community of scholars" permeates the text.[34] It should not come as a surprise, then, that Seneca encodes the sublimity of his project through other figures who are more straightforwardly connected with the topics that he investigates and who allow him to draw attention to the importance of intellectual dialogue and cooperation.

In book 5, Seneca pauses his account of winds that emerge from underground to tell a story that he came across in Asclepiodotus.[35] According to Asclepiodotus, Philip of Macedon once sent an expedition into an old mine to examine its condition and to ascertain what riches were left for them to harvest (5.15.1–2):[36]

> descendisse illos cum multo lumine et multos duraturo dies, deinde longa uia fatigatos uidisse flumina ingentia et conceptus aquarum inertium uastos, pares nostris, ne compressos quidem terra superimminente, sed liberae laxitatis, non sine horrore uisos. cum magna hoc legi uoluptate. intellexi enim saeculum nostrum non nouis uitiis sed iam inde antiquitus traditis laborare, nec nostra aetate primum auaritiam uenas terrarum lapidumque rimatam in tenebris male abstrusa quaesisse.
>
> They descended with a plentiful supply of light that would last many days. Then, exhausted by the long journey, they saw massive rivers and boundless reservoirs of stagnant water, just like our own, not even compressed by the earth hanging over them but having free range, and they saw them not without a shudder. I read this with great pleasure. For I understood that our age was suffering not from novel vices but from ones handed down from long ago, and it was not in our time that greed had first peered into veins of earth and rocks and searched in the shadows for things that had been poorly buried.

[34] The phrase comes from Hine (2006, 53–60). For our purposes here, it does not matter how many of these sources Seneca consulted directly; he cultivates the impression of entering into dialogue with all of them.

[35] Seneca calls Asclepiodotus an *auditor* of Posidonius (2.26.6, 6.17.3), and many identify him with the Asclepiodotus who authored the extant treatise on military theory. For discussion, see Vottero (1989, 28–32).

[36] For the identification of this Philip with Philip II, Alexander's father, see Williams (2012, 81 n.95).

The passage points to the opening of *De rerum natura* 3, where *horror* and *uoluptas* seize Lucretius as he realizes that the underworld does not exist and has a vision of atoms moving through the void (3.25–30).[37] As we saw in Ch. 1.3, this scene encodes the sublimity of studying the natural world through texts. Seneca's narrative encodes the same experience, but with a twist. Philip's men have discovered "a new arrangement of nature" (*nouam rerum positionem*, 5.15.4); their discovery, which Seneca encounters in Asclepiodotus's text, corresponds to Epicurus's exposition of the atomic system. But the members of the expedition experience only *horror*, just one half of the dual experience of the sublime. The problem lies in the actors' moral compass, the fact that greed motivates their sordid inspection of things pointedly not on high. Such individuals are not open to wonder and cannot experience the sublime. It is Seneca, by contrast, for whom *uoluptas* accompanies *horror* as he reads a text about the natural world and discovers a profound truth about human nature: greed is a foundational and perhaps ineradicable vice, the degrading consequences of which long predate Seneca's own avaricious age.[38] Seneca rounds off the episode by circling back to Lucretius: despite the fact that this expedition discovered a new reality, the explorers still fear the underworld (*deinde cum ista fecerunt inferos metuunt*, 5.15.4), the very fear that Lucretius's vision had dispelled. Philip's men may not have made the ethical progress signaled by the sublime, but perhaps Seneca and his readers have.

It is significant that Seneca so clearly ties his sublime experience to a textual encounter: reading is one of the primary ways by which he gains access to those aspects of the natural world that interest him, and it is through reading that both he and his own readers learn about those unfamiliar arrangements of nature. The story about the expedition into the mine dramatizes the sublime experience of reading texts about natural phenomena and venturing into the unknown. Put differently, it dramatizes the sublime experience of engaging the community of scholars in an intergenerational dialogue about the natural world. At the same time, this commitment to intergenerational dialogue gives thematic weight to Seneca's ongoing conversation with anonymous interlocutors (emphasis on the plural), which might otherwise be written off as a mere stylistic convention. Taken together,

[37] Porter (2016, 455) observes the allusion.

[38] Seneca's indignation is topical; see Glauthier (2011, 190–193). For more on Seneca and mining, see Zientek (2020, 78–82).

these conversations show the reader how to engage with texts, Seneca's own above all. Meteorological inquiry emerges as a fundamentally cooperative and, by and large, textual enterprise that, as Seneca's experience of Asclepiodotus demonstrates, potentially unleashes the electrifying current of the sublime.

At other points in the *NQ*, Seneca returns to this nexus of ideas. The fragmentary book 4a originally explored two longstanding questions about the Nile that, by Seneca's time, had come to represent quintessentially unanswerable problems: where does the Nile rise and why does it flood during the summer? (For the history of these questions and their symbolic value, see Ch. 4.6.) As he recounts the river's known course, Seneca relays the following information (4a.2.6):

> inter miracula fluminis incredibilem incolarum audaciam accepi. bini paruula nauigia conscendunt, quorum alter nauem regit, alter exhaurit. deinde multum inter rapidam insaniam Nili et reciprocos fluctus uolutati, tandem tenuissimos canales tenent, per quos angusta rupium effugiunt; et cum toto flumine effusi, nauigium ruens manu temperant, magnoque spectantium metu in caput missi, cum iam deploraueris mersosque atque obrutos tanta mole credideris, longe ab eo in quem ciciderunt loco nauigant, tormenti modo missi. nec mergit illos cadens unda, sed planis aquis tradit.
>
> Among the river's wonders, I have heard about the unbelievable boldness of the locals. They mount tiny boats in twos; one of them steers, the other bails out bilge water. Then, after getting tossed about a great deal among the Nile's rushing madness and back-and-forth waves, at long last they reach very tight channels through which they escape the narrow crags. Swept along with the entire river, they guide the rushing boat by hand, and, to the great fear of the onlookers, they are tossed on their heads, and when you have despaired of them and believe them sunk and overwhelmed by such a mass of water, they are sailing far off from the spot where they fell, sent forth as if from a catapult. The falling wave does not submerge them but rather passes them along through calm waters.

The narrative, explicitly characterized as a *miraculum*, encodes the sublime: the fate of the boat produces great fear in the spectators (*magnoque spectantium metu*), the reader is invited to experience the same fear (*deploraueris, credideris*), and that fear is ultimately resolved into exaltation when the boat dramatically reemerges on the other side of danger (*longe ab eo in quem cediderunt loco nauigant*) and continues its journey in calm waters (*planis aquis*). At the same time, the boat's sublime journey offers a compelling image for the sublime journey of natural inquiry. The expedition demands excessive boldness (*incredibilem... audaciam*) and is dangerous and exhilarating, full of rapid-fire ups and downs, complete with a headlong fall (*in caput missi, ceciderunt*); in all of this, the *incolae* resemble Phaethon. The fact that the *incolae* work in pairs, though, adds an important twist: cooperative dialogue. Insofar as they navigate the Nile before a crowd of spectators, the *incolae* dramatize Seneca's textual navigation of the Nile before the reader, and their cooperative engagement symbolizes Seneca's dialogue with the community of scholars. To engage in such an adventure or simply to read about it sparks the affective intensity of the sublime, which, again, is an enticement to study.

This reading of the story about the Egyptian *incolae* suggests the extent to which Seneca's project not only relies on textual dialogue but is also inescapably embedded in complex social networks that reach across the Empire. Seneca, at various levels of remove, draws on local traditions and knowledge-systems in his attempt to survey the world, and he can experience these encounters as sublime.[39] A good example occurs in *NQ* 6.8, where Seneca tries to demonstrate that subterranean waters must exist. His culminating piece of evidence comes from a story about two centurions whom Nero reportedly sent to find the source of—once again—the Nile (6.8.3–5). Seneca himself has heard the centurions narrating the long journey (*audiui narrantes longum ipsos iter peregisse*, 6.8.3) on which they were assisted by the king of Ethiopia and given recommendations to the neighboring rulers, all of which allowed them to "penetrate to what lay beyond" (*ad ulteriora penetrassent*, 6.8.3). The centurions, working in pairs and assisted by networks of power, cross boundaries and reach into the unknown. What they found resembled a kind of primordial chaos (6.8.4):

> "inde" aiebant "peruenimus ad inmensas paludes, quarum exitum nec incolae nouerant nec sperare quisquam potest: ita implicatae aquis herbae sunt et <herbis> aquae, nec pediti eluctabiles nec nauigio, quod nisi

[39] In general, see Hine (2006, 50–53). In the case of the Nile, Seneca might be working in dialogue with his earlier treatise *De situ et sacris Aegyptiorum* (Serv. ad *Aen.* 6. 154 = fr. 19 Vottero). See Williams (2012, 117), with ample bibliography, on Seneca's many connections to Egypt.

> paruum et unius capax limosa et obsita palus non fert. ibi" inquit "uidimus duas petras, ex quibus ingens uis fluminis excidebat."
>
> "Next," they said, "we arrived at boundless marshlands. The locals did not know where they ended, nor can anyone hope [to know]: the weeds are so entangled with the waters and the waters <with the weeds>. They are impassable, both for one on foot and for a boat; the muddy, overgrown marsh does not support a boat unless it is small and carries only one person. There," he said, "we saw two rocks from which the mighty force of a river was falling."

The pair journeys to the point beyond which journeying in pairs is no longer possible. At the heart of this limitless expanse, they discover a mighty force, and they bring back their news to Rome. Cast in this light, the narrative resembles Lucretius's account of Epicurus's expedition beyond the *moenia mundi* (1.73), whence he conveys to others what is possible and what is not and the limits to everything's power (*potestas*, 1.76). Without claiming that Seneca alludes to Lucretius here, I suggest he aims to create a similar experience of sublimity. While topography hampers the reach of flesh-and-blood explorers, the intellectual quests of Epicurus and Seneca know no bounds. Seneca, though, sets the expedition in the real world, and we can infer that one of the preconditions for undertaking such an enterprise is cooperation with others.

It is noteworthy that this expedition appears to be historical. Other sources mention Nero dispatching troops to the Upper Nile and Ethiopia in anticipation of launching a military invasion, and Seneca's centurions may have formed a part of this mission.[40] Whether or not there was more than one expedition, Seneca frames this excursion as purely intellectual ("Nero Caesar, a great lover of the other virtues and especially of truth, had sent them to track down the source of the Nile," *quos* [sc. *centuriones*] *Nero Caesar, ut aliarum uirtutem ita ueritatis in primis amantissimus, ad inuestigandum Nili caput miserat*, 6.8.3), thereby suggesting Nero's personal investment in meteorological inquiry. At this point, we should ask under what circumstances Seneca is supposed to have heard the centurions

[40] Other sources: Plin. *HN* 6.181, 184–186; 12.19; Cass. Dio 63.8.1–2. The relationship between these expeditions and the one that Seneca mentions has generated considerable scholarly interest; see De Vivo (1998, 165–181), Gauly (2004, 198–199), Hine (2006, 63–65), De Vivo (2012, 102–104), Williams (2012, 235–236), Doody (2013, 297–299), and Wilson (2014, 175–176).

narrating their journey. Presumably, it was in the presence of the emperor, but we can easily imagine a larger and more public forum for disseminating the centurions' tale. From the perspective of the *NQ*, then, the study of *sublimia* is a popular pursuit, one that interests the emperor personally and can attract an audience. This brings us back to the question of success and failure. Do the centurions find the proverbially unknowable source of the Nile? Seneca refuses to decide, pointing out only that the water must have emerged from underground (6.8.5). Not infrequently, Seneca's brand of inquiry cannot offer conclusive answers to the problems that it sets itself, and this inconclusiveness is both confounding and exhilarating. At least some of those who heard the centurions' report along with Seneca are likely to have felt the same.

One of Seneca's primary collaborators has yet to be mentioned—namely, the reader. In an important article, Courtney Ann Roby (2014) explores how Seneca creates models and stages experiments through rich descriptions that engage and absorb the reader. For instance, she draws attention to Seneca's account of a "halo" (in modern terminology, a corona; 1.2.2):

> cum in piscinam lapis missus est, uidemus in multos orbes aquam discedere et fieri primum angustissimum orbem, deinde laxiorem, ac deinde alios maiores, donec euanescat impetus et in planitiem inmotarum aquarum soluatur. tale quiddam cogitemus fieri etiam in aëre: cum spissior factus est sentire plagam potest; lux solis aut lunae uel cuiuslibet sideris incurrens recedere illum in circulos cogit.
>
> When a stone is thrown into a pool, we see the water recede in many circles: at first, the narrowest circle is produced, then a broader one, and then other greater ones, until the force fades away and is resolved into the flat surface of unmoved waters. Let's suppose that something like this happens in air as well: when it is made thicker, it can experience a blow; the light of the sun or moon or any star at all, striking it, compels it to withdraw in circles.

While Seneca relies on analogical thinking here, Roby underscores that he is also doing something remarkably innovative:

> He simulates a viewing experience shared with the reader (*uidemus, cogitemus*), turning the text itself into a kind of virtual experimental space,

> realized by his vividly detailed description. He is not satisfied to assert that water currents and air currents sparked by a point impact have the same shape. Instead he sets the water flowing before our eyes, lets the ripple subside, and shifts the quieted substrate to air as smoothly as the ripple died down: now it is the impact of a ray of light that strikes the substrate, and the "halo" of a star that ripples outward.[41]

Seneca's thought experiments are provisional, collaborative, and dramatic, and they saturate the treatise. To quote Roby again, "Seneca creates opportunities for the reader to become involved in this kind of experimental theatre throughout the text."[42] Based on the paradigms considered above, we should conceptualize the experience of participating in such theatrical philosophy as sublime. Through the *NQ*, then, we become not just audience members who passively listen to the reports of others but practicing philosophers who relentlessly probe the properties of nature's elements and adventurous explorers who boldly chart the remotest regions of the earth; Seneca's vibrant and engaging prose turns the words of scholars and centurions alike into an interactive spectacle.

3.4. Comets and *magnitudo mundi*

Because certainty often remains elusive, Seneca is prepared to take intellectual leaps of faith—to push back against commonly accepted ideas about nature and reality, to propose theories that logic may be able to challenge, but that throw into relief the undeniable greatness of the cosmos (*magnitudo mundi*). In such cases, the motivation for accepting an explanation of a particular phenomenon is grounded less in physics or meteorology than in the conviction that the world ought to expand and ennoble the reader's consciousness. Seneca's treatment of comets in *NQ* 7 is the best example of this mode of excitedly elastic thinking. At the beginning of the book, Seneca asks the reader to consider whether comets might be planets (7.1.5–2.1).

[41] Roby (2014, 163). A key part of Roby's agenda in this article is to push back on outdated views of analogy. The standard account of Seneca's analogies in the *NQ* is Armisen-Marchetti (1989, 283–308).

[42] Roby (2014, 163). Gunderson (2015, 56–73) repeatedly conveys the dynamic and often dramatic quality of the interactions between different voices in the text. Lehoux (2012, 102–105) argues that we should think of Seneca as performing a courtroom speech.

The proposition has a long history, and Seneca returns to it at several points, rehearsing various arguments that one might use to undermine it.[43] Two-thirds of the way through the book, the reader is left with the impression that comets cannot be planets, and so it comes as an electrifying surprise when Seneca argues in the end that comets are in fact planets because planets are not really what we think they are: comets are extra-zodiacal planets that move beyond the confines of the ecliptic (7.22–25). The rationale for accepting his own proposition is clear: it befits the *magnitudo mundi* (*uide ne hoc magis deceat magnitudinem mundi*, 7.24.2), and contemplation of the *magnitudo mundi* can unleash our own *magnitudo animi*.[44]

Seneca's critical doxography surveys three main explanations for comets: comets are generated by whirlwinds in the atmosphere; they are optical illusions that appear when planets are in conjunction; they are themselves planets whose orbits are simply unknown.[45] The account of the optical illusion theory contains a forceful rebuke of a shadowy Artemidorus.[46] According to Seneca, Artemidorus held that while we have observed only five stars that move erratically (planets), countless others exist that are simply unknown due to the dimness of their light or the orientation of their orbits, and when these stars appear, their light blends with the light of the fixed stars, which produces the optical illusion that we call a comet (7.13.1). After quoting Artemidorus, Seneca denounces his optical illusion theory as the least important of the lies that he tells (*hoc ex his quae mentitur leuissimum est*) and condemns Artemidorus's entire system as a shameless falsehood (*tota eius enarratio mundi mendacium impudens est*, 7.13.2). Although Seneca's indignation is immediately directed at Artemidorus's insistence that the outer sphere of heaven is a solid mass, his scorn attaches to the very idea that multiple unknown planets exist, for he later asks, with incredulous disbelief, how one might prove that there are more than five planets and that they move in different regions of the universe (*quid est deinde quo*

[43] Aristotle (*Mete.* 1.6.342b–343a) ascribes a planetary theory of comets to unnamed Pythagoreans, as well as to Hippocrates of Chios and his pupil Aeschylus. Aetius 3.2.9 (Mansfeld and Runia 2023, 254) adds Diogenes (of Babylon), whom Kidd (1988, 495) tentatively identifies with the anonymous Stoic *quidam* at *NQ* 7.19.2. Although the planetary theory is sometimes characterized as an outlier (e.g., Gigon 1991, 326), it appears in Manilius (1.867–873, see below) and Pliny (*HN* 2.94) and is implied by Ovid when Caesar's comet-soul flies beyond the moon (*luna uolat altius*, *Met.* 15.848); the idea holds special appeal at Rome.

[44] The phrase *magnitudo mundi* occurs also at 4b.3. Cf. 2.7.6; *Ep.* 65.9, 19; *Dial.* 5.6.1.

[45] For ancient theories and classifications of comets, see Keyser (1994).

[46] Presumably to be identified with the Artemidorus of Parium cited at 1.4.3 and otherwise unattested; see Vottero (1989, 244).

probes non quinque tantum stellas moueri, sed multas esse et in multis mundi regionibus, 7.14.4). From Seneca's perspective, the optical illusion theory rests on no compelling evidence (*sine ullo probabili argumento*, 7.14.4), but one senses that the same verdict might be passed on the idea that unknown planets move along undiscovered trajectories.

Seneca records only one fully articulated version of the planetary theory, that of the otherwise unknown Apollonius of Myndus (7.17).[47] Apollonius holds that many comets are planets, that a comet is its own star, just like the sun and moon, and that they only become visible when they reach the lowest point of their orbits. Apollonius's position is met with three hypothetical objections (*respondetur*, 7.18): on Apollonius's theory, comets ought to get bigger the closer they approach, but in reality comets are always biggest when they first appear; if a comet were a planet, it would move along the ecliptic, where all planets move, but comets do not follow this pattern;[48] human vision cannot see through a star or planet, but we can see through comets; "therefore, it is clear that a comet is not a planet but a faint and haphazard fire" (*ex quo apparet illum non esse sidus sed leuem ignem ac tumultuarium*, 7.18.2). Although Seneca does not explicitly endorse the rebuttal, Apollonius's position looks untenable.

The point about the ecliptic is significant since it showcases Seneca's familiarity with basic planetary theory—and Seneca's own explanation of comets hinges on the existence of extra-zodiacal planets that do not move along the ecliptic. Seneca has already observed that the planets move along the ecliptic, while comets do not (7.12.8), and he initially introduces Artemidorus in precisely this context (7.13.1). Seneca's rejection of Artemidorus's position begins and ends with appeals to what appear to be incontrovertible astronomical fact. At other points, Seneca similarly presents planetary theory as straightforward and unproblematic. For instance, when he argues against the optical illusion theory, he takes for granted that the number of planets is fixed (*his illud respondebimus, certum esse numerum stellarum mobilium*, 7.12.2) and that both they and comets have appeared at the same time, which makes plain that comets are not the result

[47] Apollonius of Myndus is known only from Seneca, who mentions his expertise in casting horoscopes and his claim to have studied with the Chaldaeans; see *NQ* 7.4.1 with Vottero (1989, 671–672). Apollonius's familiarity with a comet that appeared in the reign of Nero (*NQ* 7.17.2) makes him appear to be a contemporary, and his interest in astrology, where planets loom large, perhaps explains his embrace of a planetary theory of comets.

[48] This is the first point that Aristotle raises when he critiques the cometary theories of his predecessors (*Mete.* 1.6.343a).

of two planets in conjunction. He then points out that certain planets regularly come into close contact with one another without producing comets and argues that if the theory were true, the high speeds at which planets move would mean that comets would only appear for a brief period of time, which is not the case (7.12.4); indeed, the "law" that governs stellar motion (*lex celeritatis suae*, 7.12.4) does not allow it. Seneca's language and thought process make the premises of planetary motion appear incontrovertible. At the end of his account of Stoic cometary theories (7.21.4), Seneca appeals to the observation that planets move in two directions simultaneously, both with the rotation of the cosmos (from east to west) and slowly in the opposite direction (from west to east); the uniformity of planetary motion does not square with the movements of comets. Even when he addresses the planets' retrograde motion, the thorniest of planetary problems, an explanation lies ready to hand: it's merely an illusion, caused by the approach of the sun (7.25.7).[49] Again, there is a clear appeal to the simplicity and straightforward nature of the planets. In his brief account of the history of planetary theory that immediately precedes the critical doxography (7.3.2), Seneca presents planetary motion similarly. There, he asserts that Democritus supposed that there are more planets but did not record their names or number because their orbits had not yet been recognized.[50] It was Euxodus who first introduced knowledge of the planets' orbits to Greece, having learned their movements from the Egyptians. In his critiques of Artemidorus and Apollonius and his general approach to planetary theory, Seneca appears to leave little room for doubt about the possibility that comets might be planets or that planets might behave in ways that the astronomers have failed to explain.

Given the survey of arguments that undermine the planetary theory of comets and its basic assumptions about planets, on the one hand, and the consistent representation of the nature of the planets as settled fact, on the other, it comes as a shock when Seneca ultimately argues that comets are extra-zodiacal planets whose orbits have not been discovered. The initial proposition and its subsequent defense implicitly characterize Seneca as a bold innovator who crosses intellectual boundaries and dares to imagine the unimaginable. We need to look at the details.

[49] For ancient approaches to retrograde motion, see Kuhn (1957, 45–77).

[50] This report is not necessarily part of Democritus's view on comets (cf. D100 *EGP*), although we know that he held them to be the conjunction of planets (D99 *EGP*).

Seneca introduces his position by setting himself in opposition to Stoic tradition: comets are not sudden fires that come out of nowhere (*subitaneum ignem*) but rather one of the eternal works of nature (*aeterna opera naturae*, 7.22.1). A positive argument follows. First, taking for granted that comets are fiery, Seneca observes that the only fires that move in a circle are stars. Although he does not know whether other comets have moved in a circle, he claims that two in his own lifetime have (*hoc an cometae alii fecerint nescio: duo nostra aetate fecerunt*, 7.23.1).[51] Next, Seneca points out that atmospheric fires, such as lightning, are short-lived, whereas stars burn eternally and always remain consistent in their appearance and trajectory. On this score, comets clearly have more in common with stars than with atmospheric fires. But Seneca wishes to go further (*nunc amplius adicio*, 7.23.3). Atmospheric fires cannot stay put or linger in one place, whereas celestial fires do just this. From here, Seneca concludes that a comet has its own location and orbit and that when one disappears from view, it is not extinguished but simply departs; in short, comets are planets.

By setting himself in opposition to the Stoics, Seneca marks himself as transgressive. The content of his initial formulation accomplishes the same task. The assertion that comets are one of nature's eternal works runs counter to the tentative conclusion, which is not formally endorsed, that a comet is a haphazard fire (*ignem... tumultuarium*, 7.18.2). The prominence of the contradiction shows that Seneca is happy to try out one line of argument here, another there, all in the pursuit of something grand or boundless—in this case, eternity (*aeterna opera naturae*, 7.22.1).[52] Indeed, the elaboration of a planetary theory of comets after an extended, if conditional, rejection of that very theory and its basic premise vividly demonstrates what the wild ride of meteorological inquiry looks like. To be more precise, meteorological inquiry itself now verges on or even becomes a new form of experimental astronomy; through Seneca's brand of study, intellectual categories start to break down.

From this perspective, several aspects of the argument and its subsequent defense are noteworthy. Seneca advances a claim that runs contrary to lived experience; on the evidence of our eyes, comets do come out of nowhere and do not behave like stars. A basic principle of the treatise, though, is that

[51] These are the comets of 54 and 60 CE; see Vottero (1989, 676) on 7.6.1.

[52] Seneca returns to this point later when he declares, once again in opposition to Stoic thinking, that comets are not chance fires but ones that are fundamentally connected to the universe (*illos non fortuitos esse ignes, sed intextos mundo*, 7.30.2).

it is a mistake to try to understand nature with the eyes rather than with *ratio* ("we grasp nature with our eyes, not reason, and we don't think about what she can do but only what she has done," *naturam oculis non ratione comprendimus, nec cogitamus quid illa facere possit sed tantum quid fecerit*, 6.3.2); too often, lived experience hampers our ability to imagine what nature is capable of accomplishing, to see the all with the mind (3 praef. 10). And yet, it is precisely philosophical *ratio* that has already undermined the theory that comets are zodiacal planets. Here, then, Seneca embraces a mode of experimental thinking that will not allow itself to be restrained by the evidence of the senses or the conclusions of previous investigations. He makes the necessary adjustment (*extra*-zodiacal planets) and moves on; his *ratio* knows no bounds.[53]

While Seneca does use visual evidence to establish that comets move in a circle, he stresses that he knows that this applies only to two recent comets, and signals that he has not bothered to seek out corroborating reports. This constitutes a conspicuous failure to do due diligence since Seneca insists at the beginning of the book that it is essential to consult records of past appearances of comets (7.3). Instead of allowing himself to be constrained by records and research, Seneca throws caution to the wind and constructs a theory that is bigger, greater, more impressive (*amplius*, 7.23.3) than a meticulous argument that requires a trip to the archives. Like Manilius, Longinus, and the Younger Pliny, Seneca is perfectly capable of disregarding the culturally sanctioned insistence on *akribeia* (Ch. 1.5, "Acrobats, Heroes, and Flawed Sublimity," and Ch. 2.3).

When an interlocutor objects that planets move along the ecliptic and comets do not (7.24.1)—the very objection to the planetary theory that has appeared twice before (7.12.8, 18.1)—Seneca offers responses that appeal not to logic but to the reader's susceptibility to the sublime: stars, including planets, are divine, and the divine accepts no limits; it might better suit "the magnificence of the universe" (*uide ne hoc magis deceat magnitudinem mundi*, 7.24.2) to have multiple orbits for planets, rather than just one that is situated along the ecliptic; no one can possibly believe that in this surpassingly great and most astonishing cosmos with its infinity of stars only five are allowed to wander, while the rest stand still, a motionless and inert multitude (*credis autem in hoc maximo et pulcherrimo corpore inter*

[53] Williams (2012) also stresses the experimental quality of Seneca's thinking in this book (esp. 258–259, 261, 275–276, 290).

innumerabiles stellas... quinque solas esse quibus exercere se liceat, ceteras stare fixum et inmobilem populum, 7.24.3).[54] The astronomer's supposedly expert knowledge about planetary motion constrains our ability to think expansively—why should planets behave only in the way that astronomy tells us? The incomprehensible vastness of the universe overwhelms the objection of the interlocutor who argues from established fact as Seneca discards the astronomical truths to which he had seemed previously to hew.[55] Seneca's position may or may not describe reality (cf. 7.24.2); what matters is that it challenges our basic assumptions about the world and forces us to cross intellectual boundaries. As Seneca subsequently writes, nature delights in her own multiplicity (*ipsa uarietate se iactat*, 7.27.5), and it is through her multiplicity that she enhances her own grandeur (*uoluit et his* [sc. the variety of comets] *magnitudinem operis sui colere*, 7.27.6); to study comets through the imaginative landscape of the *NQ* is to bring into sharp focus the *magnitudo mundi*.

This is how Seneca operates throughout the treatise. Book 7 is unique, though, in that Seneca frames his project in these very terms. At the outset, he laments that familiarity tends to blind us to the true greatness of nature (*magnitudinem rerum consuetudo subducit*, 7.1.1); only when something unusual happens do we pay attention and ask questions.[56] Comets are a case in point, but when one appears the casual observer does not know whether he should feel wonder or fear (*ignarus utrum debeat mirari an timere*, 7.1.5), since comets are often thought to be bad omens but might actually be heavenly bodies. The contrast is significant: *mirari* indicates the dual experience of the sublime, connected with a philosophical view of nature, while *timere* describes a simple, superstitious state of being overwhelmed. Through the treatise, Seneca attempts to lead the reader to this sense of wonder, to an experience of *magnitudo mundi* that overwhelms the faculties and opens the mind to new possibilities, an experience of wonder without fear.[57] At the opening of book 7, though, Seneca wishes us to

[54] Seneca's use of *populus* here suggests to Housman (1903–1930, 5.94) the image from the end of the *Astron*. 5: "greatest is the group that is carried along at the highest pinnacle" (*maximus est populus summo qui culmine fertur*, 5.742). For this passage, see Ch. 2.6.

[55] Interestingly, Gell. *NA* 14.1.11–12 attributes to Favorinus a similarly incredulous skepticism of astronomical dogma that manifests itself in speculation about unknown planets.

[56] We have already seen (Ch. 1.6, "Plato and Aristotle") that wonder can lead to philosophical inquiry. Aristotle also seems to have understood that familiarity can inure us to the sublimity of the natural world (Cic. *Nat. D.* 2.95 = Arist. fr. 838 Gigon).

[57] For wonder without fear in the *NQ*, cf. Toulze-Morisset (2004).

wonder at and to be attuned to the *magnitudo mundi*, in all its cacophonous complexity, when we confront not only uncommon or devastating phenomena but also all those aspects of nature that are entirely ordinary and familiar: we take no notice, he laments, of the infinity of stars or the movements of the sun and moon as long as regularity is maintained (7.1.4).[58] Seneca will disrupt that soothing sense of order and flood the natural world with aporia, a bold and daring act. In this context, aporia symbolizes the recognition of epistemological limitations, a commitment to cast aside traditional categories and modes of thought, a willingness to jettison the familiar cast of characters in favor of virtual unknowns, like Artemidorus and Apollonius.[59] Exuberant and unapologetic aporia, not without hope of success (*nec sine spe*, 7.29.3), is the defining attitude of Seneca's brand of natural inquiry, and it has all the quickening thrill of the sublime.[60]

To round out our consideration of book 7, it will be helpful to consider two sets of comparisons, both of which shed light on the distance that separates Seneca from Manilius. We can read Seneca's treatment of comets against Manilius' own analysis of the same topic (1.809–926). At the end of *Astron.* 1, Manilius discusses three explanations for comets: the view that comets are caused by terrestrial exhalations, a planetary theory, and a less technical hypothesis that comets are signs sent from god.[61] Interwoven with the exposition of causes is considerable extraneous information, including repetition, contrafactual speculation, and historical narrative. Overall, this narrative operates much like that of the Milky Way: it allows Manilius to accumulate a seemingly endless amount of detail, to toggle back and forth between intellectual discourses, and to treat as fact ideas that blatantly contradict the broader astrological worldview that he elsewhere espouses. This is the modus operandi of the celestial daredevil. From this angle, the exploration of comets reveals fundamental similarities between Manilius and Seneca. While Manilius's anonymized doxography simply lists competing explanations and Seneca's critical doxography forcefully interrogates the

[58] Similarly Master (2015, 347–348).

[59] Cf. Hine (2006, 61) on the symbolic value of Artemidorus, Apollonius, and Epigenes.

[60] This is an important corrective to the view of Rosenmeyer (2000, 111) that "the air of adventure that informs [the *NQ*] rests on a pessimistic sense of surprise and randomness." Donini (1979, 214) attributes disorder in the *NQ* to role of humans. This view is essentially correct, since human imperfections make it all but impossible to perceive the order of the cosmos, but Donini is thinking about Seneca's castigations of morally bankrupt figures who act contrary to nature.

[61] For discussion and bibliography, see Volk (2009, 46–48) and Glauthier (2017, 291–296).

views of specific individuals, both authors test the demands of logical consistency and construct narratives that animate and thrill.

But another aspect of the planetary theory complicates this comparison. As Seneca continues to defend his theory, an interlocutor again objects: Why has the path of comets not been observed in the same way as that of the zodiacal planets (7.25.1)? The question leads to a bold analogy that likens comets to the human mind: we grant that the human mind (*animus*) exists even if we do not know what or where it is.[62] As if to dramatize the point, Seneca lists a variety of theories about the nature of the mind before drawing a conclusion: the mind is so far from having clarity about other things that it is still investigating itself (7.25.2). Now, despite the confident assertion of the interlocutor, and elsewhere of Seneca himself, that the paths of the zodiacal planets are well-known, their orbits constituted a paradigm of uncertainty from a very early date, requiring ever more complex models and calculations, none of which ever offered lasting resolution.[63] Moreover, Emma Gee has convincingly shown that planetary motion functions as a symbol of disorder in imperial Latin literature, including in Neronian Rome.[64] Manilius has little to say about the planets, almost certainly because their unpredictability threatens to inject chaos and disorder into his well-oiled celestial system (Volk 2009, 48–57). Seneca, by contrast, embraces and even amplifies the unknowability of the planets by positing the existence of additional extra-zodiacal planets with even more fantastical and inexplicable orbits. The very idea that only five stars wander while an infinite number of stars stands fixed and immovable is an affront to Seneca's conception of the universe. Elsewhere, he goes so far as to suggest that perhaps some of the fixed stars are not actually fixed—perhaps they move as quickly as the cosmic sphere and so only appear motionless (*propter uelocitatem uniuerso parem inmotis similia*, 2.32.8). Seneca wants more erratic mobility and less confident understanding; that is what befits the *magnitudo mundi*. Indeed, Seneca's energetic reflections on the limits of human understanding lead him to exclaim that god lies hidden and implicitly cannot be known, at least not on the first try, nor perhaps on the second; as with certain mystery

[62] For Seneca's bold analogies, see Williams (2012, 220–225). This example is unprecedented in ancient speculation about comets, even if it is based on the standard comparison of the universe to a living being and appears in doxographical and skeptical sources. On this last point, see Runia (2002, 307). Porter (2020, 273–275) connects this passage with Sen. *Ep.* 121.12.

[63] For ancient planetary theory, see Kuhn (1957) and Evans (1998, 289–443).

[64] Gee (2013, 110, 118–129), focusing on Manilius, Lucan, and Statius.

cults, one must be initiated repeatedly into the study of nature (7.30.4–6). "Philosophy as mystery cult" is a well-worn trope, but here the conceit is particularly apt: progress is slow and requires multiple attempts, and the entire experience is wildly thrilling.[65] Whereas Manilius flies through the sky with inexhaustible confidence of one who has already communed with god, Seneca's encounter with the *magnitudo mundi* results in the ecstatic aporia of a low-level initiand, one poised on the brink of a new revelation and eager to learn more, if only he can. Eventually, we are told, the community of scholars will untangle all the various questions that perplex us in the here and now. A generational view of intellectual progress, however, does not obscure the significance of a quasi-religious aporia in the present, and too often, critics focus on the optimistic implications of the progress narrative. It is this experience of aporia, in all its sublimity, that draws us into the study of the natural world.[66]

The elevation of aporia, in contrast to the worldview of Manilius, reflects the changing circumstances of virtually every aspect of public life as one emerges from the landscape of the late Augustan Principate into the final years of Nero's regime. In a world of theatrical play and radical uncertainty, it no longer makes sense for any discipline to claim a privileged relationship to truth and reality, including the truth and reality of the fixed stars. Even in a moment where Seneca seems more confident in humankind's abilities, a close reading suggests profound ambivalence. Originally, book 7 was followed by book 1. In the preface to book 1, Seneca presents the acquisition of true celestial knowledge as a sublime experience that remains tantalizingly out of reach (1 praef. 6–13). "That virtue after which we strive," writes Seneca, "is magnificent... because it expands the mind and prepares it for the study of celestial things and makes it worthy to come into partnership with god" (*uirtus enim ista quam adfectamus magnifica est... quia animum laxat et praeparat ad cognitionem caelestium, dignumque efficit qui in consortium <cum> deo ueniat*, 1 praef. 6). The thought of realizing *magnitudo animi* is electric: "It is thrilling [for virtue] to wander among the stars themselves and to laugh at the paved floors of the wealthy and the whole

[65] For "the extraordinary experience" of the mysteries, which includes a confrontation with the divine and constitutes a form of transcendence, see Burkert (1987, 89–93).

[66] For Seneca's conception of progress, see Parroni (2000, 441–442), La Penna (2001, 548–550), Hine (2006, 57–58), and Williams (2012, 263–268). On the tension between belief in progress and a more pessimistic skepticism, see Runia (2002, 306–330) and Limburg (2007, 349–356). Ozanam (1990, 287–288) argues that Seneca (and Lucan) respond to the rationalist tendencies of contemporary Stoicism with "un certain mysticisme."

earth with all its gold" (*tunc iuuat inter ipsa sidera uagantem diuitum pauimenta ridere et totam cum auro suo terram*, 1. praef. 7). Virtue's leisurely stroll looks back to the opening of Manilius's first book, where it is similarly thrilling to roam aimlessly about the heavens ("it is thrilling to go through the air itself and to live roaming about in the boundless sky," *iuuat ire per ipsum / aera et immenso spatiantem uiuere caelo*, 1.13–14).[67] But whereas Manilius relishes the ability to live on high (*uiuere*), Seneca can only picture an abstract *uirtus* soaring into the heavens, a *uirtus* that we only ever pursue, and one that delights not in viewing the heavens but in laughing at those on earth who have not realized their own *magnitudo animi*. This *uirtus* is not an attainable subjectivity; it represents the perspective of the sage.

Later in the preface to book 1, Seneca returns obliquely to the difficulty of attaining such a perspective and its fleeting nature. He imagines the celestial journey of the mind (*animus*) that leaves behind the impediments of earthly life and watches, free from anxiety, the settings and risings of the stars and their fundamentally divergent and yet simultaneously regular paths (*secure spectat occasus siderum atque ortus et tam diuersas concordantium uias*, 1 praef. 12). At first, the *animus* appears to work on two levels, representing both the embodied mind of the philosopher who investigates the natural world and the disembodied soul that rises to the heavens after death.[68] But certain knowledge of the paths of the planets, including comets, is precisely what eludes Seneca in book 7, and at 7.25.2, as we have seen, it is none other than the *animus* that has no clear knowledge of either itself or the external world. The experience of the *animus* in book 1, then, is not one that is readily available to those on earth and cannot be viewed as that of the ordinary student of philosophy. To know the heavens, one must be virtuous, and we are always only striving after virtue, which makes the experience of the *animus* that of the sage, who is practically indistinguishable from a god. For such an individual, the community of scholars is irrelevant, and knowledge comes easily; this is the outlook of the *Astronomica*. In Seneca's view, the rest of us travel a much more uncertain path, constantly challenged and inspired by an inescapable sense of aporia.

[67] Note the parallels (Sen./Man.): *iuuat/iuuat, uagantem/spatiantem, inter ipsa sidera/per ipsum aera, ridere/uiuere*. For *iuuat*, "it is thrilling," see Ch. 4.2 ("Nero-Phaethon").

[68] Thus Leitão (1998, 139 n.17); cf. Setaioli (2007, 354–355) and Williams (2012, 59–60, 293–294).

3.5. A Vital Calmness

Williams has persuasively shown that in *NQ* 6 Seneca deploys the sublime as part of a strategy of consolation.[69] Throughout the book, the destructive power of earthquakes terrorizes the reader. Philosophical analysis, however, demystifies the phenomenon, removing causation from the divine sphere, and demonstrates that death is not to be feared—in fact, the very act of studying the natural world strengthens the mind, opens it up to the cosmic viewpoint, and allows it to find tranquility.[70] In what follows, I offer two important caveats to this reading. First, we should not locate instability and danger only in descriptions of natural disasters. The world of the *NQ* is experienced as fundamentally volatile, at the elemental level, which means that danger is omnipresent. Theoretically, the universe might be harmoniously organized, but that is not how it appears in our day-to-day lives. Second, we need to keep in mind the distinction between the sage and the *proficiens* when we consider how Seneca utilizes the sublime to comfort and to fortify. The sage, like the *animus* in the preface to book 1, calmly surveys the *magnitudo mundi*, shielded from the hubbub of life on earth by the interiorization of the cosmic viewpoint; the sage does not need Seneca's *consolatio*.[71] Those who are only progressing toward wisdom and virtue, by contrast, experience that hubbub quite differently. Writing of Seneca's prose oeuvre as a whole, Jonathan Mannering (2013, 191) notes, "By virtue of its discomfiting paradoxes, its graphic, sensual, often violent imagery... and its obsessive ruminations on mortality and dying, Seneca's rehabilitative philosophy could rightly be branded 'shock therapy.'" Mannering's formulation suits the present context quite well. Seneca is aware that such shock therapy is itself a traumatizing experience that leaves the *proficiens* in a paradoxical state that is both still and agitated, what I will call a vital calmness. This state of mind differs significantly from that of the sage.

The elemental volatility of Seneca's world comes to view at every turn.[72] For our purposes, one example will suffice: the tension of air (2.6–11).

[69] Williams (2012, 213–257). Although Williams is primarily concerned with the sublime in book 6, its role at the end of book 3 receives careful consideration (113–116, 225, 255). For other approaches to book 6, often with political emphasis, see De Vivo (2012), Wessels (2016), and Wiener (2016, 205–208).

[70] Seneca articulates this goal clearly at 6.32.1–3. Images of cosmic dissolution serve a similarly consolatory function at *Dial.* 6.5–7.

[71] For the *magnitudo animi* of the sage in Seneca's prose works, see Williams (2016).

[72] Waiblinger (1977, 23) describes the focus of the treatise as a world where "Veränderung, Werden und Vergehen, Unruhe, Unordnung und Zufall herrschen"; Vottero (1998, 296–299) lists

Throughout this passage, Seneca writes interchangeably of *aër* and *spiritus*.[73] The argument is convoluted, but this is how he frames the matter: while it is easy enough to prove that certain violent (*uehementia*) and powerful things (*magnique impetus*) exist in nature, nothing is very violent without tension, and nothing possesses tension in and of itself more than *spiritus* (2.8). What does the violence of air in tension look like? When *spiritus* has violently roused and unleashed itself, like a missile (*cum uehementius concitatus ipse se torsit*), it tears apart trees and woods, rips up and smashes whole buildings, and stirs up the sea (2.6.4).[74] *Spiritus* also exercises its might below the level of sense perception and over imperceptibly long intervals of time (2.6.5):

> consideremus quae ingentem uim per occultum agunt: paruula admodum semina, et quorum exilitas in commissura lapidum locum inuenit, in tantum conualescunt ut ingentia saxa deturbent et monumenta dissoluant; scopulos interim rupesque radices minutissimae ac tenuissimae findunt. hoc quid est aliud quam intentio spiritus, sine qua nihil ualidum et contra quam nihil ualidum est?
>
> Let's consider things that exercise great force in secret: very tiny seeds, slight enough to find space in the crevices of rocks, grow powerful to such a point that they dislodge huge stones and break apart monuments, and sometimes the most miniscule and slender roots split cliffs and crags. What else can do this but the tension of breath, that without which and against which nothing is powerful?

Powerful seeds operating in empty spaces that we cannot perceive—Seneca's air mimics Lucretius's atoms. Despite the occasional celebration of nature's divinity and order, Seneca's universe is every bit as chaotic and violent as Lucretius's atomic battlefield. Moreover, Seneca's *spiritus* wreaks havoc on our mental state (*quid est aliud quo animus noster agitetur* [sc. *quam spiritus*], 2.6.6),

some of the many places in the text where "la natura stessa ci assale"; Rosenmeyer (2000, 109) emphasizes that, beyond the prefaces, "the accent is often on the unexpected and the exciting, on disasters and catastrophe, with little effort to accommodate their strangeness to a divinely constituted whole." Porter (2020) offers a broad reevaluation of the Stoic worldview in the imperial period that emphasizes similar qualities; for Seneca, see esp. 238–243, 265–275.

[73] See Hine (1981, 122–123, 139) and Williams (2012, 189–191, 247).

[74] The violence of nature here draws on familiar images for the sublime; cf. Long. *Subl.* 1.4, where the sublime tears up everything like a bolt of lightning.

letting loose the disruptive forces of nature deep within us; air is an awesome force.[75]

Seneca's reflections on air, however, subtly point in another direction. Although he presents 2.6–11 as a preliminary examination of the nature of *aër*, both elemental "air" and "atmosphere," that sets the stage for the book's discussion of thunder and lightning, it is *spiritus* and its tension that steal the spotlight. This emphasis suggests that we ought to read *spiritus* as Stoic breath (*pneuma*), a meaning that Seneca gives *spiritus* elsewhere in the *NQ* (e.g., 6.16).[76] The Stoics held that breath is composed of a total blending of fire and air, which exists in a state of tension, simultaneously expanding from the force of fire's heat and contracting from the pull of air's cold.[77] Breath itself pervades or is totally blended with the substance of the world, and its vital tension imparts tension to the inert elements water and earth. David Furley (*CHHP* 440) succinctly describes what became the common Stoic position: "By mixing with the whole world it exercises control over everything.... As the active ingredient in all things in the world, *pneuma* is responsible for the 'tension' that holds all the world and everything in it together." If, in *NQ* 2.6–11, we identify a slippage between elemental air and Stoic breath, the result is to impart to air a power and significance generally reserved for *pneuma*, and to characterize all-pervading *pneuma* as a threatening force of nature.[78]

Seneca, then, does not focus on the constructive power of breath's tension, the idea that it sustains the world and makes the existence of inanimate objects, plants, and animals possible. On the contrary, breath in tension spins out of control, and its dynamic force rips apart everything with which it comes into contact. Later in the book, Seneca reflects on the many terms by which we can accurately name god: fate, providence, nature, the

[75] This characterization of air recurs in other books (e.g., 6.16.4). For the sublime spectacle of violent wind, see Ch. 1.4 on Virtuv. 1.6.2. This is a pervasive theme in the *Aetna* as well (e.g., *spectacula uenti* at *Aetna* 384). For an early analysis of air's potency and might, see Hippoc. *Flat.* 3.

[76] See, again, Hine (1981, 122–123, 139) and Williams (2012, 189–191, 247).

[77] For the origins and development of the idea, see D. Furley at *CHHP* 440–441 and L-S 47. Wildberger (2006, 1.61, 2.575–576) challenges the traditional understanding of *pneuma/spiritus* as mixture; her argument does not affect my point here.

[78] In 2.6.6, Seneca explicitly states that *spiritus* holds (*teneret*) our bodies together and emphasizes its unity and tension (*intentio, unitas*); this is Stoic *pneuma*. At 2.9.4, he describes the ability of *spiritus* not simply to surround objects but to pass through (*transit*) and pervade (*permeat*) them; this is total blending, as Hine (2010, 212–13 n.14) suggests. Both passages push the reader to take *spiritus* as Stoic *pneuma* and to read the entire discussion (2.6–11) as an exploration of that concept; cf. Donini (1979, 210), Gigon (1991, 330), and Gunderson (2015, 64).

universe (2.45.2–3).[79] But, more specifically, he is also the breath of the cosmos (*spiritum mundi*, 2.45.1). The reader who remembers the invincible might of *spiritus*, so vividly described earlier on, may find the concept disturbing.

It is against this broad background of omnipresent instability and destruction that Seneca offers his readers relief in *NQ* 6. At the beginning of the book, Seneca draws attention to the fact that while he had promised *solacium* against infrequent dangers, he actually creates an overwhelming of sense of catastrophe (6.2.1). The motivation is consolatory. Seneca imagines three kinds of readers: the foolish (*stulti*); the inexperienced (*imperiti*), who are implicitly advancing toward wisdom and virtue; and the wise (*prudentes*).[80] The foolish will simply succumb to fear, and Seneca seems not to care (*sine remedio timor stultis est*, 6.2.1).[81] For the others, something different lies in store: "Reason strikes fear out of the wise, and for the inexperienced a great freedom from anxiety results from the abandonment of hope" (*ratio terrorem prudentibus excutit, imperitis magna fit ex desperatione securitas*, 6.2.1). Seneca's goal is to stage the dual logic of sublime for these two groups: the wise and the inexperienced will feel terror and then will overcome that terror, the former through pure *ratio*, and the latter by learning to let go of false hopes and steeling themselves to the idea that everything is to be feared (*cogitate omnia esse metuenda*, 6.2.3). The experience of the sage is one to which we all aspire, but that of the inexperienced is more realistically attainable.

When the destructive power of nature confronts the wise, they emerge serene, dispassionate, and emotionally unscathed. Seneca's account of the learned man (*eruditissimo et grauissimo uiro*, 6.31.3) who was bathing at the time of the Campanian earthquake is a good example: he affirms that he saw (*uidisse se adfirmat*) the tiles in the bath separate and come back together, and that water entered the cracks when the tiles drew apart and was forced out again when they were compressed. In addition, Seneca heard the same individual relating that he had seen (*eundem audiui narrantem uidisse se*)

[79] See Wildberger (2006, 1.21–48) and Kreuzwieser (2016, 30–33). D. Furley at *CHHP* 448–451 offers a succinct account of the cosmology and teleology that this passage implies.

[80] Unnecessary editorial interventions have sometimes obscured the threefold classification here. Alexander (1948, 309–311), however, argues cogently for the transmitted text. He cites, inter alia, *Ep.* 76.34–35, where Seneca contrasts *stulti*, *imperiti*, and the *sapiens*, and *Ben.* 7.17.1, where Seneca includes himself among the *imperiti*. On Alexander's view, the *imperiti* are "persons of good natural quality who have not…attained to anything near the insight of the sages," "those who have not yet completed the course" (i.e., the *proficientes*).

[81] Hine (1996a, 234; 1996b, 94–95) emends the text to *sine remedio timor stultis <salutaris> est*. This creates only two groups of readers, the *stulti*/*imperiti* and the *prudentes*.

walls shaking with more flexibility and with greater frequency than the nature of matter allows (*macerias mollius crebriusque tremere quam natura duri sinit*). There is no trepidation, no anxiety, even though, for all intents and purposes, this figure has discovered a new configuration of nature; unlike the explorers whom Philip sent into the mine (Ch. 3.2), Seneca's acquaintance does not experience *horror*. This is the subjectivity of the sage, and it is an aspirational goal for the less-than-perfect reader. By the end of the book, it may be possible for those who are progressing toward wisdom and virtue to experience the effects of the quake calmly, without trepidation, from a philosophically informed perspective.

The emotional state of the inexperienced (*magna... ex desperatione securitas*) is something very different. We should note, first, that *securitas* is entirely positive, and in a book so heavily indebted to Lucretius, it suggests Epicurean *ataraxia*.[82] But there's a paradox here. To illustrate his meaning, Seneca cites Aeneas's words to his troops in the midst of a burning Troy: we should tell ourselves that "there is one safety for the conquered, to hope for no safety" (*una salus uictis nullam sperare salutem*, *Aen.* 2.354 at *NQ* 6.2.2).[83] Although Seneca leaves it unsaid, the effect of Aeneas's words on his men is important—they unleash madness (*sic animis iuuenum furor additus*, *Aen.* 2.355). In the *Aeneid*, then, the abandonment of hope leads to dehumanizing delirium.[84] For Seneca, by contrast, to surrender to nature results in a form of serenity that is free from fear and simultaneously aware of nature's might. It constitutes a transformation of Virgilian *furor* into vital calmness, serene tension, or inspired *ataraxia*, a state that not a few readers have associated with Lucretius himself and that encapsulates the dual logic of the sublime. It is a mode of being in the world that keeps us alive to nature's volatility and tuned in to her power and that differs essentially from the peaceful stillness of the sage toward which we ought to aspire. This is the *solacium* for the *imperiti*, those of us who have not yet attained wisdom and virtue.

Seneca's account of the survivors of the universal flood in book 3 (3.27–30) describes a related mental state.[85] Stranded on isolated mountain peaks, unable to mingle, the remnants of humanity had only one *solacium* (3.27.12):

[82] Although Lucretius cannot use *securitas* for metrical reasons, his gods live a life that is *securus* (5.82, 6.58), the repose of death is *securus* (3.211), a personified *rerum natura* describes death as *securus* (3.939), and Lucretius himself uses the epithet in response to *natura*'s speech (3.977).

[83] De Vivo (1992b, 21–33) discusses the quotation in depth.

[84] Aeneas immediately compares his troops to ravenous wolves (*Aen.* 2.355–360).

[85] Delarue (2006) reads this scene through Ovid, Quintilian, and Longinus. For more on Seneca's engagement with Ovid here and throughout the flood narrative, see Mazzoli (1970, 238–247),

> quibus in extrema perductis hoc unum solacio fuit, quod transierat in stuporem metus. non uacabat timere mirantibus; ne dolor quidem habebat locum, quippe uim suam perdit in eo qui ultra sensum mali miser est.
>
> Brought to the brink, they had this one fact as a source of comfort: that their fear had passed into stupefaction. In their awe, there was no time to be afraid. Even grief had no place, since it loses its power in one who is too wretched to feel anything bad.

The stupefaction and awe of the survivors of the flood encode the sublime: the overwhelming power of nature has crushed them, and they have transcended their fear and grief.[86] Unlike Seneca's wise bather, who floats unperturbed above the din of earthly life, these subjects live immersed in the chaos that Seneca creates on every page of the *NQ*. To a certain extent, then, they resemble the *proficiens*. But in contrast to the *proficiens*, these figures are in no position to investigate nature's causes and consequently have no hope of spiritual elevation. The comfort of their sublime subjectivity constitutes a bleak counterpoint to the aspirational *securitas* that Seneca offers to the inexperienced in *NQ* 6.

Williams (2012, 113–116) argues that through the riveting and expansive account of the flood (3.27–30) Seneca's own voice models a form of sublimity. In contrast to the "helpless astonishment" of the survivors of the flood, Williams explains, Seneca himself plays the role of "another kind of spectator, serene and detached," whose mind-set constitutes "a revelatory departure from the smallness of everyday perception, and attunement instead to the rhythms of cosmic decay and rebirth." The Senecan voice, in other words, momentarily plays the role of the sage.[87] Can the reader interiorize the enlightened perspective that Seneca models and that the wise bather exemplifies? With any luck, but only for brief stretches of time. The gaze of the survivors offers a more accessible subjectivity that is no less free from fear, no less elevated, and that simply does not understand the mechanics of the phenomena that are unfolding. This contrast in perspectives explains a peculiarity of the narrative structure that emerges when we try to plot how

Degl'Innocenti Pierini (1990), Limburg (2007, 159–163), Berno (2012, 57–68), and Garani (2021; 2022; 2023). Star (2021, 140–153) situates Seneca's flood in the context of ancient apocalyptic narratives.

[86] In the following sentence (3.27.13), Seneca quotes part of *Met.* 2.264, from Ovid's Phaethon narrative. The sideways glance toward Phaethon picks up on the importance of that figure at the opening of the book and confirms that Seneca is thinking in terms of sublimity.

[87] For such Senecan role playing, see Edwards (1997, 33–34).

it dramatizes the sublime. A narrative of overwhelming historical disaster (3.27) leads into philosophical analysis, which locates the disaster within a cosmic framework that assuages and uplifts (3.28). And yet, calm detachment gives way to causal overdetermination, reflections on the violence of water, and the vehement insistence that destruction will come, in the future, from all sides at once, that the day of reckoning is not far off (3.29.1–30.6). When Seneca finally relents, affirming that the waters will eventually recede and the cosmic cycle will begin anew, he once again pivots to vice's easy and inevitable spread (3.30.7–8), which the reader knows will lead to more wholesale devastation since the flood itself constitutes a punishment for humanity's moral failings and transgressions (3.29.5–6, 30.7–8).[88] The drama, then, takes the reader from the overwhelming to the uplifting and then back again to that which overwhelms. The narrative arc of the conclusion poses a challenge, as Seneca vehemently pushes the reader toward the same immersive state of stupefaction that engulfs the survivors of the historical flood. The devastation at which they marvel is simply a vividly mythologized panorama of the turbulent uncertainty that surrounds us all, and it is a worthwhile *solacium* to behold it with wonder and awe.

Closely related to the strategy of consolation that Williams analyzes is the so-called *praemeditatio malorum*, the therapeutic practice of imagining future hardships in an effort to recognize their true status as indifferents and thereby to free oneself from mental disturbance.[89] The flood narrative of book 3 and book 6 as a whole operate along these lines.[90] I suggest, however, that we read the *NQ* itself as an extended and experimental use of the *praemeditatio*. Throughout the text, Seneca creates a volatile and dangerous world, not a future world, but the very world that we inhabit right here, right now. The sense of instability and the awareness of nature's might, which ricochet throughout the work like emotional pinballs, are not to be eradicated. Seneca's world is energetic, unpredictable, and overwhelming, and to see the all with the mind (3 praef. 10) is to be alive to those very

[88] Mader (1983, 65–66) convincingly argues that, with respect to Stoic tradition, Seneca innovates when he emphatically ties the deluge to moral decline. If so, the effect on the reader is all the more devastating. For Berno (2019, 86–90), Seneca's flood stands in for an account of the great fire that ravaged Rome in July 64 CE; the desolation of a post-flood *orbis terrarum* figures the desolation of a post-fire *urbs*.

[89] Armisen-Marchetti (1986) discusses Seneca's use of the *praemeditatio*. Hadot (1995, 81–125) considers the concept of "spiritual exercises" more broadly. For indifferents in Stoic ethics, see B. Inwood at *CHHP* 690–697 and L-S 58. Seneca discusses the topic explicitly at *Dial.* 1.3.1.

[90] Book 3: Garani (2021, 172–175; 2022, 146). Book 6: Armisen-Marchetti (1986, 189 with n. 23) and Limburg (2007, 319–321); cf. Wessels (2016, 209).

qualities, without succumbing to fear. While Seneca may raise some of his readers to a momentary state of joyful wonder, a vital calmness waits around the bend, and this may be our only *solacium*, the best outcome of the *praemeditatio* and Seneca's own brand of shock therapy; anything more is temporary at best.

3.6. Delusions of Grandeur

For Seneca, progressing toward knowledge and virtue is a sublime journey. The investigation of nature is a part of that journey, and while we might catch a fleeting glimpse of the ultimate destination, where the mind lives in fortified tranquility and dispassionately examines the cosmos, that is not a place to which ordinary human beings have regular access. In contrast to this narrative, the *NQ* hints at parallel experiences of sublimity that are problematic, even dangerous: a sublime experience that takes as its object not true greatness but something masquerading as greatness, and that leads not to spiritual elevation but to moral degradation. Such bursts of perilous sublimity unfold in a contemporary landscape where individuals are obsessed with bodily pleasure and political power, where exploitation and domination are the norm, and where things are not what they appear to be. Seneca sketches the contours of these dangerous modes of sublimity in clear contrast to the sublimity of natural inquiry, and his indictment of the former is a condemnation of the shortcomings of Neronian culture. Indeed, we might compare such sketches to the topsy-turvy world of Petronius's *Satyrica*, which dates to roughly the same time and is similarly anxious about the ability of the inhabitants of Nero's Rome to achieve true sublimity.[91] In opposition to this moral rot, Seneca holds out hope that natural inquiry itself might be able to protect us with its own brand of the sublime.

The Mullet Eaters

In 3.18, Seneca castigates a new kind of extravagance (*luxuria*): diners who insist not just on a fresh mullet but on one that they can watch expire on the

[91] For the *Satyrica*'s anxiety about the sublime, see Conte (1996, 37–72), who draws a connection (p. 44) with Long. *Subl.* 44. I return to this nexus of texts in the Epilogue.

table before their very eyes. Repeated exclamations convey the affective intensity of the experience, as the quoted voices insistently plead with the reader to share their enthusiasm (*uide... uide... ecce... quam...!* 3.18.5). Such aficionados view the phenomenon as a wildly splendid spectacle (*tam pulchro spectaculo*, 3.18.1), and three times—at the beginning, middle, and end of the passage—the sight is characterized as supremely exquisite (*nihil est... mullo expirante illo formosius*, 3.18.1; *nihil est moriente* [sc. *mullo*] *formosius*, 3.18.4; *nihil est enim illa* [sc. *morte*] *formosius*, 3.18.7). Seneca's contemporaries are enthralled by the sight of such beauty. Subtle confirmation of the sublimity of the experience comes in the form of an unstated contrast: Seneca used to be amazed (*mirabamur*, 3.18.2) that diners insisted on eating mullets that had been caught the same day—his wonder, one infers, has been outdone, for this madness has crossed all boundaries, smashed all norms (*furor usitata contemnens*, 3.18.3).[92]

The sublime experience of the glutton is a travesty of the sublime experience of the investigator of nature, and it serves as a cautionary tale, prominently placed in the first book of the treatise.[93] Like the investigator of nature, the glutton is a careful observer of the natural world: as the mullet's death throes obstruct its breathing, the glutton carefully marks the progression of colors that suffuse the fish and the changing appearance of its scales (3.18.1). Like the investigator of nature, the glutton questions the reports of others about the natural world. When told that a mullet was caught today, he refuses to take the statement at face value and insists on seeing the evidence firsthand (3.18.3). The glutton, however, does not seek to understand nature's causes or to make progress toward wisdom and virtue. On the contrary, he consumes nature and rejoices in suffering and death. Whereas the investigator of nature privileges reason over the evidence of the eyes (see 6.3.2, cited in Ch. 3.4), the gluttons are gluttonous with an array of body parts, including their eyes (*dentibus et uentre et ore... oculis quoque*, 3.18.7), and appear to reason not at all. The result is not the serenity of the sage or even the vital calmness of the inexperienced, but unrestrained *furor* in the hopeless service of *luxuria*. Seneca's point, though, is not merely the

[92] Similar shifts in tense and markers of time underscore the contemporaneity of the episode: Seneca "used to wonder at" (*mirabamur*, 3.18.2) certain excessive behaviors, but "now" (*iam*, 3.18.3) things are even worse; "we used to hear" (*audiebamus*) one sort of extravagant claim, "but now we hear" (*at nunc audimus*, 3.18.4) something even more incredible, etc.

[93] Williams (2012, 77) and Roby (2014, 156, 178) similarly contrast the normative investigator of nature, without, however, framing the issue through the sublime.

folly of the gluttonous fans of mullets. Rather, he condemns his contemporaries who abuse their senses and refuse to apply their rational faculty to the study of nature. The journey ends in moral depravity; this is the cultural milieu in which Seneca studies the natural world.

Hostius Quadra

Seneca's condemnation of Hostius Quadra (1.16) works similarly. Hostius fills his bedroom with funhouse mirrors that make objects appear bigger than reality, a setup that allows him to be both producer and spectator of all his lurid sexual adventures. The vocabulary of spectacle saturates the passage, inviting us to gaze at the extravagant show and to share in Hostius's emotional rollercoaster ride.[94] Affective intensity, both uplifting and overwhelming, is very much the point, since the story illustrates that sexual passion (*libido*) rejects no means of arousing pleasure (*inritandae uoluptatis*) and is exceptionally clever at exciting its own frenzy (*ad incitandum furorem suum*, 1.16.1). The object of Hostius's gaze, however, is not something truly magnificent but something that only masquerades as greatness: the mirrors make his body and those of his male and female partners appear far greater than they really are (*imagines longe maiores reddentia* [sc. *specula*]), and he rejoices in this false bodily greatness as if it were real (*ipsius membri falsa magnitudine tamquam uera gaudebat*, 1.16.2). Hostius's own words vividly encode the sublimity of this deceitful grandeur (1.16.8–9):

> quo nequitiam meam, si ad naturae modum pecco? id genus speculorum circumponam mihi quod incredibilem magnitudinem imaginum reddat. si liceret mihi, ad uerum ista perducerem; quia non licet, mendacio pascar. obscenitas mea plus quam capit uideat, et patientiam suam ipsa miretur.
>
> What is the point of my depravity if I commit wrongs within nature's bounds? I will surround myself with the kind of mirror that produces an

[94] The story is a drama (*fabellam* 1.16.1); Hostius puts his foulness on stage, with a clear pun on *obscenitas/scaena* (*obscenitatis in scaenam usque perductae*, 1.16.1, cf. Varro *Ling.* 7.96); he is a *spectator* (1.16.3) and repeatedly spectates (*spectabat* three times in 1.16.5); and he makes his foulness a spectacle (*obscenitatem suam spectaculum fecerat*, 1.16.6). For connections between Hostius and the theater, see Walters (1998, 362–363) and Berno (2002, 224–228; 2003, 53–61). Lehoux (2012, 101–102) approaches Hostius's morally problematic gazing from the perspective of how we evaluate witnesses and testimony in philosophical inquiry.

unbelievable greatness of images. If I could, I would make them a reality. Since I can't, I will feed on the lie. Let my foulness see more than it is capable of, and let it wonder at its submission!

Hostius presents himself as a transgressor of boundaries who revels in images of false *magnitudo*, and he summons the reader to share his experience of intoxicating awe. Ultimately, he comes to a spectacular end: Seneca has already told us that this metaphorical slave (*sestertii milies seruum*) was killed by his own slaves (1.16.1), and his rousing exhortation to wonder is met with a curt reminder of his undignified demise (*facinus indignum! hic fortasse cito et antequam uideret occisus est*, 1.16.9).

Hostius experiences the sublime; his speech encodes it (*miretur*), and he himself even attains a form of transcendence, with the boundaries of his body dissolving in his mirrors and giving way to the penetrations of his partners, as his slaves liberate his soul from his body.[95] What to make of this situation? A chorus of twenty-first-century critics read Hostius as an antihero, the inversion of the sage, of the philosopher, even of Seneca himself.[96] These categories, though, are not interchangeable, and we should be more careful. To be sure, numerous verbal echoes set up a contrast between Hostius and the sage, represented by the *animus* that surveys the heavens at the beginning of the book (1 praef. 11–13).[97] While both Hostius and the *animus* are spectators of sublime shows, their locations and the objects of their gazes differ radically—one occupies the heavens and looks at its divine surroundings, the other lives at ground level and scrutinizes a labyrinth of illusions. But what, exactly, is the opposite of the sage? One compelling answer is simply this: everyone who is not a sage, those of us who fall short of perfection and simply dwell, to adopt Platonic terminology, in the world of becoming rather than being. On this reading, through the enormity of his distorted perceptions and the intensity of his passions, Hostius brings to view and magnifies the distorted perceptions and uncontrolled passions of ordinary people.[98]

[95] On the dissolution of Hostius' body, see Leitão (1998, 143–144, 152–153).

[96] Berno (2002, 221–224 and 227–228; 2003, 45–50), Toulze-Morisset (2004, 215), Limburg (2007, 273–274), Williams (2012, 55–60; 2016, 177–179), Wilson (2014, 178), and Gunderson (2015, 71); cf. Scott (1999, 60). Bartsch (2006, 103–114) explores Hostius's perversions of philosophical traditions without framing him as an anti-philosopher. Gauly (2004, 129) reads the story as an allegory about the overturning of social order in contemporary Rome. Toohey (2004, 261–282) focuses on Hostius's self-alienation, drawing on Lacan.

[97] See Leitão (1998, 136–146) and Berno (2002, 218–219; 2003, 50–52).

[98] Bartsch (2006, 108) suggests reading Hostius as an "Everyman."

A significant consideration undermines this line of interpretation. In the *NQ*, Seneca does not follow the older Stoic tradition that divides the human world into sages and fools, with nobody in between. Instead, as we have seen, he places great emphasis on the role of the *proficiens*, and it is against the *proficiens* that we should read Hostius. Of course, there are many from which to choose. At one level, Hostius acts as a counterweight to Phaethon. His rousing speech at the end of the penultimate book sets off the same emotional fireworks as Phaethon's rallying cries at the opening and in the middle of the treatise, and the awesomeness of Hostius's implicit tragedy nicely balances the weight of Phaethon's epic. Both Hostius and Phaethon are larger-than-life figures who dream big, lose control, and come to cinematic ends, and we should not doubt that Hostius very much leaps.

And yet, Seneca unequivocally condemns Hostius; this is where his similarities to the philosopher and to Seneca come into play. Like the philosophically minded student of nature, exemplified by Seneca as the author of the *NQ*, Hostius undertakes an innovative and experimental journey that changes the way he and others see and experience the world. Mirrors have an important role to play in that journey: they are tools that allow us to study celestial phenomena (1.17.2–3) and, ultimately, to know ourselves (1.17.4), since natural inquiry, properly conceived, ultimately leads to wisdom.[99] And just like Seneca's ideal investigator of nature, but in contradistinction to Phaethon, Hostius works with others—many others, as Seneca makes clear—in a fundamentally theatrical undertaking. Hostius's wild inventiveness, however, seeks not to know nature but only to manipulate it in the service of sexual gratification, and rather than strive to wonder at true *magnitudo mundi*, he loses himself in a false *magnitudo* (cf. *ipsius membri falsa magnitudine tamquam uera gaudebat* at 1.16.2, *incredibilem magnitudinem imaginum* at 1.16.8). So, too, his collaborative impulse is problematic since his eagerness to play the passive role in his sexual encounters (1.16.2–5, 7, 9) runs contrary to Seneca's active interrogation of others and the world. From this perspective, Hostius represents the degradation of natural inquiry, what happens when the investigator of nature mistakes appearances for reality, forgets the provisional status of the undertaking, and fails to test and probe the theories and phenomena at issue. Hostius, then, is not the anti-philosopher but the philosopher *manqué*, one who falls short in a fun-

[99] For the point about mirrors, see Berno (2002, 220–221; 2003, 43–45) and Bartsch (2006, 103–106).

damentally spiritual journey even while thinking that he is progressing toward virtue and truth. After all, how else does the philosopher construe the erotic ecstasy of the sublime? We can experience the sublime in different contexts, in ethically opposed scenarios and activities. Hostius's story offers the reader a choice: to ignite and embrace this experience through the study of the natural world properly understood, or to set it aflame in narcissistic and self-destructive servitude to our own *libido*, which is nothing more than a sad parody of the passion that is supposed to animate philosophy.[100]

Hostius died while Augustus was still alive (1.16.1), which means that his story does not belong to the distant past. Taken together, he and the mullet eaters stage the same degradation of the sublime, and both narratives indict ongoing threats to Rome's moral and intellectual fabric. Instead of the sublimity of sexual showmen and fanatical gluttons, Rome needs a sublime that aims for what is virtuous and true, and the study of nature, carefully conducted and closely monitored, provides just such an experience; the alternative is not just luxurious but deadly.

Lucilius, Sicily, and Roman *imperium*

Seneca's exhortation to Lucilius at the opening of book 4a suggests that yet another sublime experience threatens contemporary Rome: the sublimity of Roman *imperium* itself. At the end of the preface, Seneca explains why he has dedicated his account of the Nile to Lucilius, his friend and the current governor of Sicily (4a praef. 20–21). Seneca hopes to provide a distraction from the ethically compromised realities of Lucilius's day-to-day life, to lessen Lucilius's loneliness through philosophical conversation, and to keep him from becoming arrogant in his awareness of the fact that he governs a province (Sicily) that has a particularly thrilling political history (*ne forte magnam esse historiis fidem credas et placere tibi incipias quotiens cogitaueris...*, 4a praef. 21). This last motivation prompts Seneca to imagine Lucilius contemplating the grand role that Sicily has played in Rome's imperial evolution (4a praef. 21–22). The monologue exploits a contrast between elevation and oppression, dramatic reversals of fortunes, and the greatness of

[100] Leitão (1998, 158–160), by contrast, concludes that Hostius's "passion" represents a path to knowledge of and communion with the divine for inhabitants of a fallen world.

everyone involved. During the First Punic War ("a massive war," *ingentis belli*), Sicily both raised up and smashed (*et sustinuit et fregit*) the armies of Carthage and Rome, the greatest cities (*maximarum urbium*). It elevated Sextus Pompey's fortune and exhausted Octavian's (*Pompeii fortunam erexit, Caesaris fatigauit*). Through the battle of Naulochus ("a massive spectacle," *ingenti spectaculo*), mortals learned how swift is the fall to the bottom from the top (*quam uelox foret ad imum lapsus e summo*) and by how many different paths fortune ruins great power (*quamque diuersa uia magnam potentiam fortuna destrueret*). All at once, the island saw both Sextus Pompey and Lepidus cast down, in different ways, from the highest peak to the lowest depths (*ex maximo fastigio aliter ad extrema deiectos*).

Sicily is a virtual theater of the sublime, but the source of that experience is very particular: contemplation of Roman political and military might, the spectacle of empire. Indeed, as a prize (*praemium*) to be won, Sicily represents the origins of Roman *imperium*, and the last days of the Republic saw the forces (*uires*) of the entire Empire (*totius imperii*) converge on this one spot, making it a distillation of the electrifying potency of imperial power—this is one of the many *mirabilia* that the island contains (4a.1.1).[101] The danger, then, lies in thinking that one can actually possess this locale and what it represents. If Lucilius says to himself "I have this province under my jurisdiction" (*hanc ego habeo sub meo iure prouinciam*, 4a praef. 21), the sublimity of Roman *imperium* will fill him with delusions of grandeur that will be impossible to counteract—he will forever gaze at the grandeur of Rome, thinking that it belongs to him, rather than contemplate the true *magnitudo mundi*, within which he is but an infinitesimal speck of no significance (cf. 4b.11, 1 praef. 8, and Ch. 1.4 on Cic. *Rep.* 6.16). The sublimity of empire distorts our perceptions and warps our sense of proportion, and Seneca is apprehensive that it may already be too late, that he will have to forcefully drag Lucilius away from his captivating post. Seneca begins the book by warning his friend not to treat his governorship as an empire (*nec efficere imperium quod est procuratio*, 4a praef. 1). At the end of the preface, the sublimity Rome's empire, symbolized by Sicilian history, explains why that is dangerous.

At a broader level, Sicily's seductive power represents the sublime allure of public life, and Seneca's antidote is the equally seductive power of natural

[101] For Sicily's long-standing connections with wonder and the sublime, see Ch. 5.5 ("*Aetna* and the Paradoxographical Tradition").

inquiry.[102] Although Sicily contains many wonders (*mirabilia*), Seneca will pass over all of Lucilius's questions (*quaestiones*)—that is, questions about Sicily's natural phenomena (4a.1.1), with the implication being that such a project would itself be sublime. Instead, Seneca will launch an investigation along with Lucilius (*quaeram enim tecum*, 4a.1.1) into the Nile's source and the causes of its summer flood, an equally exhilarating topic the treatment of which, we can only surmise, will have downplayed Egypt's political status.[103] For the dangerous sublimity of Sicily, imbued with all the arrogance and swagger of empire, Seneca will introduce his friend to the community of scholars and take him on a far-flung adventure that aspires to truth and virtue. Once again, the contrast with Manilius is telling. The *Astronomica* exhibits an unapologetically triumphalist approach to the investigation of the stars and human life, whereby Roman *imperium* and the *mundus* converge, and Manilius's poem is seen to contain and to be coextensive with both. But whereas Manilius's exuberant cultivation of the infinite allows the reader to revel in the vastness of cosmos and empire, Seneca dissociates cosmic from imperial greatness, conspicuously marginalizing Rome and the emperor.[104] Contemplation of the *magnitudo mundi* becomes an opportunity to elevate oneself by turning inward, shutting out the noise of *imperium*, and searching, however provisionally, for virtue, wisdom, and god.

As I will discuss in the following chapter, the questions that Seneca proposes to investigate with Lucilius are fundamentally unanswerable. But concrete facts are not the point; it is the journey that matters. And yet, for Seneca's nephew, Lucan, even the journey is problematic, a deceptive lie, an irresponsible charade. In the grinding brutality of Lucan's universe, unadulterated *imperium* is what counts, not whether one studies the stars or atmospheric phenomena—all such pursuits are vanity and madness, deadly distractions from the political turmoil that engulfs the entire world. Seneca's articulation of the scientific sublime will come in for the harshest of criticism.

The Scientific Sublime in Imperial Rome: Manilius, Seneca, Lucan, and the Aetna. Patrick Glauthier, Oxford University Press. © Oxford University Press 2025. DOI: 10.1093/9780197787588.003.0004

[102] Williams (2012, 48–53) discusses Stoicism in general and the *NQ* in particular as responses to the encroachment of the principate on traditional aristocratic claims to power. For Senecan contrasts between natural philosophy and empire, note Tracy (2014, 120–121) and Garani (2020a, 113–122). Maruotti (2019, 179–181), by contrast, reads 4a praef. in the context of sublimity and the possibility of failure in Seneca's moral journey.

[103] For the history of these questions and their relationship to the sublime, see Ch. 4.6.

[104] For the tension between marginalization and presence of Rome, see Hine (2006).

4
Lucan
Natural Inquiry in a Time of Civil War

4.1. Introduction

This chapter examines the sublimity of natural inquiry in Lucan's *Civil War*. While Manilius daydreams about the destruction of the cosmos and Seneca forces the reader to experience his ordered world as dangerous and unpredictable, Lucan relentlessly insists that the universe itself is in the process of disintegration, unraveling before our very eyes.[1] In the midst of such material breakdown, the study of the natural world does not lead to moral progress or bring the student closer to god—the sublimity of such study constitutes a distraction from the all-consuming focus of civil war. Paradoxically, Lucan cannot refrain from plunging the reader into the physical minutiae of nature's mechanics. A pathological urge to narrate and to speculate about nature and natural phenomena, divorced from any higher moral purpose, saturates the text, potentially elevating but ultimately oppressing the reader with theoretical detail and open-ended questions. At the same time, those characters who understand how the material world functions cease to inspire wonder and simply crumble into political and historical irrelevance—other sources of sublimity, including Caesar's military might, take precedence and render the scientific sublime a shadow of its former self. Seneca's concerns about the seductive allure of *imperium* are fully realized in Lucan's Caesar, while his anxieties about the perversion of philosophical inquiry give way to the realization that such inquiry is entirely disengaged from the exigences of day-to-day life. For Lucan, the investigation of nature is a rarefied object whose electric vitality short-circuits in a time of civil war/*Civil War*.

[1] For cosmic dissolution and world-ending catastrophe in Lucan, see Lapidge (1979), Roche (2005), and Star (2021, 169–190). Instability and breakdown have been major themes of Lucan scholarship for decades; two especially influential studies are Henderson (1987) and Masters (1992). Throughout this chapter, "Lucan" refers to the narrator of the *Civil War*.

Lucan's interest in physics and natural philosophy is well documented.[2] Astronomical knowledge is especially conspicuous in the poem, but Lucan makes use of or alludes to ideas from a wide range of fields, including cosmology, meteorology, and medicine. Numerous studies track down sources or attempt to map Lucan's knowledge onto the intellectual landscape of the twentieth or twenty-first century. Only relatively recently, however, have critics begun to interpret such passages, to ask how theory and observation relate to the broader concerns of the text.[3] The present chapter contributes to this scholarly trend. Lucan's deprecation of the sublimity of natural inquiry parallels his rejection of other traditions and institutions, and the inability of characters to make effective use of sublime knowledge contributes to the overarching sense of breakdown and collapse that permeate the text. By looking at how both characters and Lucan himself, as narrator, think about natural inquiry and try to deploy natural knowledge, we can better appreciate how such passages participate in the poem's thematic and ideological economies.

One aspect of source criticism may seem especially relevant in the present context. Nineteenth-century critics, and some of their twentieth- and twenty-first-century descendants, made a concerted effort to establish Lucan's familiarity with Seneca's *Natural Questions* (*NQ*), which dates to roughly the same period as the *Civil War*.[4] I am not concerned to establish Lucan's dependence on Seneca, whether it be the *NQ* or any other work. What emerges from my reading is a clear sense that Lucan and Seneca are thinking about natural inquiry and the sublime in similar ways. Just as Seneca darkens and destabilizes the Manilian cosmos, Lucan accentuates and magnifies the precariousness and aporia of the *NQ*, but whereas Seneca holds out hope for transcendence, Lucan immerses the reader in relentless anxiety and doubt. Potentially worse, several episodes suggest that Seneca's

[2] See most recently Meyer (2023). Important treatments include Pichon (1912, 165–216), Schotes (1969), Beaujeu (1979), Migliorini (1997, 95–125), Loupiac (1998), Schrijvers (2005), and the essays in Landolfi and Monella (2007). For Lucan and Manilius, see Schwemmler (1916). For Lucan and Germanicus, see Magnavacca (2020).

[3] See esp. Barrenechea (2004; 2010), Walde (2007), Tracy (2010; 2014, 99–273), Manolaraki (2013, 45–117), Taylor (2020), and Meyer (2023). Schrijvers (2005) takes the literariness of Lucan's technical passages seriously and helpfully analyses how he embeds such material in the text.

[4] See Diels (1886) and Hosius (1892, esp. 337–339, 352–353). Manolaraki (2013) and Tracy (2014) have taken this study in new and exciting directions; for discussion, see Ch. 4.6. According to the *Vitae*, Lucan recited (some of) the poem and/or published three books of it (presumably the first three) at some point after the Neronia of 60 CE; see Griffin (1984, 157–158), Fantham (1992a, 1–3), with Rostagni (1979, 143–144, 183). Ahl (1976, 333–353) argues for a relatively late date (62/63–65 CE, the year in which Lucan died). The *NQ* dates to the early 60s; see Ch. 3.1.

entire physical project, with its promise of moral and spiritual advancement, is hopelessly naïve or comically absurd, if not downright deadly.[5]

The relevance of the sublime to the *Civil War* has also received significant attention. Numerous critics associate Lucan with the sublime in passing.[6] More particularly, Matthew Leigh and Francesca D'Alessandro Behr partially frame their readings of Lucan through the expositions of *enargeia*, *phantasia*, and apostrophe that Longinus and other ancient literary critics develop.[7] Henry Day's (2013) *Lucan and the Sublime* takes this kind of approach much further, consistently reading the entire poem through the analyses of Longinus, Burke, Kant, and more recent theorists of sublimity.

Day makes clear that the charged energy of the sublime is the poem's "primary aesthetic mode."[8] Indeed, it is appealing to read Lucan's exuberant transgressions and inversions and the poem's pervading ethos of excess—from the very first line, this civil war/*Civil War* is always multiple, always "more than" (*bella...plus quam ciuilia*, 1.1)—as fundamentally sublime. But Day is more specific, identifying two conflicting currents of sublimity in the work, the Caesarian and the Pompeian. "Lucan," writes Day, "portrays Caesar as a larger-than-life, hyper-kinetic, awe-inspiring source of destruction, a literally superhuman force" who, as a sublime subject, grinds the world into submission.[9] Caesar's sublimity is firmly rooted in the overwhelming power of nature—he is, at his core, the sublime thunderbolt that terrifies the world (1.151–157). Moreover, Day convincingly demonstrates Lucan's use of Lucretius in constructing this mode of sublimity. Not only does Lucan deploy Lucretian imagery and language, especially from *De rerum natura* 6, but he also models his Caesar on Lucretius's Epicurus: "Fearlessly surmounting all obstacles... [Lucan's Caesar] puts himself on a par with the divine, supplanting the traditional epic gods who are so conspicuously

[5] Narducci (2002, 162–164) connects Lucan with a broader "processo...di dissoluzione della teodicea stoica." My position is related, but I am not concerned with the gods and fate per se.

[6] Although he does not name the sublime, Morford (1967, 28) connects Lucan with Lucretius, Manilius, Seneca, and the *Aetna* poet as examples of a strain of Roman culture that was attuned to the wonders of nature.

[7] Leigh (1997, 10–15) and D'Alessandro Behr (2007, 165–166) on *enargeia* and *phantasia* (*Subl.* 15). D'Alessandro Behr 1–2 on apostrophe (*Subl.* 16). Note, too, the entry for "sublime" in the index of de Nadaï (2000).

[8] Day (2013, 17). Seitz (1965) reads the poem as a virtual distillation of *pathos* that I would call sublime; at 214, he remarks "Das an sich Große...genügt Lucan noch nicht: er steigert es zum Kolossalen."

[9] Day (2013, 106). Narducci (1979, 105–107) makes Caesar qua tyrant the embodiment of *megalopsuchia*. Less relevant approaches to Caesar and the sublime are de Nadaï (2000, 188–214) and Alston and Spentzou (2011, 55–64).

absent from Lucan's poem."[10] Contrarily, Day identifies a mode of sublimity in the poem that attempts to recuperate past trauma and to create wholeness out of historical loss. It is a state "whereby defeat appears as, indeed becomes, triumph and greatness manifests itself amid disaster."[11] This is the Pompeian sublime, which stands in opposition to the Caesarian mode. As a form of the fall, it is short-lived: "Rome's greatness can only block out the rupture inflicted by Caesar's victory for the very briefest of moments before the poem's affective balance tilts back towards the pain of defeat."[12]

Day's analysis provides an essential point of departure for our investigation. And yet, despite the centrality of Lucretius to Day's position, the sublimity of natural inquiry does not receive sustained attention in his study. Elaine Fantham, however, explicitly connects Lucan's interest in natural phenomena with sublimity, writing that "instead of the *deorum ministeria* Lucan appealed to the glories of the natural world... to produce the sense of awe" that is the Longinian sublime.[13] Fantham's point is legitimate but only tells part of the story. Lucan actively engages with the tradition of the scientific sublime, and it is clarifying to read the *Civil War* from this perspective against the *Astronomica* and, especially, Seneca's *Natural Questions*. With this in mind, the conflict between different modes of sublimity that Day stresses will be crucial. We have already seen such conflict in Seneca, and we will chart further tensions below. In addition to the Pompeian sublime, the sublimity of natural inquiry attempts to resist Caesar's brilliant onslaught, and it struggles accordingly.

4.2. Sublime Distraction

For Lucan, civil war is all-consuming and demands our full attention. In such a setting, the study of the natural world constitutes a dangerous distraction, a sublime daydream about unanswerable questions during which we momentarily let our guard down and disengage from what truly matters.

[10] Day (2013, 106). Schrijvers (2005) gives more evidence of Lucan's familiarity with Lucretian physics. In the context of sublime ruins, Weiner and Benz (2018, 286–297) connect Lucan with Lucretius and Longinus.

[11] Day (2013, 203); cf. Schrijvers (2006, 101) on "the historical/political sublime" in Lucan.

[12] Day (2013, 206). Chaudhuri (2014, 181–192) builds on Day's reading, adding further dimensions to Pompey's sublimity and drawing Cato into the mix.

[13] Fantham (1992a, 19). For the idea that natural philosophy constitutes one of the grandest themes for poetry, see Innes (1979, 165–171).

The study of natural world may promise to bring us into contact with the divine, but that lofty goal comes at an unacceptable cost if it is even attainable at all.

The Gallic Tides

Early in book 1, a programmatic passage establishes the sublimity of natural inquiry and urges the reader to turn a blind eye. At 1.392–465, Lucan catalogs the Gallic tribes whom Caesar leaves behind as he mobilizes his troops and begins his march on Rome. The passage upends the traditional epic catalog of fighters and frustrates the reader's generic expectations in multiple ways—it is, notably, a catalog of places and peoples unconnected with the looming war rather than of heroes and their contingents.[14] The breakdown of epic norms indicates the extent to which the genre can no longer function properly in the midst of civil war and in Neronian Rome; under such pressures, traditions begin to founder. This is the basic context against which to read Lucan's reflections on natural inquiry.

One of the areas that Caesar's men leave behind is a stretch of the Gallic coast that exists alternately as land and sea, according to the movements of Ocean (1.409–411). This prompts Lucan to consider the causes of the tides (1.412–419):

> uentus ab extremo pelagus sic axe uolutet
> destituatque ferens, an sidere mota secundo
> Tethyos unda uagae lunaribus aestuet horis,
> flammiger an Titan, ut alentes hauriat undas,
> erigat Oceanum fluctusque ad sidera ducat,
> quaerite, quos agitat mundi labor; at mihi semper
> tu, quaecumque moues tam crebros causa meatus,
> ut superi uoluere, late.

Whether wind from the furthest clime rolls the sea in this way and, while carrying it, abandons it, or whether the wave of wandering Tethys, set in motion by the second star [i.e., the moon],[15] surges with the lunar hours, or whether

[14] For the catalog's generic violations, see Green (1991, 243–251) and Roche (2009, 277–278).

[15] The Bern scholia ad 1.413 comment: *primum sidus est solis, secundum lunae.* The order in which ancient authors arrange the planets varied, but Hannah (1996, 179) observes that before 150 CE

flame-bearing Titan, so that he may drink the nourishing waves, raises Ocean and leads the billows to the stars—investigate, you whom the toil of the world excites. But you, whatever cause you may be who sets in motion such frequent movements, may you always lie hidden from me, as the gods above have wished.

This is precisely the kind of passage that Fantham (1992a, 19) has in mind when she observes that Lucan's descriptions of the workings of the natural world infuse the *Civil War* with sublimity. In support of her proposition, Fantham cites Hermogenes, *On Types of Style* 1.6 (pp. 243–244 Rabe), where the investigation of the causes of certain natural phenomena, including the movements of the sea, produce solemnity (σεμνότης), a byword for the sublime.[16] The historical Lucan was undoubtedly familiar with such thinking, but we can be more precise. The heightened emotional state of the imaginary student of nature (*quos agitat mundi labor*) encodes the sublime. Such individuals are both exited and agitated, inspired and shaken, and their tumultuous state of mind (*agitat*) is caused by the intense and unrelenting turmoil of the sea (*uolutet, aestuet, moues tam crebros... meatus*). The passage also stages the sublime's twofold movement. Through the tumult of the tides, Lucan moves the reader from the furthest reaches of the sky up to the moon and finally to the sun, passing from a purely terrestrial explanation for tidal movement to theories that demonstrate the interconnection between heaven and earth and the physical forces that literally draw that which is down below up toward the divine. Unexpectedly, however, Lucan cuts the inquiry short (*at mihi semper... late*). The direct address to *causa* is extraordinary and arresting. To determine the answer to this kind of question is to cross the boundary between the human and that which is more than human, a movement that is a priori sublime and which we have seen both Manilius and Seneca configure as such; Lucan countenances no such surge of transcendence.[17]

Lucan's feigned ignorance dramatizes the breakdown of poetic norms and intellectual traditions in the chaos of civil war and contemporary Rome.[18]

the "normal" order of the planets in Greek horoscopes is Sun, Moon, Saturn, Jupiter, Mars, Venus, Mercury. Barrière (2019, 77), by contrast, writes that the moon "peut avoir une influence parce qu'il s'agit d'un astre favorable à l'eau (*secundo*)."

[16] For Hermogenes, see Ch. 1.6 ("Longinus and the Rhetorical Tradition"). For solemnity as sublimity, see Long. *Subl.* 12.3 and Porter (2016, 181).

[17] Ozanam (1990, 287) connects Lucan with the proposition that "le mystère crée la grandeur." From this perspective, the scientific sublime here yields to a different mode of sublimity.

[18] At 6.479 and 10.204, Erictho and Acoreus, respectively, take the moon's influence on the tides for granted; the historical Lucan is not actually uncertain.

More specifically, Lucan's activation and subsequent rejection of the pursuit of the sublime through natural philosophy characterize the scientific sublime itself as a distraction. At one level, this constitutes a rejection of the tradition of natural inquiry that, in a Latin context, is closely connected with Lucretius.[19] Lucretius regularly writes of Epicurus or reason itself as banishing darkness and obscurity and shedding light on the true nature of the world,[20] and those working in this tradition often describe their subject matter as something hidden that they themselves must uncover or reveal.[21] By emphatically ordering the *causa* of the Gallic tides to lie hidden, Lucan appropriates this rhetoric and turns it upside down, thereby configuring himself as the exact opposite of Lucretius and the Lucretian-inspired investigator of the cosmos. As we have seen, this tradition also includes Seneca, and an inquiry into the causes of the tides is exactly the kind of meteorological puzzle that Lucan's uncle could have treated in the *NQ*.[22] But Lucan's aporia is not Seneca's—there is no hope that the community of scholars might one day arrive at the answer, that just by setting our minds to the task we might come into momentary contact with god. Lucan fleetingly raises the possibility of undertaking such a journey only to shut it down, and his rejection of the ethical or spiritual promise of this kind of project implies a disavowal of Seneca's philosophical aspirations; the gods wish to keep us in the dark, and in the dark we will remain.

At another level, we should remember that this passage occurs within the broader context of Caesar's march on Rome. As Caesar's troops invade Italy and throw the world itself—not just the waters of Ocean—into confusion, the Lucretian/Senecan pursuit of *ataraxia* becomes a kind of idealistic and irresponsible escapism. Under such circumstances, no Roman can afford to linger on the outskirts of the known world and to play the part of Lucretius's Epicurus, a wanderer beyond the *moenia mundi*—or, in this case, a wanderer along the Gallic coast—who willfully neglects the concerns of state. Lucretius, of course, acknowledges the political violence that threatens the Republic in his own day, and conspicuously prays for peace (1.38–43),

[19] For the Lucretian associations of multiple explanations, see Schrijvers (2005, 36–39, on our passage at 37) and Hardie (2009, 231–263, on our passage at 250–251). Green (1991, 245–246), de Nadaï (2000, 69–70), and Domenicucci (2013, 9–10) read this passage similarly.

[20] See West (1969, 79–93) and note the use of *latere* at Lucr. 3.273–281. Leigh (1999, 175) observes that this kind of rhetoric is common from Plato onward.

[21] See, e.g., Ov. *Met.* 15.72, 147 (Pythagoras), and Man. 3.47, 4.883.

[22] Arist. *Mete.* 2.1–3.353a–359b discusses the nature of the sea; cf. Sen. *NQ* 3.28, [Arist.] *Mund.* 4.396a.

only to follow Epicurus's lead and devote himself to philosophical speculation. In contrast to this kind of figure, Lucan takes sides in the ongoing conflict, regularly represents himself as an active participant in the narrative that he creates, and repeatedly attempts to conscript the reader into the fight.[23] The Lucretian tradition is a seductive distraction, and the sublime experience that it offers only allows the reader to ignore more urgent concerns.

Lucan's address to the Bardi and the Druids at the end of the catalog (1.447–462) supports this reading. The Bardi, poet-priests (*uates*) who memorialize the deeds of the war dead (1.447–449), play the traditional role of the epic poet, which Lucan finds unavailable. Once again, the perverse nature of civil war and Neronian Rome is incompatible with the norms of Homeric and Virgilian epic.[24] The Druids are more complicated. Now that the Romans have left their posts, the Druids lay down their weapons (*positis... ab armis*, 1.451) and return to their ominous religious rites (*barbaricos ritus moremque sinistrum / sacrorum... repetistis*, 1.450–451). At this point, attention shifts to their purported knowledge of the gods and the heavens (*solis nosse deos et caeli numina uobis / aut solis nescire datum*, 1.452–453) and their belief in the doctrine of transmigration (1.454–458), about which they sing (*canitis*, 1.457). Although this teaching is mistaken, it is a great blessing for the Druids' fellow citizens (*felices errore suo*, 1.459) since it means that they are not afflicted by the fear of death (*quos ille timorum / maximus haud urguet leti metus*, 1.459–460). In light of the poetic credentials of the Bardi, we should read these singing priests as poet figures, too. As poets who claim privileged knowledge about the gods, the heavens, and the soul, and whose songs prevent the *leti metus*, their most obvious counterpart is Lucretius.[25] Given the Druids' conspicuous promotion of transmigration and the soul's immortality, doctrines that Lucretius fundamentally opposes, we can think about other ways in which the Druids differ from the author of *De rerum natura*. Lucretius, modeling himself on Epicurus, stares into the void and fights a philosophical battle. As he aimlessly traverses

[23] Studies that emphasize Lucan's active participation in the narrative and/or the reader's resulting engagement include Seitz (1965), Marti (1975, 84–90), Henderson (1987, esp. 149), Masters (1992), Bartsch (1997), Leigh (1997), Narducci (2002, 88–106), Roche (2005), and D'Alessandro Behr (2007). Like Cato (2.289–297), Lucan does not want, and does not want the reader to be able, to watch the end of the world with the serene detachment of the Stoic sage.

[24] Cf. Feeney (1991, 277–279) and Green (1991, 247).

[25] Green (1991, 248) claims that the Druids "are indistinguishable from Stoic philosophers." Roche (2009, 298) remarks that freedom from the fear of death is a core ethical concern of both Stoics and Epicureans. Note the historical Caesar's account of the Druids' interest in *rerum natura* at *BGall.* 6.14.6.

the pathless mountains of the Muses (1.921–930, 4.1–5), he remains unengaged in the turmoil that engulfs contemporary Rome.[26] The Druids, by contrast, fight actual battles and prepare their fellow citizens to do the same. They are practically minded and civically engaged, and their songs inspire a headlong rush into certain death (*inde ruendi / in ferrum mens prona uiris animaeque rapaces / mortis*, 1.460–462). This is that state of Caesarian sublimity, and it contrasts radically with the awestruck gaze of the student of nature who merely contemplates the causes of the tides. If the unspeakable crime of civil war explodes the norms of Homeric and Virgilian epic, it renders cowardly and irresponsible the sublime pursuit of philosophical truths about nature; better to rush boldly into war, eyes fixed on the ground (*prona*), psychologically fortified by philosophical falsehoods (*felices errore suo*).

Pompey's Soul

Lucan's approach to the tides establishes one side of his attitude toward the sublime journey of natural inquiry—it is a distraction in which no responsible Roman reader can afford to indulge. His deployment of the flight of the mind topos at the opening of book 9 adds a slightly different perspective. Here, Pompey's soul (*manes* at 9.1, *umbram* at 9.2) leaves its pyre and flies into the heavens (9.1–4). According to Lucan, the semi-divine shades (*semidei manes*, 9.7) of those virtuous individuals who lived blameless lives occupy the border-zone where air and ether meet (9.5–11),[27] and this is where Pompey's spirit lingers (9.11–18):

> illic postquam se lumine uero
> impleuit, stellasque uagas miratus et astra
> fixa polis, uidit quanta sub nocte iaceret
> nostra dies risitque sui ludibria trunci.
> hinc super Emathiae campos et signa cruenti
> Caesaris ac sparsas uolitauit in aequore classes,
> et scelerum uindex in sancto pectore Bruti
> sedit et inuicti posuit se mente Catonis.

[26] This, I suggest, is how Lucan characterizes Lucretius. For the politics of the Lucretius's poem, see Fowler (1989).

[27] Narducci (2001, 73–75; 2002, 337–39) connects this location with Manilius's final Milky Way hypothesis (1.758–804).

There, after it filled itself with true light and wondered at the wandering planets and the stars fixed in the sky, it saw the magnitude of the night beneath which lies our day and laughed at the indignities inflicted on its headless corpse. Next, it flew over the fields of Emathia and the standards of blood-stained Caesar and the ships scattered on the sea, and, as an avenger of crimes, it took a seat in the sacred breast of Brutus and placed itself in the mind of invincible Cato.

Pompey's soul undertakes a literal flight of the mind.[28] Once it perceives the light of heaven unobscured and gazes in awe at the paths of the stars and planets, it suddenly recognizes the profound ignorance that envelops those on earth and grasps just how misguided our values and priorities are. The combination of awe (*miratus*) in the face of cosmic greatness and laughter (*risit*) at what is literally mundane encodes the sublimity of the experience and indicate Pompey's *magnitudo animi*. In particular, this scene of sublime spectatorship acts out the process of natural inquiry and philosophical enlightenment that Seneca attributes to the *animus* in the preface to *NQ* 1 and that leads to precisely the same kind of ethical realignment, complete with sublime laughter.[29] Both souls attain a cosmic point of view that remains all but inaccessible to ordinary human beings, but whereas this vantage point remains a goal for the Senecan *proficiens*, Pompey's soul refuses to accept the life of the mind. From its semi-divine perch among the celestial spheres, the soul leaps headlong back to earth, planting itself firmly in the bodies of two heroes who are destined to fall in their turn. In choosing political engagement, in the here and now, over the awesome spectacle of celestial knowledge, it is as if the soul echoes Lucan's command that the cause of the Gallic tides remain hidden. In rushing unafraid into a battle that it is doomed to lose, it is as if the soul has been listening to the songs of the Druids.

There is, however, another way to approach this text. W.R. Johnson (1987) reads it as sarcastic ("ridiculous by design" at 72, a "cartoonlike apotheosis" at 83): "It is a peculiar picture, this, and a still more peculiar shift in mood and tone. Pompey smiling at his headless corpse is Stoic enough.... But we are, I think, amazed and not wholly persuaded to find this rarest and richest of Stoic prizes so abruptly bestowed on this least Stoic of

[28] This is an example of what Leigh (1999, 176) calls "a practice typical of Lucan: the concretization of other writers' metaphorical repertoires."

[29] See Ch. 3.4. Laughter pervades *NQ* 1 praef. 6–13 (note esp. *iuuat...ridere* at 1 praef. 7). Narducci (2001, 79; 2002, 344–345) compares, inter alia, Lucian, *Icar.* 17–18, and stresses the connection with diatribe and Cynicism.

men" (70). In rebuttal, Emanuele Narducci argues that the various elements that Johnson finds ludicrous in Lucan's text have well-established and high-minded parallels in the ancient world: "Che Lucano possa... apparire «ridicolo» al lettore del tardo Novecento... non significa che egli apparisse tale ai lettori ai quali si indirizzava" (2001, 71; cf. 2002, 336). Narducci's point is well taken but disregards Johnson's emphasis on the radical "shift in mood and tone" that the opening of book 9 represents. Taken in isolation, Pompey's adventure performs all the expected maneuvers of a straightforward or entirely serious flight of the mind, a text along the lines of Cicero's Dream of Scipio (Ch. 1.4). But we cannot be content to read the episode in isolation. Given the chaotic uncertainty and grinding brutality of the *Civil War*, Pompey's burst of celestial insight is surely out of place.[30] Nobody in the poem exhibits this kind of knowledge or consciousness, and Pompey himself, early in the previous book, has shown himself to be entirely incapable of appreciating the wonders of heaven and the significance of the stars (see 8.159–201 with Ch. 4.4, "Pompey Questions his Helmsman"). Pompey does not make progress in philosophy and thereby ascend to heaven; he stumbles blindly into death and emerges a parody of the *proficiens*.

Nero-Phaethon

The *Civil War* contains another significant moment of celestial travel that sheds further light on Lucan's undermining of the scientific sublime. In the invocation of Nero at the opening of book 1, Lucan imagines the (soon to be) deceased emperor ascending to the stars and gives him advice about where to make his eternal abode (1.45–59).[31] The first half of the passage, in particular, exploits both the flight of the mind topos and the Phaethon paradigm of natural inquiry (1.45–52):

te, cum statione peracta
astra petes serus, praelati regia caeli
excipiet gaudente polo: seu sceptra tenere

[30] Cf. Sklenář (2003, 126–127), Wildberger (2005, 80–82), and Chaudhuri (2014, 189). Contrast Manolaraki (2013, 57–58) and Tracy (2014, 103).

[31] The bibliography on this passage is vast; Nelis (2011, 254–259) helpfully surveys the major contributions. Note, too, Dinter (2012, 79–80) on the absence of *fama* and Ambühl (2015, 163–165) on possible Callimachean influence.

seu te flammigeros Phoebi conscendere currus
telluremque nihil mutato sole timentem
igne uago lustrare iuuet, tibi numine ab omni
cedetur, iurisque tui natura relinquet
quis deus esse uelis, ubi regnum ponere mundi.

When you have completed your guard duty and at a late date make for the stars, the palace of your preferred heaven will welcome you and the celestial sphere will rejoice. Whether it is thrilling to hold the scepter or to mount Phoebus's flame-bearing chariot and with wandering fire to circle an earth that has no fear, now that the sun has been transferred, every divinity will yield to you, and nature will leave in your power what god you wish to be, where you wish to establish your reign over the universe.

The two roles that Lucan initially envisions for Nero are ostensibly those of Jupiter and Apollo.[32] But to take control of the chariot of the sun is also to play the role of Phaethon, and several contemporary texts appear to connect the emperor with this very figure, an association that Nero himself may have actively cultivated.[33] At this point, Phaethon does not appear to lose control, and this impression creates an implicit contrast with the final lines of Virgil's first *Georgic*, where the image of the out-of-control charioteer anxiously construes Octavian as Phaethon (*G.* 1.511–514). Given the favorable comparison with Octavian and the fact that Nero was an avid charioteer, Lucan's image will have likely pleased the princeps.[34]

What matters for us, though, is that this Phaethon is an investigator of nature (cf. Ch. 1.5, "Phaethon"). The key lies in *iuuet*. Lucretius programmatically uses this verb, two times in successive lines in two different places, to describe the metapoetic experience of approaching new springs and plucking new flowers (1.927–928, 4.2–3). In the context of poetic initiation and vatic inspiration, the experience is not just "pleasing" but "thrilling." Those writing in the tradition of Lucretius use the same verb to encode the sublimity of celestial exploration. At the beginning of the *Astronomica*,

[32] Nero connected himself with both Jupiter and Apollo; see Roche (2009, 140–141).

[33] See Duret (1988); cf. Champlin (2003, 134–135). A scholion to line 49 (Endt 1969, 8) already points out the allusion to Phaethon.

[34] After the great fire and Nero's demise, any connections between Nero and Phaethon will have landed quite differently; see Newlands (2002, 315–316) and Heslin (2007, 18). The emperor's near-fatal crash during the chariot race at Olympia (Suet. *Ner.* 24.2, Cass. Dio 62.14.1) will have similarly affected the reception of the Nero-Phaethon construct.

Manilius declares that it is thrilling (*iuuat*, 1.13) to go through the air and to stroll and live in the boundless sky and to come to know the paths of the planets, and that it is even more thrilling (*iuuat*, 1.17) to discover the laws of astrology (Ch. 2.3).[35] As Ovid's Pythagoras prepares to explain the nature of things in *Metamorphoses* 15, he asserts that it is thrilling (*iuuat*, 15.147) to go among the stars and that it is thrilling (*iuuat*, 15.148) to travel on a cloud, to stand on Atlas's shoulders, and to look down on the fools below and to enlighten them.[36] In the *Natural Questions*, it is similarly thrilling (*iuuat*, 1 praef. 7) for Senecan *uirtus* to wander (*uagantem*, 1 praef. 7) among the stars and to laugh at the foolish priorities of those on earth (Ch. 3.4). When Lucan speculates that it might thrill (*iuuet*) Nero to mount the Sun's chariot and to circle the earth with wandering fire (*igne uago*), he activates this topos.[37] Here, a contrast emerges. Lucretius and Manilius mount celestial chariots at programmatic moments near the end of their poems (*insignem conscendere currum* at Lucr. 6.47 and *aetherios... conscendere currus* at Man. 5.10). What Lucretius and Manilius do in their final books, chariot-loving Nero boldly does in Lucan's first (*flammigeros... conscendere currus*). The allusion to Phaethon, who figures the investigator of nature in the opening books of the *Astronomica* and *Natural Questions*, and the presence of the imagery and vocabulary of the flight of the mind topos collectively hold out the possibility that Nero's soul might become a student of the stars, a natural philosopher, a model of Stoic *uirtus*.

But again, we cannot read in isolation. Paul Roche (2009, 9), for one, cogently argues that while the invocation of Nero reads as straight panegyric on its own, fulfilling all the generic expectations of such praise, its insertion into a poem that is relentlessly hostile to the imperial system underscores the extent to which such a panegyric is out of place. So, while the image of Nero-Phaethon itself might qualify as unproblematic panegyric, when we set it alongside Pompey's flight of the mind, we get a clearer sense of Lucan's verdict on such study. At best, it is a modus vivendi available only to the dead and viable only in some alternative universe, when all wars,

[35] For Manilius's dependence on Lucretius here, see Bailey (1947, 3.1562–1563), Wilson (1985, 283), and Volk (2002, 225).

[36] Landolfi (2003, 19) connects Man. 1.13–14 and Ov. *Met.* 15.146–149. For Ovid's Pythagoras and the sublime, see Hardie (2009, 136–152).

[37] Note parallels with *NQ* 1 in particular: both Seneca and Lucan use *iuuat*/*iuuet* one time only, while Lucan's *uago* picks up on Seneca's *uagantem*. In other contexts, *iuuat* is not necessarily repeated. For instance, at Virg. *G.* 3.291–293 and at *Octauia* 385–387, it occurs only once. In the second instance, the speaker is Seneca's character; see Ch. 5.3.

including civil, will have ceased (1.60–62). From here, it does not take much imagination for Nero-Phaethon to be seen to devolve into the same sort of satirical joke as Pompey's Stoic soul.

If we situate Nero-Phaethon within the context of the opening of book 1, something different happens: as the text continues, Phaethon's symbolic value shifts from that of an avatar of natural inquiry to that of the architect of universal destruction.[38] When Lucan warns the emperor to maintain the balance of heaven and to locate himself in the middle of the sphere (*librati pondera caeli / orbe tene medio*, 1.57–58), the reader is reminded of the advice that Ovid's Sol gives to Phaethon: maintain a middle course (*medio tutissimus ibis*, *Met.* 2.137). A few lines later, after the panegyric ends, in the middle of the simile that equates civil war with cosmic collapse, Lucan declares that the world's final hour will "again seek ancient chaos" (*antiquum repetens iterum chaos*, 1.74), and the reader hears Ovid's Earth, who laments that, thanks to Phaethon's fire, "we are plunged into ancient chaos" (*in chaos antiquum confundimur*, *Met.* 2.299).[39] With these allusions in the background, we should revisit Nero's "wandering flame" (*igne uago*, 1.50). Ancient sources tend to regard the sun as a planet (hence, "wandering"), but the sun does not "wander" in the same (apparently) unpredictable ways as Saturn, Jupiter, Mars, Venus, and Mercury. The Ovidian echoes and the epithet *uagus* raise the possibility that Nero-Phaethon, as the bearer of the sun, may be headed off course from the very beginning (cf. Hinds 1987, 28). At this point, we should also reconsider the image of an earth that feels no fear (*telluremque nihil mutato sole timentem*, 1.49). In isolation, we might take this to mean that the earth has nothing to fear because Nero, unlike Phaethon, will not unleash universal disaster. Similarly, a lack of fear in the face of cosmic transformation (*mutato sole*) might indicate greatness of soul, and perhaps Nero inspires true *magnitudo animi* among his earthly spectators. But given the broader context, especially the "wandering flame," an earth that feels no fear might just be an earth that is dumbstruck as it watches the charioteer of the sun lose control and the

[38] Duret (1988, esp. 147–148) claims that Lucan and his contemporaries Seneca and Vagellius create an alternative mythology wherein Phaethon does not crash. My reading of *NQ* 3 praef. (Ch. 3.2) and the present discussion push back against Duret's conclusions. Lucan himself is familiar with the traditional story (2.410–415), and there is no reason to suppose that readers are not meant to think of it here.

[39] For this allusion, see Hinds (1987, 28–29). For the simile itself, see Johnson (1987, 14–18), Narducci (2002, 42–50), Sklenář (2003, 3–10), and Roche (2005). Wheeler (2002) explores Lucan's reception of the *Met.* more broadly.

cosmos catch fire. On this reading, the earth resembles the survivors of Seneca's cosmic deluge, those wretched few whose fear turned into stupefaction and whose mindset constituted a sad counterpoint to the vital calmness of the *proficiens* (Ch. 3.5 on 3.27.12).

In isolation, then, Nero-Phaethon soars through the heavens, a virtual carbon copy of Lucretius's Epicurus, while in the broader context of book 1 he ignites the great conflagration that constitutes the subject matter of the poem and that will never lead to the regeneration that such fires are supposed to inaugurate.[40] Either way, Phaethon does not offer Lucan a meaningful model to shape his own identity or to structure his worldview, and this failure is only partially a function of genre. Just as Nero coopts the epic Muse (1.63–66), thereby corrupting Lucan's connection to the literary past (Feeney 1991, 278), he seizes the identity of Phaethon, thereby foreclosing the possibility that anyone in contemporary Rome might undertake the journeys of Manilius or Seneca. In the ongoing emergency of civil war, there is no time for that kind of adventure; it must wait for an impossibly distant future, and even then, it might look hopelessly self-indulgent and irresponsible.[41]

The account of the Gallic tides suggests that the investigation of nature is simultaneously sublime and a distraction from the urgent necessity of civil war. The celestial journeys of Pompey and Nero further underscore the incompatibility of the investigation of nature, on the one hand, and civil war/*Civil War*, on the other. Collectively, these passages identify Rome's imperial system, including the Neronian moment, as the cause of the breakdown of the scientific sublime. We should read other passages in which Lucan is unable to offer an explanation for natural phenomena from this same perspective and as similarly reaffirming the corrosive impact of Roman *imperium* on the study of nature and the experience of sublimity. When Lucan does not know why the sun rises obscured by clouds (1.233–235), cannot explain the nature of the Syrtes (9.303–318), or remains mystified about the origins of Libya's snakes (9.619–623), he casts a sidelong glance at the sublime and quickly turns away; to pursue the matter further would be a dereliction of duty.

[40] Roche (2005, 59–71) discusses the reader's shocking realization that rebirth will not follow this conflagration.

[41] For a complementary reading of the rejection of Manilius's worldview in the opening of Luc. 1, see Galli Milić (2016).

4.3. Knowing Too Much

Philip Hardie (2013b, 236–237) writes of the "cloud of unknowing" that hangs over Lucan's poem, and it is undeniable that both Lucan and his characters are traumatized by deep and disquieting uncertainties. But despite this "cloud" and the poem's own insistence that there is no time for the study of the natural world, Lucan repeatedly recounts the physical processes according to which natural phenomena operate, and he digs deep into their technical details. Such passages—from a brief allusion to the circulation of air within the human body to the description of the strange spectacle of the night sky at a southern latitude—collectively insist that we conceptualize the natural world in philosophically rigorous terms, on all levels, at all times. On Fantham's reading (1992a, 19), these passages animate the poem with the electricity of the sublime, but the text itself suggests that while at first such detailed knowledge might ennoble and uplift, it ultimately overwhelms and incapacitates.[42] To suffer through Lucan's encyclopedic knowledge of nature is to feel crushed beneath the weight of all there is to know. Whereas Manilius revels in the infinite accumulation of ideas and competing theories, even when such accumulation leads to no satisfying resolution, as in the case of the Milky Way (Ch. 2.2), Lucan construes this kind of experience differently. It is a feeling reminiscent of the vital calm that sometimes marks Seneca's *proficiens* (Ch. 3.5), and it contributes to the sense of oppression that rattles the poem from its foundations. This experience is darkly, unsettlingly sublime.[43]

Lucan encodes the dark sublimity of natural knowledge through the Pythian priestess Phemonoe (book 5) and in the context of the Thessalian witch Erictho (book 6). Multiple studies have shown the metapoetic status of these characters, both of whom are *uates* (5.165, 6.651), and I take those conclusions as read.[44] More recently, Day has argued that the sublime contributes in significant ways to the metapoetic functions of Phemonoe and

[42] Cf. the astute remarks of Bramble (1982) regarding the storm in book 5: "Scientific theory is invoked to make us more aware of the breach in natural law" (554), "scientific observation…helps once more…[to] further our fears on nature's behalf" (555).

[43] This dark sublimity has something in common with the theories of Burke and Kant, for whom terror is so central (see the Introduction). The well-known prayer for ignorance of the future (2.14–15) points to the oppressiveness of certain kinds of knowledge; the phenomenon I discuss here is closely related.

[44] Phemonoe: O'Higgins (1988, 211–217) and Masters (1992, 133–149). Erictho: O'Higgins (1988, 218–225), Masters (1992, 205–215), Dinter (2012, 62–75), and Celotto (2018). Quint (2020) discusses Erictho's fascinating reception.

Erictho.[45] In the case of Phemonoe, Day draws attention to *Subl.* 13.2, where Longinus compares the inspiration that comes from close imitation of great writers of the past to the divine inspiration of the Pythia. Since Longinus makes the priestess' experience a model of sublime literary production, the argument goes, we can read her literal possession and ecstasy in the *Civil War* through the same lens. Erictho's sublimity, according to Day, lies primarily in her sinister ability to induce "a sickening vertigo" through spells that upend the natural world (Day 2013, 103):

> Under her spells the sea rises up though the winds are still, waterfalls hang from cliffs, rivers flow uphill, 'mountains dip their peaks' and, instead of towering on high, Olympus looks up at the clouds (6.469–477); such inversions offer a perverted literal counterpart to the radical changes wrought upon (our perception of) the world by the sublimity of Lucretian atomism.

The following discussion builds on Day's perspectives. The passages that focus on Phemonoe and the Thessalian witches carefully encode the sublime, which frees us from relying on Longinus. In addition, both texts vividly thematize the terrifying experience of knowing too much. When read in this light, the priestess and the witches of Thessaly, not just Erictho, provide essential interpretive frameworks for making sense of the overabundance of natural knowledge in the whole poem. Their stories shed light not just on the experience of the poet but on that of the reader and suggest that the *Civil War*'s excess of erudition systematically enacts a dark sublimity.

Day is certainly right to read Phemonoe's experience as sublime. Setting aside *Subl.* 13.2, we should note that Phemonoe comes into contact with the god at a literal level, and the result is explosive. Lucan initially compares the process of inspiration to a volcanic eruption (5.97–101):

> hoc ubi uirgineo conceptum est pectore numen,
> humanam feriens animam sonat oraque uatis
> soluit, ceu Siculus flammis urgentibus Aetnam

[45] Day (2013, 98–105); cf. Seitz (1965, 215–216). Day also connects Phemonoe with the *matrona* who appears at 1.673–695. Since the *matrona* does not have any connections with natural knowledge, I have not focused on her here.

undat apex, Campana fremens ceu saxa uaporat
conditus Inarimes aeterna mole Typhoeus.

When this power is received in the virgin's breast, it strikes her human spirit, sounds forth, and undoes the prophetess' mouth, just as the Sicilian peak undulates when the flames put pressure on Aetna, just as Typhoeus, buried by the eternal mass of Inarime, roars and heats the Campanian rocks.

Aetna has longstanding associations with the sublime, as we will see in Ch. 5. What matters here is that the destructive force of Phemonoe's encounter with the god packs all the punch of an archetypical moment of sublimity. Significantly, though, Lucan's mythological aition subtly hints at a more technical explanation of the mountain's volcanism: the presence of underground steam (*uaporat*) suggests the subterranean wind theory of seismic and volcanic activity.[46] The simile, then, gestures toward a philosophically grounded rationale for Aetna's combustible nature and thereby characterizes Phemonoe's inspiration through the scientific sublime.[47]

Lucan pays close attention to how Phemonoe experiences this process. At first, she is terrified (*conterrita uirgo*, 5.161). The initial moment of inspiration (5.161–169) emphasizes the physical irruption of *numen* and *spiritus* into her body (*concepit pectore numen, / quod...spiritus ingessit uati*, 5.163–165), the forceful displacement of her mind (*mentemque...expulit*, 5.167–168), and the god's command that the human part of her withdraw entirely from her breast (*hominem toto sibi cedere iussit / pectore*, 5.168–169); by submitting her body to the god, she transcends her own mortality. By the end, however, the situation has evolved. While a volcanic fire still suffuses her complexion (*rubor igneus inficit ora / liuentesque genas*, 5.214–215), Phemonoe is no longer frightened: "She has a pallor, not the kind that is characteristic of one who is afraid, but one that causes fear" (*nec qui solet esse timenti, / terribilis sed pallor inest*, 5.215–216). The movement from terrified to terrifying encodes the dual experience of the sublime, the tension between simultaneously being overawed and inspiring awe; this is

[46] For the subterranean wind theory, see Ch. 5.1. Schrijvers (2005, 33–36) discusses Ovid's penchant for, and Lucan's occasional experimentation with, treating a phenomenon as simultaneously both "natural process and supernatural being."

[47] In subsequently describing Phemonoe's mental state as a form of orgiastic madness (*bacchatur demens*, 5.169), Lucan invokes a different tradition of sublime transport, one that is present already in Plato (e.g., *Symp.* 215a–216b, 218b) and is widespread in the imperial period (e.g., Dion. Hal. *Dem.* 22, Long. *Subl.* 3.2). This tradition characterizes not only the affective response to a work of literature but also the emotional upheaval of encountering philosophical truth.

magnitudo animi as embodied by Lucan's Caesar or Virgil's Hercules (Ch. 1.4) And once the god leaves her, she immediately falls (*uixque refecta cadit*, 5.224). Lucan has already made clear that to burn up in such a volatile encounter can be viewed as either a punishment or reward (*numinis aut poena est mors immatura recepti / aut pretium*, 5.117–118); like Phaethon, Phemonoe can celebrate her divine incineration.[48]

The most central part of Phemonoe's experience is the knowledge to which she gains access (5.177–182):

> uenit aetas omnis in unam
> congeriem, miserumque premunt tot saecula pectus,
> tanta patet rerum series, atque omne futurum
> nititur in lucem, uocemque petentia fata
> luctantur; non prima dies, non ultima mundi,
> non modus Oceani, numerus non derat harenae.

All time comes together into one mass, and so many ages oppress her wretched breast, so great a chain of events is exposed, everything that is to come presses toward the light, and fates struggle in their search for a voice. The world's first day was not absent, nor its last, nor the size of Ocean, nor the number of the sands.

The initial emphasis falls on historical knowledge, which gradually expands to encompass both the origins and end of the *mundus*. The size of the Ocean, however, and the number of the sands do not qualify as historical knowledge at all. In Herodotus, the Pythia herself affirmatively declares that she knows these facts (οἶδα δ' ἐγὼ ψάμμου τ' ἀριθμὸν καὶ μέτρα θαλάσσης, 1.47.3), which grounds her authority in a claim to universal knowledge. Phemonoe's knowledge is similarly universal, equally infinite.[49] But rather than establishing her unquestionable authority, all that data crowds her mind: boundless knowledge oppresses and makes wretched (*miserumque premunt…pectus*), it is a jumbled heap (*congeriem*) in which undifferentiated facts struggle and fight (*nititur*, *luctantur*), and Phemonoe can only find the

[48] Phemonoe's experience also resembles cosmic dissolution, which further justifies a comparison with Phaethon: at 5.119, her human "frame" totters (*compages humana labat*), which recalls the collapse of the cosmic "frame" in passages such as 1.72 (*compage soluta*) and 5.633 (*motaque poli compage laborant*). For the language, see Lapidge (1979, 360–370).

[49] For the proverbially uncountable number of sands and the sea, see McCartney (1960, 81–83). On Lucan's use of the trope here, note Graziano (2021).

answer she's looking for with difficulty and after an arduous investigation (*uix inuenit... inter fata diu quaerens tam magna latentem* [sc. *Appium*], 5.188–189). Couched in the language of philosophical inquiry (*inuenit, quaerens, latentem*), Phemonoe's knowledge of Ocean and the sands looks more meteorological than historical, and her sublime experience of that knowledge provides a lens through which to read Lucan's own use of similar knowledge. Her fall is not that of the investigator of nature who experiments with different hypotheses or modes of discourse, but of one whom the very mass of natural knowledge, symbolized by the god of prophecy himself, crushes and asphyxiates, even at the very moment of transcendence (*cetera suppressit faucesque obstrinxit Apollo*, 5.197).

Day is also on the right track when he suggests a connection between Erictho and Lucretius, but further links are worth considering. Day's discussion of the vertiginous effects of Erictho's songs examines a few lines from an account of the effects of the songs of Thessalian witches in general (6.438–491). At a simple level, the text encodes the sublimity of the witches' power: "Once the spell is heard, the *mundus*, in its headlong spin, is dumbstruck, and as Jupiter pushes and impels the celestial sphere on its swirling axis, he marvels at the fact that it does not go" (*torpuit et praeceps audito carmine mundus, / axibus et rapidis impulsos Iuppiter urguens / miratur non ire polos*, 6.463–465). The responses of the *mundus* and Jupiter can be read into one another: the *mundus* is paralyzed both physically and psychologically (*OLD* s.v. *torpeo* 2), while Jupiter is immobilized with awe; something exists that is even greater than the *mundus* and Jupiter himself, and they have both experienced it firsthand.

A more subtle process of encoding unfolds a few lines later (6.481–484):

> terra quoque immoti concussit ponderis axes,
> et medium uergens titubauit nisus in orbem.
> tantae molis onus percussum uoce recessit
> perspectumque dedit circum labentis Olympi.

Earth, too, shook the axis of its immobilized weight, and the thrust that tends toward the middle of the world faltered. The burden of so great a mass, shattered by the voice, withdrew and gave a clear view of Olympus gliding all around.

The song of the Thessalian witches physically shakes the earth's axis, which decenters the *mundus*, disrupting the gravitational pull that draws objects toward the center.[50] What happens next? The earth breaks apart (*recessit*), allowing one—who, exactly?—to see clear through to the cosmic sphere spinning on the other side (*perspectumque dedit circum labentis Olympi*).[51] The movement from the violent disruption of a worldwide earthquake to a disembodied gaze that peers calmly through the earth to the fixed stars opposite encodes the sublime.

Two important comparanda are missing from this analysis, though. First, the action of 6.683–684 is best understood as a reimagining of the opening of *De rerum natura* 3. There, as we have seen (Ch. 1.3), Lucretius explains that when he hears the voice of Epicurean philosophy (*simul ac ratio tua coepit uociferari*, 3.14), the walls of the *mundus* withdraw and he sees things moving through the void (*moenia mundi / discedunt, totum uideo per inane geri res*, 3.16–17). More specifically, the earth falls away and allows everything that moves through the void beneath his feet to be discerned clearly (*nec tellus obstat quin omnia dispiciantur, / sub pedibus quaecumque infra per inane geruntur*, 3.26–27). The verbal echoes with our passage are not especially close (*uoce* ~ *uociferari*, *recessit* ~ *discedunt*, *perspectum* ~ *dispiciantur*), but the basic idea—a voice is heard, the earth opens, a cosmic panorama comes into view—is virtually identical in both cases and highly unusual. Moreover, in addition to atoms falling through the void, Lucretius sees the dwelling place of the gods (3.18–24). This passage draws on the description of (mythological) Olympus at *Od.* 6.42–46, and it is precisely (philosophical) Olympus (= the heavens, *OLD* 1c) that the Thessalian witches bring to view in the poem (*perspectum... Olympi*). Lucretius's experience is programmatically sublime, that thrilling combination of *uoluptas* and *horror*, "a rapturous earthquake," "a catastrophic pleasure." This is the experience that the Thessalian witches make available to Lucan's readers; it is both an earth-shattering shudder and the calm detachment of a view of heaven. In this, the witches cast the reader in the role of Lucretius, forcing

[50] A scholion to 6.482 (Endt 1969, 228) describes the phenomenon simply as an earthquake. The same verb makes civil war, and perhaps the poem itself, a universal earthquake at 1.5 (*concussi... orbis*); cf. Gauly (2018, 83).

[51] The phrase *perspectum... Olympi* is tricky. Housman (1926, 170) glosses "terra perforata dari sic dicitur" and compares Liv. 40.22.3, where a dense network of crisscrossing branches makes it almost impossible to see through to the sky (*ut... perspici caelum uix posset*). A scholion to 6.484 (Endt 1969, 228) understands the passage similarly: "the earth is surrounded by *aer* and has now revealed the *aer* and given it to sight" (*terra aere cingitur, quae nunc aerem proditum dedit aspectibus*).

us to see the *mundus* from a shockingly different perspective, to feel the disorienting jolt of discovering a new reality.

The second relevant background text comes from Longinus. In *Subl.* 9.6, Longinus praises Homer's depiction of *theomachia* in a passage that combines *Il.* 21.388 and 20.61–65: the heavens and Olympus resound, and Aidoneus shouts aloud in fear that the earth may split open, on account of Poseidon's violent shaking, and reveal the underworld to the gods. In Lucan's poem, a powerful voice sets in motion a cosmic earthquake, and depending on one's perspective, either that earthquake realizes Aidoneus's fears, as Erictho prepares to bring the underworld quite literally to the surface, or the serene view of heaven demonstrates the foolishness of mythological thinking. In the first instance, what initially appears to be a calm look at the cosmic sphere conceals a terrifying glimpse into the underworld. In the latter case, Lucan sides with Lucretius, who may himself be thinking of *Il.* 20 when the earth recedes at the start of *De rerum natura* 3: in pointed contrast to Aidoneus's worries, Lucretius declares that the underworld is nowhere to be seen (*at contra nusquam apparent Acherusia templa*, 3.25).[52] Even if we disallow the allusion on Lucan's part, Longinus demonstrates the dark sublimity of a seismic event that splits the world apart. A view of heaven, discerned through a rift in the earth, is ominously sublime.

As the reader continues, the shockingly gruesome details of the Erictho narrative itself quickly subsume any sense of serenity that the disembodied view of Olympus might have produced. The *uoluptas* that the reader feels in the physical details of the *Civil War* is always combined with a disorienting and oppressive *horror*—it is all too much, even if it is pleasurable. From this perspective, the experience of the revivified corpse also symbolizes the reader's experience of the haunted house that is Lucan's political and natural histories. After it gives its speech, the corpse begs for death. It requires help to fall (*carminibus magicis opus est herbisque, cadauer / ut cadat*, 6.822–823), and Erictho kindly sets it ablaze before she departs (*accensa iuuenem positum strue liquit Erictho*, 6.826). Like Phemonoe, the corpse is exhausted from all that it has learned, and after such an ordeal, the fires of death are greeted as a welcome reprieve. In Lucan's poem, too much knowledge is darkly, oppressively sublime, and knowledge of the natural world is no exception. The contrast with the infinite accumulation of knowledge in both Manilius and Seneca is striking—we have moved from the cheerful

[52] The commentaries of Bailey (1947) and Kenney (2014) cite no parallels for this line.

exuberance of Manilius through the wide-eyed and transformative aporia of Seneca into the crushing terror of Phemonoe and those who hear the songs of the Thessalian witches.

Two examples will suggest some of the ways in which Lucan's perspective can play out. Frequently, Lucan's obsession with natural phenomena overwhelms by exposing the reader to an overload of unfathomably vast panoramas, even in just a few lines. Take the prominent comparison of Caesar to the thunderbolt (1.151–157). The sudden brilliance, earth-shattering crash, and swift unstoppable might of thunderbolts figure sublime speech from Aristophanes through Longinus and Quintilian.[53] All of these qualities are operative in the expositions of Lucretius (6.96–422) and Seneca (*NQ* 2) and appear in Lucan's simile, which, against this background, functions "as a programmatic statement of the future dictator's sublimity" (Day 2013, 111). Specifically, Lucan encodes the dark sublimity of the thunderbolt through the terrified trembling and stunned vision of those who witness its explosive power (*populosque pauentes / terruit* [sc. *fulmen*] *obliqua praestringens lumina flamma*, 1.153–154); this is what it means to be thunderstruck in the world of the *Civil War*. Crucially, however, the thunderbolt is not Jupiter's but one that operates according to philosophical theory: "Forced out by winds through clouds, accompanied by the sound of shattered ether and the crash of heaven, the thunderbolt flashed forth" (*expressum uentis per nubila fulmen / aetheris impulsi sonitu mundique fragore / emicuit*, 1.151–153). The physics of the line are well-established: multiple philosophical traditions explain that thunder and lightning are produced by the collision of clouds and/or the violent movement of air within and bursting out of clouds.[54] In particular, Lucan's *expressum uentis per nubila* resembles Seneca's *exprimi* [sc. *spiritum*] *conlisibus nubibus* at *NQ* 2.12.5.[55] The terror

[53] Day (2013, 107–111) collects the evidence. Note esp. Ar. *Ach.* 530–531; Long. *Subl.* 1.4, 12.4, 34.4; Quint. *Inst.* 12.10.23–24, 64–65. Schrijvers (2005, 31) connects Lucan's simile with the natural sublime.

[54] The main sources are collected by Getty (1940, 49), Roche (2009, 193), and Day (2013, 111 n.17). Numerous early Greek natural philosophers also explain thunder and lightning through the collision of clouds and the violent movement of air: see Anaximander D33 *EGP* (A23 D-K), Anaximenes D22 *EGP* (cf. D21 and 23 *EGP*, A17 D-K), and Democritus and Leucippus ATOM. D117 *EGP* (A130 and A25 D-K, respectively). The evidence for Anaxagoras is conflicting, but Sen. *NQ* 6.9.1 (D63 *EGP*, A89 D-K) seems to make him hold that thunder and lightning are produced by the collision of clouds and the friction of air being forced out (cf. D53 *EGP*, and A1.9, A84 D-K). Diogenes appeals to wind, among other causes (D23 *EGP*, A16 D-K). For more, see the extensive doxography at Aetius 3.3 (Mansfeld and Runia 2023, 256–263).

[55] Seneca's passage forms part of a translation of Arist. *Mete.* 2.9; for a comparison, see Hine (1981, 246–247).

that shakes and blinds the reader is rooted, from the opening line of the simile, in an accurate knowledge of the physics of the *fulmen*, and it is through that knowledge that the reader's gaze is directed to the cosmic dimensions of the thunderbolt's destabilizing and destructive reach. The thunderbolt is darkly sublime because of its overwhelming visual and aural effects and, at the same time, because of the inescapable materialism and relentless logic through which it bursts into life.

Often, Lucan incorporates excessive amounts of technical information within a larger narrative. Thus, when Cato's army reaches the temple of Jupiter Ammon, Lucan dedicates over fifteen lines to the peculiar astronomical sights that one encounters at a latitude far south of Rome, beginning with the odd shadows cast at midday (9.528–543): "Here, too, nothing blocks Phoebus when day stands balanced at the zenith; with difficulty, the tree protects its trunk, so short is the shadow driven by the rays toward the center" (*hic quoque nil obstat Phoebo, cum cardine summo / stat librata dies; truncum uix protegit arbor, / tam breuis in medium radiis compellitur umbra*, 9.528–530). The sun's power is unchecked, and its untrammeled centripetal brilliance threatens to overpower the natural world. From here, Lucan expands our view: this is the tropic of Cancer, and whoever lives beyond this boundary, your shadows fall to the south, whereas ours fall to the north (9.531–532, 538–539);[56] for you, Cynosura and the Wain set, there are no circumpolar stars, both poles of the celestial axis are equidistant (9.540–543); at this latitude, the constellations of the zodiac rise perpendicular to the horizon and in equal periods, Scorpio and Taurus are equally erect, Aries and Libra move at the same speed, as do Astraea and Pisces, Chiron and Gemini, Capricorn and Cancer, Leo and the Urn (9.543, 533–537). The passage may be riddled with technical errors,[57] but the emotional impact cannot be denied: Phoebus's unrelenting power, the spectacle of the constellations of the zodiac rising in unfamiliar ways, and "you," the reader, witnessing it all seemingly at firsthand. By compelling readers to explore the strange sights of an extreme southern latitude, Lucan plunges them into the unknown, and we are inevitably overwhelmed by the dizzying experience of that excess of knowledge. The night sky, a sight that seems fixed and unchanging, is revealed to be contingent—other skyscapes exist, other ways of perceiving the world are possible. The very idea expands the mind and is potentially electrifying, even if the details are factually

[56] The printed text runs 531–532, 538–543, 533–537, 544, etc.
[57] See Housman (1926, 329–333) and Wick (2004, 2.196–198).

inaccurate.[58] In this, Lucan resembles both Manilius, who delights in transporting the reader to distant latitudes where the familiar becomes unrecognizable (Ch. 2.5), and Seneca, who is unafraid to dare the reader to imagine a heaven full of unfamiliar planets and stars that only appear to be fixed (Ch. 3.4). For all three authors, technical accuracy counts for less than the vital and relentless accumulation and display of theoretical possibilities. But when they are read through the lenses of Phemonoe and those who hear the spells of the Thessalian witches, these kinds of hyper-displays ultimately crush the viewer, throwing our world into disarray. If the pursuit of infinite knowledge brings Manilius and Seneca into contact with the divine, in the world of the *Civil War* it is possible to know too much. The strange sight of the stars rising at a right angle to the horizon creates chaos and is positively terrifying; to behold such a show is to be directly exposed to the devastating power of the sun's rays as they hurtle relentlessly toward the center.

Such passages are not digressions or virtuoso displays of learning, nor are they mere reflections of the recommendations of rhetors such as Hermogenes. They harness the awesome power of natural knowledge to heighten the sense of inescapable and all-pervading crisis that dominates the *Civil War* and are manifestations of *bellum ciuile* itself.[59]

4.4. Spectacular Irrelevance

Throughout the text, the reader encounters characters who possess precisely the kind of natural knowledge that, under regular circumstances, triggers the sublime. In the narrative setting of the *Civil War*, however, such knowledge turns out to be entirely detached from reality and/or incapable of inspiring the thrill to which ancient readers are accustomed. Through such miniature dramas, the scientific sublime itself begins to break down, to cease to function properly, to become spectacularly irrelevant, just like so much else in Lucan's crumbling universe.[60]

[58] Beaujeu (1979, 213), writing of this and similar passages, rightly declares that they create "une impression d'exotisme surprenante, parfois saisissante."

[59] Perhaps the most spectacular of such passages are those that envision the dissolution of the *mundus* itself and equate civil war with cosmic disintegration. These passages draw heavily on Stoic physics and the language of Latin Stoicism to reveal to the reader a new reality that is terrifyingly sublime. Again, see Lapidge (1979), Roche (2005), and Star (2021, 169–190).

[60] Cf. Feeney (1991, 291): "All public institutions in the poem become irrelevant, leaving the solitary individual as the focus, living on posthumously, surviving the death of the *res publica*."

The "Earthquake" at Massilia

An illuminating example occurs at Massilia. In preparation for his siege of the city, Caesar orders his troops to fell a grove that the locals consider sacred (3.399–452). Leigh has shown that Caesar's confrontation with the grove takes as its model Epicurus's confrontation with *religio* in Lucretius's opening book (Ch. 1.3): the dark and terrifying grove fills the locals with superstitious awe, but Caesar's bold daring (he is the first to swing an axe) reveals the irrationality of those fears as he lets in the light of reason and the barbarous gods fail to exact vengeance for a supposedly sacrilegious act.[61] We should read what happens next against this very Lucretian background of philosophical enlightenment.

Caesar, ever impatient, departs for Spain, leaving his troops to conduct the siege. A giant ramp is built, with two towers that are as high as the city walls. The towers were not fixed in place, "but crept along their lengthy path by a hidden cause" (*sed per iter longum causa repsere latenti*, 3.458). Commentators remark that the towers move on rollers or wheels (Braund 1992, 256; Hunink 1992, 189), but this distracts from the nature of the drama that is unfolding: the combination of *causa* and *lateo* pushes the reader to think about philosophical explanations for obscure natural phenomena, such as the Gallic tides, whose rationale Lucan dramatically orders to remain concealed (*causa...late*, 1.418–419). Next, the massive towers sway as they move along, and the Massilians think they know what has happened: "The soldiery believed that a wind had shook the empty hollows of the earth in its search to break free, and they wondered at the fact that their walls were still standing" (*telluris inanis / concussisse sinus quaerentem erumpere uentum / credidit et muros mirata est stare iuuentus*, 3.459–461). The Massilians misidentify the cause of the towers' movement as an earthquake, the physical mechanisms of which they understand perfectly well—Lucretius and Seneca, for instance, explain earthquakes through the violent movements of air trapped inside the earth.[62] The philosophical idiom of

[61] Leigh (1999); Chaudhuri (2014, 159–165) builds on Leigh's arguments. For other influences and readings, see Masters (1992, 25–29) and Kersten (2018, 68–97). In the conclusion to an important article on Seneca and the sublime, Michel (1969, 257) points in passing to this passage in Lucan.

[62] Lucr. 6.535–607, Sen. *NQ* 6.12–21.1. This account of seismic activity is closely connected with the subterranean wind theory of volcanism; see Ch. 5.1.

ancient meteorology and the language of wonder encode the sublimity of the soldiers' experience.

But no earthquake has occurred. The philosophical analysis of the Massilians bears no relationship to their present reality, and their state of sublimity is triggered not by a direct confrontation with nature's might or a striking insight into the physics of an earthquake, but by a delusion. In Lucan's topsy-turvy world, it is possible to mistake a man-made weapon of war, a symbol of Caesar's own indomitable power, for a force of nature. Such confusion undermines any attempt to champion the scientific sublime. The soldiers' experience, then, suggests that natural philosophy cannot or, at the very least, does not consistently lead to enlightenment and transcendence.

Caesar, Amyclas, and the Aratean Tradition

The inability of natural knowledge to make good on the promise of transcendence plays a key role in two episodes that juxtapose Caesar and Pompey. Through Caesar's encounter with Amyclas in book 5 and Pompey's exchange with his helmsman in book 8, Lucan probes the different ways in which celestial and meteorological knowledge fail to expand and elevate the mind. The episodes are closely linked. Both Amyclas and the helmsman are described as *rector ratis* (5.515, 8.167), both take their passengers on nighttime voyages, and both deliver speeches that reveal their knowledge of the natural world, including the heavens. In addition, as we will see, their stories display the shortcomings of traditional sources of sublimity in Lucan's fractured landscape.

After transporting his army to Epirus and encamping at Dyrrachium, Caesar slips away in the middle of the night to search for a way back to Italy. Immediately, he stumbles upon a small boat belonging to a certain Amyclas and demands that Amyclas convey him across the sea. In response, Amyclas lists the sun signs, moon signs, and sublunar portents that make him reluctant to sail (5.540–556). The speech draws heavily on the second half of Aratus's *Phaenomena* and Virgil's adaptation of those lines in the first *Georgic* (1.351–463).[63] Given the prestige of the literary background and

[63] See Esposito (2007), Matthews (2008, 115–117), and Meyer (2023, 78–80). Kersten (2018, 137–152) discuses influence of the *Georgics* more broadly on the entire scene. Tomcik (2021) places Amyclas's speech within the didactic tradition by considering its use of acrostics and related forms of wordplay.

the celestial/meteorological nature of the topic, one might have expected Amyclas's recitation to dazzle the general, to stop him in his tracks. Lucan's Caesar, after all, is familiar with natural philosophy and a self-stylized lover of truth, but that is not how the narrative plays out.[64] Caesar prevails upon his interlocutor, and when a storm of earth-shattering proportions arises, the reader easily understands the accuracy of Amyclas's technical knowledge. What, then, is the lesson about Amyclas's Aratean erudition?

All signs indicate that Amyclas aims for sublimity but finds only a hollow simulacrum. At the outset, Amyclas appears lowly, humble, even primitive—his hut is built of rush and reed, his bed consists of seaweed (5.515–521).[65] Amyclas's poverty lends him a certain kind of philosophical peace of mind (*securus*, 5.526), everything about him is diminutive (*scintillam tenuem* at 5.525, *angustique lares* at 5.528), his simple life is easy (*tuta facultas*, 5.527), and he is not predisposed to feel the affective intensity that a godlike Caesar can activate (*nullo trepidare* [sc. *potuit*] *tumultu / Caesarea pulsante manu*, 5.530–531). Nevertheless, Caesar tries to inspire in him that greatness of spirit that describes the mindset of the sublime subject: "Expect more than your modest wishes and expand your hopes! Do not refrain from entrusting your fate to a god who wants to fill your cramped home with sudden wealth!" (*expecta uotis maiora modestis / spesque tuas laxa... ne cessa praebere deo tua fata uolenti / angustos opibus subitis implere penates*, 5.532–533, 536–537). In response to Caesar's invitation to greatness, Amyclas deploys Aratus and the Aratean tradition, which affects Caesar not at all. Like Longinus (*Subl.* 10.6), Caesar is unimpressed with the *Phaenomena*.[66]

The beginning and end of Amyclas's speech, however, suggest that he is indeed striving for the exaltation that Caesar challenges him to seize, and that he falls short (5.540, 557–559):

[64] For Aratus, the Aratean tradition, and the sublime, see Ch. 2.2 and Glauthier (2022). For Caesar's familiarity with natural philosophy in the poem, see Meyer (2023, 57–63). At 10.189, Caesar boasts of his own *tantus amor ueri*.

[65] On the "primitive" quality of Amyclas's hut, see Matthews (2008, 92). For his *paupertas*, see Narducci (1983; cf. 2002, 258–261).

[66] Note Narducci (1979, 106–107) on Caesar's *megalopsuchia* here and Seidman (2017, 73–82) on the grandeur and violence of his speech. In a slightly different but equally helpful way, Narducci (2002, 254–256) connects Caesar's "monstruoso egocentrismo" with passages from Sen. *NQ*. One reason Caesar can summon Amyclas to greatness is that the general plays the role of a god in a theoxeny narrative. For the importance of this tradition, esp. of Ovid's Philemon and Baucis, see Narducci (1983), Matthews (2008, 87–89), Ambühl (2015, 36–39), and Reitz (2020).

> multa quidem prohibent nocturno credere ponto....
> [The interpretation of sun, moon, and sub-lunar signs follows.]
> sed, si magnarum poscunt discrimina rerum,
> haud dubitem praebere manus: uel litora tangam
> iussa, uel hoc potius pelagus flatusque negabunt.

To be sure, many things prevent me from trusting the sea tonight.... But if a crisis of great events demands it, I would in no way hesitate to lend a hand. Either I will touch the shores as you have commanded, or this sea and the winds, rather than I, will deny it.

Although the evidence and cultural weight of the Aratean tradition forbid Amyclas to sail, he throws caution to the wind. To abide by the recommendations of the experts in the face of catastrophic necessity is anything but sublime—Amyclas will leap into the void. It is, however, a timid leap. Caesar calls on him to entrust his fate to a god (*praebere deo tua fata*, 5.536), and Amyclas answers that he will lend a helping hand (*praebere manus*). Later, engulfed in the cosmic dissolution of the storm, Caesar declares that if the glory of his death has been given to the sea and he can no longer wage war, he will accept unflinchingly whatever death the gods give him (5.656–659). This is the voice of a Phaethon, one whose unfathomable daring unleashes an unimaginable disaster and who rejoices in the fact that world comes crashing down around him.[67] Amyclas can only halfheartedly ventriloquize such sentiments—either he will make it to Italy, or at least it won't be his fault if the boat founders. It is unsurprising that when the storm immediately erupts, vindicating Amyclas's predications and reluctance, he recommends that they simply turn back (5.568–576).[68]

The contrast between Caesar and Amyclas, from the perspective of the sublime, constitutes a judgment on the Aratean tradition and its astronomical and meteorological expertise. In Lucan's world, the sublime aspirations of Aratus and his Latin translators and imitators go nowhere. Aratus may teach you to decode the language of god, which is written in the stars, but it is Roman *imperium*, which finds its distillation in Caesar's every word and deed, that summons you to true greatness and brings you face to face with

[67] Cf. 1.150: Caesar "delights in having made forward progress through ruination" (*gaudensque uiam fecisse ruina*). For Caesar's sublimity amidst the storm, see Day (2013, 149–153).

[68] Amyclas is thus a parody of Virgil's Palinurus, who recognizes the signs of dangerous weather and confidently tells Aeneas that they cannot make it to Italy under such conditions (*Aen.* 5.13–25); cf. Quint (1993, 137–138), Narducci (2002, 253–254), and Meyer (2023, 81–82).

the divine. To predict the weather is a bookish pastime that shelters you from and makes you unequal to the kinds of experience that truly overwhelm and uplift, that plunge you to the ocean floor and raise you to Olympus all at once (cf. 5.638–653). If you hope to leap into the void, you must become Phaethon, you must embrace Caesar, you must lose yourself in the majesty of empire.[69]

So much, at any rate, is the lesson that Caesar himself surely takes from his expedition, and we may well agree. But those who read between the lines might be tempted to see in Caesar's refusal to heed Amyclas's Aratean advice nothing more than obstinate stupidity. Rather than lose his life in a deed of great daring, Caesar hops gingerly back to shore and takes comfort in the lamentations of his men (*non ingratis…querellis*, 5.681). It is *dictu mirabile* (5.672), or it is fate's cruel joke at the expense of freedom and the Republic, to have fed Caesar's delusions and to keep the madman alive to try and try again. In this case, the Aratean tradition's claim to the sublime looks weaker, sadder, even more desperate.[70]

Pompey Questions His Helmsman

Pompey's exchange with his helmsman offers a complementary critique of celestial inquiry and astronomical knowledge. Jonathan Tracy (2010) reads the episode against two related philosophical traditions, both of which he connects with Seneca, whereby the contemplation of the stars consoles the distressed and/or leads to spiritual improvement: a defeated Pompey, in dire need of consolation and growth, asks his helmsman about the stars, but the helmsman's response fails to comfort or improve the dejected general, whose mental state only continues to deteriorate. This perspective allows Tracy to contrast Lucan's heavens with those of Manilius and Seneca: "When he looks upward in the hope of a pure realm beyond the reach of terrestrial conflicts and vices, as depicted by Seneca and Manilius, Pompey finds instead all the treachery of his own plane staring back down at him" (2010, 655). The elevation and transcendence that are supposed to derive

[69] The connection with empire is reinforced by the traditional ship-of-state metaphor: Caesar in the boat quite simply is the storm-tossed *res publica*.

[70] This is not to deny the potential humor of the scene. Seidman (2017, 73–74) sees comedy in the incongruous pairing of mighty Caesar and poor Amyclas, and Johnson (1987, 104–112) and Quint (1993, 137–140) find comic deflation in the storm's hyperbole.

from the study of the stars—Tracy's primary example is the preface to *NQ* 1—do not materialize: Pompey "has received no salutary broadening of perspective from his brief encounter with the heavens" (2010, 656). Tracy's reading is persuasive, but it remains to situate the episode within Lucan's broader approach to the investigation of the natural world and the problematic sublimity thereof.[71]

In the wake of Pharsalus, Pompey is directionless, "seeking out detours" (*petens dispendia*, 8.2), "confusing with uncertainty the steps of his flight and the paths that he has entangled with wandering" (*incerta fugae uestigia turbat / implicitasque errore uias*, 8.4–5). His disorientation is the result of his precipitous fall from a great height (*summo de culmine lapsus*, 8.8), and that whirling change in fortune leaves those who meet him stupefied and thunderstruck (*occursu stupuere ducis uertigine rerum / attoniti*, 8.16–17). In his total and almost instantaneous defeat, Pompey becomes an object of sublime wonder, at least for others. But the memory of his past greatness crushes him (8.21–23), for such a reversal of fortune is a source of shame unless it ends in death (8.29–32); Pompey wishes his fall could have been that of a Phaethon.[72]

By the time Pompey sets sail from Lesbos, the sublime aura of his vertiginous collapse has faded. Relentless anxiety about his destination keeps him up at night (*uigiles…curae*, 8.161), crushing the exhausted surge of his uncertain breast (*fessos incerti pectoris aestus*, 8.166). In this state of mind, Pompey turns to his helmsman and asks him about the stars (*rectoremque ratis de cunctis consulit astris*, 8.167), "by which ones he distinguishes the lands, what is his method in the sky for cutting through the sea, by what star he sets a course for Syria, or how small a fire in the Plow guides him to Libya" (*unde notet terras, quae sit mensura secandi / aequoris in caelo, Syriam quo sidere seruet / aut quotus in Plaustro Libyam bene derigat ignis*, 8.168–170). The set-up prompts the reader to anticipate that the grandeur of the stars and the magnificence of celestial knowledge will recuperate Pompey's aimless and abject degradation.[73]

The helmsman, a learned observer of the quiet expanses of heaven (note *doctus* at 8.171), seems to oblige (8.172–176):

[71] Meyer (2023, 101–114) offers a complementary reading that adds more astronomical and geographical detail to Tracy's position. For a different approach, see Barrière (2020, 46–54).
[72] McRoberts (2015) discusses possible connections between Pompey and Ovid's Phaethon.
[73] Pompey's indirect questions plug into the Latin didactic tradition, which may similarly raise expectations of the sublime; for the didactic associations of such indirect questions, see Ch. 4.5.

signifero quaecumque fluunt labentia caelo,
numquam stante polo miseros fallentia nautas,
sidera non sequimur, sed, qui non mergitur undis
axis inocciduus gemina clarissimus Arcto,
ille regit puppes.

We do not follow whatever stars flow and glide in constellated heaven, deceiving wretched sailors with the sphere never standing still. Rather, it is the pole star that guides ships, the one that is not plunged into the waves and that never sets, the brightest one in the twin Bears.

The helmsman's vocabulary, here and in the rest of his speech, is that of Lucretius, Manilius, and the Aratean tradition; it is the language of celestial inquiry at Rome, and it ought to ennoble Pompey.[74] At *Astron.* 1.474–482, for instance, the regularity of the fixed stars is evidence of the divine reason of the cosmos. Their movements look confused but are entirely certain ("what is so confused in appearance, what is so regular in its changes," *quid tam confusum specie, quid tam uice certum est*, 1.482), and there is nothing that is more wondrous than the orderliness of the movements of the heavens ("there is not anything in so great a mass that is more marvelous than its *ratio* and the fact that all things obey fixed laws," *nec quicquam in tanta magis est mirabile mole / quam ratio et certis quod legibus omnia parent*, 1.478–479). For Pompey's helmsman, by contrast, the constant motion of the fixed stars is deceitful and a source of pain (*miseros fallentia nautas / sidera*).[75] While his turn to the pole star and its fixity might inspire confidence, by the end of the speech the helmsman reveals that he also navigates by Canopus, "a star content to wander in the southern sky" (*Australi caelo contenta uagari / stella*, 8.182–183). Not only is this one of the *fallentia sidera*, but also the fact that it "wanders" (*uagari*) makes it a planet (Tracy 2010, 649), one of those most incomprehensible and untrustworthy of celestial bodies. The helmsman is not ignorant or confused (*doctus*, 8.171); he speaks with the same astronomical confidence as Lucan. If his words register as sublime for the reader, it is a dark and disconcerting sublimity that they convey.

[74] See the relevant notes in D'Urso (2019).

[75] Tracy (2010, 643–644) detects a very specific reworking of Man. 1.294–295 in Luc. 8.173–174, concluding that the helmsman uses Manilian language "to undermine Manilius's optimistic view of cosmic governance."

This speech, though, has no discernable effect on its internal audience. When the helmsman asks where his passenger wishes to go, Pompey's heart remains doubtful (*dubio contra cui pectore Magnus…respondit*, 8.186–187), still mired in anxious thoughts and uncertain worries. He then asks that the helmsman keep the ship far from Emathian shores and Italy, leaving everything else to the winds and to *fortuna* (8.187–192). For a moment, Pompey sounds like Caesar in the storm (cf. 5.580–583 on *fortuna*), but his words are best taken as the same kind of hollow mimicry of the sublime to which Amyclas gives voice. A contrast with Virgil's Pallas underscores his lack of heroism. At *Aen.* 10.160–161, Pallas clings to Aeneas's left side, like a dutiful student,[76] and asks about the stars as the two sail back from Pallanteum in preparation for war (*Pallasque sinistro / adfixus lateri iam quaerit sidera*). Lucan's Pompey plays the role of the young protégé, not the hero, who sails to his untimely death, and he fails to live up even to those expectations. There will be no *aristeia*, only centrifugal flight and an ignoble end; the helmsman's speech falls on deaf ears.[77]

If Caesar and Amyclas demonstrate the pedestrian nature of Aratean learning in comparison with the electric charge of Roman *imperium*, at least from the Caesarian perspective, Pompey and his helmsman show something else: the inability of celestial inquiry and astronomical knowledge to offer a way out of the grinding misery of defeat in civil war. Pompey's world, the world of the Republic, is hopelessly broken, and the heavens provide neither consolation nor transcendence. While Tracy rightly reads this as an inversion of Manilius's optimistic worldview, the relationship with the *NQ* is more complicated. We have seen that Lucan recognizes and rejects the sublimity of celestial inquiry and deploys astronomical knowledge to create an experience that is darkly sublime; all of this constitutes an inversion of the basic ethos of the *NQ*. But the *NQ* itself has already begun to upend the worldview of the *Astronomica* (Ch. 3.4–5). Overall, the heavens of the *NQ* are theoretically orderly but almost impossible to understand for those on earth, and Seneca is intent on maximizing our sense of their apparent limitlessness and unpredictability—perhaps the fixed stars only appear to

[76] See Harrison (1991, 105).

[77] Lausberg (1990, 184) and D'Urso (2019, 365–366) observe the connections between Pompey and Pallas. Tracy (2010, 650) and D'Urso discuss points of contact between the helmsman and Palinurus. Tracy (2014, 15–30) treats Pompey's persistent state of flight and his transformation, later in book 8, into "a risible parody of a Parthian king, of Alexander, and (by extension) of Caesar himself."

be fixed, perhaps there are more planets with even more extraordinary orbits than we realize. This celestial aporia can bring the *proficiens* into momentary contact with the divine, but for the foolish and the wise such an experience is of no value. Lucan's Pompey is neither sage nor *proficiens*, which leaves only the fool.[78] In the dangerous world of the *NQ*, the foolish find no remedy for their ever-present fear (*sine remedio timor stultis*, 6.2.1), a description that seems entirely fitting for Lucan's paradoxically shameful run-away leader (*profugum...ducem*, 8.259).[79] The stars that guide epic heroes, from Homer's Odysseus to Virgil's Palinurus, and that amaze the readers of Aratus and Manilius do not speak to Pompey. He asks for divine guidance, but his mind is closed to the possibilities of the sublime.

4.5. Cato and the Snakes

In book 9, Cato and his men undertake a brutal march across the Libyan desert during which they arrive at a spring that swarms with poisonous snakes (9.604–618). At this point, Lucan declares himself unable to explain the origins of the snakes according to the principles of natural history and instead offers a lengthy mythological aition (9.619–699) before cataloging the names and most salient attributes of seventeen serpents in thirty-four lines (9.700–733). When the historical narrative resumes, Cato's troops fall victim to the snakes one after another in a comically grotesque and learned massacre before the Psylli, a North African tribe immune to snake venom and equipped with magical incantations, come to the rescue and miraculously save the beleaguered army (9.734–937). Cato's seemingly mythic ordeal shows certain parallels with a variety of literary and historical adventures, and scholars have argued that Lucan models his "hero" on such diverse figures as Hercules, Jason, Perseus, Aeneas, Alexander, Ophellas, Regulus, and Marius.[80] Others have attempted to read the episode on the basic assumption that not only language but also the history and nature of

[78] Tracy (2010, 654–659) argues convincingly against reading Pompey as a *proficiens*.
[79] Johnson (1987, 79) nicely unpacks the biting implications of this phrase.
[80] Scholarly analysis of the expedition has tended to focus on either Lucan's sources or his characterization of Cato and Cato's Stoicism. Bartsch (1997, 31–35) and D'Alessandro Behr (2007, 113–116) survey the major approaches. Raschle (2001) and Wick (2004) provide detailed introductions. Meyer (2023, 143–174) is a helpful overview that situates the episode within book 9 and connects it with other aspects of Lucan's interest in nature and natural inquiry.

poetry itself constitute core thematic concerns of Lucan's poem.[81] Whatever texts and paradigms may be in play, however, Cato's battle with the Libyan snakes deploys and problematizes the scientific sublime in ways that we have already encountered—the narrative oppresses the reader with an excess of knowledge and pointedly asks whether the sublimity of natural inquiry can actually lead to transcendence. But there's a twist. Although I am not concerned with Lucan's attitude toward Cato as such, I sympathize with those who find his presentation at the very least problematic; in particular, I side with those who have detected implicit humor in the sheer excesses and improbabilities of the Stoic sage's battle with an army of poisonous snakes.[82] The absurdity of the episode deflates the grand pretensions of the scientific sublime and mocks its disingenuous calls to greatness. Lucan offers a surreal critique of an intellectual tradition that takes itself in earnest to a deadly degree.

The episode begins with display of natural knowledge that has the potential for sublimity. When the parched army reaches the infested spring, the soldiers are terrified. To save his men (*aspexit perituros fonte relicto*, 9.611), Cato dismisses their fear—they are frighted only by a mirage of death (*uana specie conterrite* [sc. *miles*] *leti*, 9.612)—and orders them to drink in safety (*ne dubita... tutos haurire liquores*, 9.613). As he explains, the serpents' poison is deadly only when mixed with blood; in other words, when they bite (9.614–616). The implication is that even if the serpents' poison has contaminated the water, there is nothing to fear. Cato's words show an unexpected degree of intellectual sophistication.[83] Celsus makes the same point about snake venom (5.27.3). In addition, he asserts that a snake can be eaten quite safely even though its bite may kill, and he pronounces it perfectly harmless both to put one's finger into a snake's mouth, provided it is not bitten, and to suck venom from a snake-inflicted wound, so long as one's own mouth contains no sores. Lucan himself takes the basic idea for granted in the simile that compares Curio to an Egyptian asp and Juba to its archenemy, the ichneumon (4.724–729). When the ichneumon provokes the asp, causing it to lunge forward, "it seizes [the serpent's] throat with safe

[81] Eldred (2000), Malamud (2003), Papaioannou (2005), and D'Alessandro Behr (2007, 113–161, 166–169). Lausberg (1990) explores the episode from a generic perspective.

[82] Johnson (1987, 46–66), Bartsch (1997, 29–30), and Leigh (1997, 267–273; 2000, 108–109); cf. Batinski (1991) and Chaudhuri (2014, 191–192). For a recent appraisal of the complicated connections between Lucan's Cato and Stoicism, see Kaufman (2020).

[83] Raschle (2001, 174) calls the idea both "ein Volksglaube" and "eine Erkenntnis der antiken Medizin."

bite, falling short of deadly gore; then the ineffectual plague is pressed out and its jaws flow with poison spent in vain" (4.727–729). In the Libyan desert, then, Cato commands his soldiers to overcome their irrational fear of death and uses theoretical knowledge to embolden them to act. For Lucan's contemporaries, this perhaps appears unremarkable. From the perspective of Cato's soldiers, however, the move surely suggests Cato's own will to sublimity. Note what happens when Cato drinks. Cato, like Caesar in the Massilian grove, leads by example: "He spoke and downed the suspected poison, and in all sands of Libya, this was the one fountain from which he was the first to demand a drink" (*dixit, dubiumque uenenum / hausit; et in tota Libyae fons unus harena / ille fuit de quo primus sibi posceret undam*, 9.616–618). The phrase *dubium uenenum* is focalized through the soldiers; they remain uncertain about the safety of the water, suspecting that it might be contaminated. In the eyes of the terrified troops, Cato's boldness shows that he has conquered the fear of death and proves that he is superior to nature's might; the singularity of the event reflects the awesomeness of the spectacle for those in attendance.[84]

Of course, this is not the end of the story. Does Cato inspire his soldiers to draw safe drinks from the spring? The reader never finds out—Lucan immediately turns to the origins of the snakes, and when the narrative resumes, the slaughter is already playing out. The unexpected and dramatic suspension of the poem's historical plot raises doubts about the efficacy of Cato's strategy and whether his confidence in contemporary medicine (and in himself) is misplaced.[85] Several factors in the aetiology of the snakes contribute to the sense that something is off. Significantly, the passage begins by invoking and immediately disavowing the scientific sublime (9.619–623):

> cur Libycus tantis exundet pestibus aer
> fertilis in mortes, aut quid secreta nocenti
> miscuerit natura solo, non cura laborque
> noster scire ualet, nisi quod uulgata per orbem
> fabula pro uera decepit saecula causa.

[84] De Nadaï (2000, 68–69) draws attention to Cato's "coeur magnanime" here.

[85] At 9.611–618, Cato does not even imagine that the snakes might go on the offensive; this, too, raises doubts about his fitness to lead. Subsequently, some snakes pose a threat that is quite separate from their fangs: a jaculus flings itself from a tree and shoots straight through the head of one soldier, and when another pierces a basilisk with his spear, its poison runs up the spear shaft to attack the soldier's hand (9.822–833). At the very least, Cato appears to have misjudged the nature of the threat.

Why the Libyan climate, fertile for death, overflows with such plagues, or what hidden nature has mixed with the harmful ground, our care and hard work are not strong enough to know, except in so far as a story that has spread throughout the world has deceived generations in place of the true cause.

The investigation of the causes of natural phenomena, especially meteorological phenomena (*Libycus...aer*), is potentially electrifying, but Lucan claims that he is not physically up to the challenge (*non...ualet*); yet another *causa* will remain hidden (cf. Ch. 4.2, "The Gallic Tides"). Rather than offering Lucretian enlightenment, Lucan will flood the reader with mythological falsehood: the snakes, we are told, originated when drops of blood that had fallen from the severed stump of Medusa's neck mixed with the desert sands and were heated by the scorching sun. Ironically, the aition concludes with an emphasis on moisture and heat that suggests the philosophically acceptable principle of spontaneous generation.[86]

If Lucan's posture casts doubt on the efficacy of Cato's pronouncement and example, the conspicuous use of didactic modes of speech deepens those anxieties. Latin didactic regularly announces its themes through series of indirect questions (Brown 1990, 315–321); a good example is the opening of Virgil's first *Georgic* (*quid faciat laetas segetes*...). On a first reading, Lucan's indirect questions (*cur exundet, quid miscuerit*) cast him as a didactic poet, apparently announcing the themes that he will subsequently treat—not until line 622 does he abruptly deny his own ability to provide satisfactory answers. More specifically, by referring to the snakes and/or their poison as plague (*pestibus*), Lucan connects the episode with didactic plague narratives, especially those of Lucretius (book 6) and Virgil (*G.* 3).[87] In particular, Lucan's professed inability to explain the origins of this plague through an analysis of the Libyan climate (*aer*) or soil (*solo*) constitutes a rejection of Lucretius's appeal to both air and earth as sources of *pestilitas* (6.1098–1102).[88] More broadly, though, of the nineteen occurrences of *pestis* in the *Civil War*, thirteen come from book 9, and all thirteen refer, directly or indirectly, to the Libyan snakes or their poison (614, 619, 630, 724, 734, 744, 779, 787,

[86] For spontaneous generation, see Osorio (2020, 35–40). Relevant readings of the Medusa narrative include Fantham (1992b), Papaioannou (2005), and Meyer (2023, 145–157). For a contrast with Lucretius here, note de Nadaï (2000, 69–70) and Wick (2004, 2.248).

[87] For the didactic plague topos, see Gale (2022). Gardner (2019, 189–206) examines plague and "quasi-plague" as part of a wide-ranging reexamination of ancient plague discourse.

[88] Cf. Fantham (1992b, 98 n.7), Raschle (2001, 178), and Wick (2004, 2.248–249).

805, 844, 922, 926, 930).[89] In a sense, Cato's battle with the snakes is simply a battle with plague.[90] In *De rerum natura* and the *Georgics*, this is a brutally terrifying battle—the suffering is horrific, the medical profession is impotent, all social institutions begin to break down. The reader who is attuned to the didactic atmosphere of Lucan's snake plague expects the narrative to present a similarly devastating scene of social implosion; Cato's sublime confidence begins to look like blithe optimism. But Lucan also has the ultimate source of this tradition in mind, Thucydides's account of the plague at Athens, where the medical profession is equally helpless and the total collapse of human *technai* and *nomoi* looms large. Lucan's feigned inability to analyze the origins of the plague represents a travesty of Thucydides's strategic decision (2.48.3) not to consider its *aitiai*, restricting himself instead to an analysis of its symptoms, which he hopes may prove helpful if another outbreak occurs. The absurd terror of the snakes threatens to deflate the inspiring thrill of Cato's daring drink just as the plague at Athens punctures the soaring rhetoric of Pericles's funeral oration.

When the historical narrative resumes, this puncturing effect is precisely what the reader experiences. At this point, the text undergoes a certain kind of bifurcation. There are now two distinct narrative planes, one on which Lucan shares his knowledge about the snakes with the reader and another wherein Cato and his men fall victim to the serpents in a grotesque spectacle.[91] At least some of Lucan's technical knowledge comes from Nicander's *Theriaca* and Aemilius Macer's Latin translation or adaptation thereof (*FLP* 295–299).[92] The text, in other words, continues to feel didactic, and that erudition is conspicuously disconnected from the realities of the historical narrative.

The bifurcation becomes clear almost as soon as the narrative picks up again (9.734–746):

> has inter pestes ducto Cato milite siccum
> emetitur iter, tot tristia fata suorum

[89] At 9.630 *pestes* refers to the plagues inside Medusa's body, which later produces the snakes. For possible models, see Schwemmler (1916, 12–13, on Manilius) and Raschle (2001, 244, on Lucretius).

[90] At Virg. *G.* 3.419, snakes are *pestis acerba boum*, which looks forward to the plague at the end of that book. At Ov. *Met.* 7.533–535, countless snakes infect the springs, rivers, and lakes of Aegina, which leads to an outbreak of plague. Instead of suggesting plague (Virgil) or explicitly causing plague (Ovid), snakes dramatically become the plague in Lucan, who concentrates Ovid's multiple sources of water into a single *fons*.

[91] Tesoriero (2005, 206–209) discusses a similar narrative split at 9.950–979.

[92] The basic studies are Morel (1928), Cazzaniga (1957), Lausberg (1990), and Landolfi (2007).

insolitasque uidens paruo cum uulnere mortes.
signiferum iuuenum Tyrrheni sanguinis Aulem
torta caput retro dipsas calcata momordit.
uix dolor aut sensus dentis fuit, ipsaque leti
frons caret inuidia nec quidquam plaga minatur.
ecce, subit uirus tacitum, carpitque medullas
ignis edax calidaque incendit uiscera tabe.
ebibit umorem circum uitalia fusum
pestis et in sicco linguam torrere palato
coepit; defessos iret qui sudor in artus
non fuit, atque oculos lacrimarum uena refugit.

Among these plagues, Cato leads his soldiers and measures out the dry journey, seeing so many sad fates of his men and unusual deaths accompanied by a small wound. Aules, the young standard bearer of Etruscan blood, was bitten by a dipsas, its head twisted back, after he had stepped on it. Scarcely was there pain or sensation of the tooth, and the appearance itself of death is free from ill-will, nor is the blow at all threatening. Behold! the venom advances in silence and a consuming fire seizes the marrow and inflames the innards with hot decay. The plague drinks up the moisture poured around his vital organs and begins to parch the tongue on his dry palate. There was no sweat to go over his tired limbs and the stream of tears fled from his eyes.

A gulf that is both epistemological and psychological separates Lucan and the reader from Cato and his troops. On the one hand, neither Cato nor Aules understands the nature of the situation. The venom produces an unbearable form of dehydration, forcing Aules to dig through the sand in search of moisture and then to swallow salt water from the sea. Like the plague victims of Thucydides and the didactic tradition, Aules cannot quench his thirst, and he soon dies trying to drink the blood from his veins. But even then, the Etruscan youth fails to "perceive" the nature of his death, "thinking" rather that this is simply thirst (*nec sentit fatique genus mortemque ueneni, / sed putat esse sitim*, 9.758–759). The standards fall to the ground and Cato orders them to be picked up in great haste: "It was not allowed for anyone to learn that thirst possessed this kind of power" (*discere nulli / permissum est hoc posse sitim*, 9.761–762). Nobody within the historical narrative of the poem learns anything at all, and Cato remains Stoically detached; he sees (*uidens*) the deaths of his soldiers and continues on his way, inhumanely unaffected.[93]

The reader, on the other hand, learns a great deal. First, Lucan names the snake (*dipsas* at 738). Then, through the interjection *ecce* at line 741 (and repeated at 789 and 822), he focuses our attention on the physical processes that begin to unravel deep within the hidden recesses of Aules's body; for the reader, at least, nature's hidden causes will come to the surface. Since we alone apprehend the profound connection between the snake, its name, and its powers, we are alone in appreciating the order and reason behind Aules's suffering: the "thirst snake" (διψάς) produces dehydration or "thirst" (δίψα). This is a lesson that the reader receives again and again as the battle continues, for Lucan takes great pains to make sure that the names of his serpents correspond precisely with their lethal powers and even modifies the physical realities of certain snakes or combines attributes from different snakes to produce a strong etymological effect.[94] Take, for example, the seps. Its Greek name (σήψ) means "putrefying sore" and in the catalog Lucan describes it as "dissolving bones along with the body" and "putrefying" (*ossaque dissoluens cum corpore tabificus seps*, 9.723). During the subsequent battle, a seps attacks Sabellus and the putrefaction begins: his skin dissolves, his limbs swim with gore, his thigh muscles melt, the membrane that holds the stomach together bursts apart, and he soon becomes nothing but a bodiless wound (*iamque sinu laxo nudum sine corpore uulnus*, 9.769); the "putrefying sore" turns Sabellus into one massive "putrefying sore." The Greek seps, though, behaves rather differently—in addition to inducing swelling, numbness, and bleeding, it causes the hair to fall out and a white rash to spread.[95] Lucan alters the physical details of the natural historical tradition to produce an exact correspondence between snake name and snake wound.

The cumulative effect on the reader is that of the same dark sublimity that characterizes the experiences of Phemonoe and those who hear the

[93] Aules's suffering echoes that of Ovid's Erysichthon, and his companions' indifference travesties the solemnity with which the Argonauts react to Mopsus's death in Apollonius; see Raschle (2001, 251–253) and Wick (2004, 2.313–314). Aules is also a caricature of Virgil's Eurydice, who inadvertently steps on a hydrus or "water snake" (rather than dipsas or "thirst snake") at *G.* 4.457–459.

[94] Cazzaniga (1957), Aumont (1968), and Migliorini (1997, 110–119, 122–123) treat the ways in which Lucan transfers or combines snake attributes. Cazzaniga 36–41, Eldred (2000), Wick (2004, 2.278–282), and Landolfi (2007) discuss Lucan's interest in the etymologies of the serpents' names; cf. Barbara (2008, 262–271).

[95] See Wick (2004, 2.324) for the ancient evidence.

spells of the Thessalian witches; we are simultaneously overwhelmed and enlightened, appalled and empowered. For the historical actors of the poem, however, no theoretical insight balances their suffering, no didactic lesson reveals the natural order that underlies their excruciating agony and situates it in a broader cosmic context. Indeed, the disheartened soldiers lament the sad fact that serpents, rather than Caesar, appear to be bringing the war to its inevitable conclusion (9.850–851). They fantasize about traveling farther south, to the torrid zone where heaven is scorched by the sun: "It is thrilling to attribute my death to celestial causes and to die by the agency of heaven" (*iuuat aetheriis ascribere causis / quod peream, caeloque mori*, 9.853–854). They mean that it is preferable to die from dehydration under the rays of a brutal sun rather than in battle with the snakes, but the clear implication is that it is sublime (*iuuat*) to perish in the midst of a cosmic upheaval, like Phaethon, an option that is conspicuously not on the table.

At the same time, a certain humor pervades the entire episode.[96] A figure like Sabellus, for instance, transcends the human, like a Homeric hero, but he does not come into contact with god—it is a parody of the sublime that he is reduced to a puddle of pus. Similarly parodic is the fate of Nasidius, a Marsian farmer who falls to a prester (*Nasidium Marsi cultorem torridus agri / percussit prester*, 9.790–791). True to its etymologically ambiguous name, the prester—ultimately derived from *pimprēmi*, both "inflate" and "burn"—inflates Nasidius's body and causes it to boil. The gore exceeds the limits of that which is human (*humanumque egressa* [sc. *sanies*] *modum*, 9.794), and when the soldiers eventually abandon the bloated corpse, it continues to grow with no limit in sight (*nondum stante modo crescens... cadauer*, 9.804). The refusal to abide by regular human boundaries or to allow oneself to be constrained is a hallmark of the sublime subject; Lucan's Nasidius deforms this intellectual drive for *megalopsuchia*.

But the reader should not miss the joke. Since the Marsians traditionally have certain immunities to and magical powers over snakes (see, e.g., Virg. *Aen.* 7.750–760, Sil. It. 8.495–510), it is not without humor that this Marsian's exotic expertise fails him utterly when he most sorely needs it. Moreover, Nicander and his Latin imitators and admirers, including Virgil, repeatedly underscore the practical applications of their didactic poems,

[96] For the tension between horror and humor here, see esp. Bartsch (1997, 29–37). Writing of the seps, Aumont (1968, 118 n.46) rightly describes a "réalisme brutal et caricatural" and identifies a likeness to the "dessin animé." Johnson (1987, 45, 57, 64) similarly highlights the cartoonish quality of Lucan's Cato and the snake episode.

and it is precisely farmers like Nasidius whom these poets claim to be able to assist with their remedies and safeguards (Nic. *Ther.* 4–7, Virg. *G.* 3.414–422). Nasidius, like his fellow soldiers, has no access to such works, and the irony only increases when we realize that the prester does not actually figure in the *Theriaca* at all—even if Nasidius had brought his copy of Nicander to Libya, he would have been totally out of luck. The battle with the Libyan snakes, then, exposes the complete divorce of didactic learning from real-world action and demonstrates the total irrelevance of the scientific sublime to those who are engulfed in the chaos of civil war. Like Amyclas and Pompey's helmsman, Cato strives for the sublime, and the historical circumstances render that effort null and void. Whatever elevation Lucan's own knowledge can impart to readers, it does no one within the historical narrative of the poem any good. What's more, the dark sublimity of Lucan's own natural historical precision contributes essentially to a radical parody of the experience of transcendence that lies at the heart of martial epic; an overabundance of natural knowledge dissolves the would-be hero into a putrid slime rather than elevating him to the level of a god.

It remains an open question whether the psychological *therapeia* that Cato eventually administers to his suffering troops (9.881–889) unleashes the uplifting surge of the sublime.[97] But even on such a reading, it is Cato's ethical fortitude, not his natural knowledge, that potentially elevates the dying. We have seen (Ch. 4.2, "Pompey's Soul") that the book opens with Pompey's soul soaring through the heavens, laughing at the injuries inflicted upon its lifeless body (*risitque sui ludibria trunci*, 9.14). Cato and his troops cannot laugh at the carnage that surrounds them, but Lucan and the reader can. Cato falls dramatically short of the Stoic sage, his exhortation to greatness lands on deaf ears, and the army of the *res publica* becomes a grotesque laughingstock.

The dramatic appearance of the Psylli only throws into relief the collective impotence of Cato, the didactic tradition, and the scientific sublime.[98] The tribe's magical incantations (*cantu* at 9.895 and 914, *carmina* at 9.927, *excantata* at 9.931) situate the Psylli outside the traditions of Greco-Roman medical and didactic learning with which Lucan has been concerned hitherto, and their procedures are truly wondrous (*magicae miracula gentis*, 9.923); it is the Psylli, rather than Cato, who have a strong claim to the sublime.

[97] For the problematic implications of even this passage, see Johnson (1987, 54–55).
[98] For the Psylli, see Phillips (1995), Migliorini (1997, 119–121), and Meyer (2023, 171–174).

Crucially, their lifesaving incantations belong to the same corrupted landscape as the poisoned *fons* with which the entire episode begins. In a sense, their *carmina* render them alien poets, foreign counterparts to the likes of Nicander, Lucretius, and Virgil. In the primordial chaos of the Libyan desert, which figures cosmic dissolution, the investigation of natural *causae* is not possible, the collective knowledge of philosophical medicine and natural history serves no purpose, and the scientific sublime, far from inspiring and uplifting, wittily deflates and degrades. All that is left is the magical voice of a people for whom poisonous snakes, the bane of the Republican army, are merely and literally child's play (*donatis lusit serpentibus infans*, 9.908). Cato and his troops suffer terribly, while Lucan, the Psylli, and the reader break into a smile that is subtly sublime.[99]

4.6. Acoreus and the Nile

Book 10 brings Caesar to Alexandria. After dining in opulence at the royal court, the general questions Acoreus, an aged Egyptian priest, about the peculiar nature of the Nile. Caesar is eager to learn the causes of the river's summer flood and the location of its source, going so far as to promise to abandon civil war altogether in return for the revelation of Acoreus's secret knowledge (10.176–193). Acoreus's lengthy response (10.194–331) contains an overabundance of astronomical, geographical, and meteorological information, including an anonymized doxographical review of theoretical explanations for the river's rise, all of which are rejected (10.219–267). After Acoreus fails to answer his interlocutor's questions, the narrative breaks off, and Lucan directs the reader's attention to Pothinus, who has been plotting Caesar's assassination.

Over the past few decades, several studies have framed Caesar's interest in the Nile as a symbol of his imperialist drive; to know the Nile is to conquer the Nile and to extend the boundaries of Roman *imperium* to the ends of the earth.[100] Without a doubt, the narrative authorizes this kind of reading; both Caesar and Acoreus characterize the inquiry with militaristic language, and Acoreus explicitly compares Caesar's query to the historical

[99] Per Leigh (1997, 234–291), Lucan configures combat as spectacle, the appropriate response to which is wonder or awe. A sublime smile, then, is not out of place in reaction to Cato's ordeal.

[100] See esp. Romm (1992, 152–156), Barrenechea (2010), Manolaraki (2013, 80–81, 103–106, 114), Tracy (2014, 114–115, 134, 271–273), and Gauly (2018).

expeditions of Cambyses, Sesostris, and Alexander. Moreover, as we have seen (Ch. 3.3), Nero dispatched at least one expedition to the Upper Nile and Ethiopia, and while Seneca (*NQ* 6.8.3) claims that the emperor was motivated by his love of truth, others (esp. Plin. *HN* 6.181) connect the adventure with a planned attack on Ethiopia; Acoreus should be on guard.

At a rather different level, the episode dramatizes the irresistible pull of natural inquiry and asks the reader to reflect on the ethical value with which Seneca, in particular, imbues it. Hermann Diels established long ago the many close correspondences between Acoreus's speech and the extant portions of Seneca's treatment of the Nile in *NQ* 4a; Seneca's voice, in a sense, pervades the passage.[101] While Diels sees Lucan's narrative as little more than a versification of the *NQ*, Eleni Manolaraki (2013, 80–115) and Tracy (2014, 99–273) argue that the episode engages with Seneca on a much more extended level, suggesting that Acoreus's interaction with Caesar reflects the inadequacy or failure of Seneca's philosophical interventions with Nero.[102] Based on my readings of the *NQ*, I develop a new framework for thinking about Acoreus, Caesar, and Seneca that integrates the episode into my overall reading of natural inquiry in the *Civil War* and shows the ways in which Caesar's encounter with the Egyptian priest expands and transforms Lucan's criticism of the Senecan project. Caesar knows that his questions have no answer, and for that very reason he sees in this intellectual adventure an opportunity for elevation and greatness; true to form, he leaps into the Nile's bottomless abyss. Acoreus, however, inundates his interlocutor with excessive erudition and irrelevant facts, and the reader is left wondering whether his many explanations spring from ignorance or cunning. Either way, the sublimity of natural inquiry leaves Caesar exposed to real-world danger and brings him breathtakingly close to his own undoing; Caesar finds no truth, Acoreus's speech is all misdirection, and Seneca's meteorological program looks laughably ineffectual. At the end of the day, the scientific sublime is a deadly distraction as civil war continues its ruinous rampage.[103]

[101] Diels (1886); cf. Hosius (1892, 337–339, 352–353).

[102] McCloskey and Phinney (1968, 85), Schmidt (1986, 39), and D'Alessandro Behr (2007, 214 n.24) similarly connect Acoreus with Seneca.

[103] There are other ways of approaching the scene that complement my own. Zwierlein (1974, 61–66), for one, compares the after-dinner song of Iopas (Virg. *Aen.* 1.740–747), with its astronomical and meteorological themes, while Ambühl (2015, 387–395) reads the episode against Callimachus and Hellenistic didactic and deipnosophistic traditions, suggesting that the scene constitutes "einer Alternative zum Bürgerkriegsepos" (397). Slightly different is Manolaraki (2013,

Before considering Caesar's dual request, we need to consider the Nile's status as an object of philosophical inquiry. By the middle of the first century CE, the cause of the Nile's summer flood and the location of its source had become paradigmatically unanswerable questions.[104] The doxographical tradition traces interest in the causes of the flood back to Thales (D9 *EGP*, A16 D-K), and Herodotus devotes considerable attention to both questions (2.19–31). Closer in time to Lucan, Diodorus Siculus characterizes the Nile's summer flood as something about which all people make inquiries (τὸ παρὰ πᾶσιν ἐπιζητούμενον, 1.37.1) but emphasizes that it is possible to argue against all those who attempt to explain it (δυνάμενοι [sc. ἡμεῖς] ποικιλώτερον ἀντειπεῖν πρὸς ἅπαντας, 1.41.10). Similarly, he declares that nobody has ever seen the river's source or received secondhand information from anybody who has (1.32.1, 37.6), and describes the question as a matter of guesswork and plausible conjecture (τοῦ πράγματος εἰς ὑπόνοιαν καὶ καταστοχασμὸν πιθανὸν καταντῶντος, 1.37.7). Seneca appears to have been of the same opinion. The fragmentary *NQ* 4a does not preserve Seneca's views on either issue, if he presented them at all. However, he states that if it were possible to identify the source, the reasons for the flood would follow (*unde crescere incipiat si comprehendi posset, causae quoque incrementi inuenirentur*, 4a.2.3); the contrafactual suggests that no positive answer was formulated.[105] In a different context (*NQ* 3.22), Seneca floats the possibility that the nature of the Nile may be inexplicable (*inenarrabilis natura est*) and that it is too great not to trace its *origo* back to the origins of the cosmos itself; like the river's geographical source, its temporal source recedes into infinity and vanishes from sight. In addition, the obscurity of the Nile's headwaters constitutes a poetic topos: Callimachus has the Nile boast that, despite its size, no mortal knows where it begins its

105–115), who argues that Acoreus/Caesar figures the poetic rivalry between Lucan/Nero. Meyer (2023, 183–223) provides a broad overview.

[104] Major ancient discussions and doxographical collections can be found at Herod. 2.19–31, Arist. frr. 686–695 Gigon, Lucr. 6.712–737, Diod. Sic. 1.37–41, Strab. 17.1.1–5 (785–791), Sen. *NQ* 4a.1–2, Mela 1.50–54, Plin. *HN* 5.51–59, Aetius 4.1 (Mansfeld and Runia 2023, 318–323), Ael. Arist. *Orat.* 48. For comprehensive lists and discussion, see Postl (1970, 11–89, 222–234). According to one tradition, Aristotle sent Alexander to discover the cause of the Nile's summer flood; see fr. 686 Gigon with Romm (1992, 151–152).

[105] Indirect confirmation comes from *NQ* 6.8.3–5, where Seneca reveals that the Nile's source is an open question for Nero (see Ch. 3.3), and John the Lydian, who, in a passage based on Seneca, represents this problem as a matter of mere speculation with no truth to be had at all (ποικίλαι μὲν οὖν αἱ περὶ αὐτοῦ δόξαι, τὸ δ' ἀληθὲς κατὰ τοὺς ἀνθρώπους τέως οὐδαμοῦ, *De mens.* 4.107 at Hine 1996a, 189). Seneca presumably treated both questions in the lost *De situ et sacris Aegyptiorum* (Serv. ad *Aen.* 6.154 = fr. 19 Vottero).

journey (fr. 384.31–32 Pfeiffer); Tibullus wonders why and where the Nile concealed its source (1.7.23–24); Horace announces that the river deliberately keeps its origins secret (*Carm.* 4.14.45–46); and Ovid explains that, when Phaethon set the world on fire, the Nile fled and hid its head, which still lies hidden even now (*Met.* 2.254–255).[106]

From the vantage point of Hellenistic and early imperial culture, the causes of the Nile's summer flood and the location of its source simply cannot be known; they are theoretical questions for which no answer exists. The unknowability of the Nile—the fact that its nature is fundamentally *inenarrabilis*, to use Seneca's formulation (*NQ* 3.22)—makes the river a potential source of sublimity. This is the intellectual background against which we should read the interaction between Caesar and Acoreus.[107]

To return to the historical narrative of the poem, Caesar does not view the investigation of nature as a distraction; on the contrary, he sees it as an integral part of the acquisition and maintenance of power. As he tells Acoreus, "in the middle of battle, I have always had time for the expanses of the stars and of heaven and for things up above, and my arrangement of the year will not be outdone by the calendar of Eudoxus" (*media inter proelia semper / stellarum caelique plagis superisque uacaui, / nec meus Eudoxi uincetur fastibus annus*, 10.185–187). Caesar presents himself as being able to successfully integrate natural inquiry with the prosecution of civil war (*media inter proelia*) and stresses that his investigations yield tangible results that even seem to put him in control of the natural world (*meus... annus*).[108] While his language hints at the sublime, a few tweaks to the translation make him name the experience. Emanuele Berti (2000, 168–169) reads *stellarum caelique plagis* as a hendiadys ("le distese stellate del cielo") and takes *superis* as neuter plural. "The starry expanses of heaven" and "things

[106] For more examples of the topos, see Postl (1970, 205–208).

[107] Note that at *NQ* 3.22, Seneca singles out the Ocean, Ister, and Nile, and that at *Subl.* 35.4, Longinus reflects on the sublimity of the Nile, Ister, Rhine, and Ocean, as well as the stars and Mount Aetna. It is tempting to see a connection here. Porter (2007, 172–74; 2016, 450–454) convincingly argues that a common source lies behind *Subl.* 35.4 and Lucr. 6.608–737, which are the only extant ancient texts that discuss the Ocean, Nile, and Aetna, and he includes *NQ* 3.22 among a list of partial parallels. Two parallels that Porter does not name are Cic. *Marcell.* 28 (Rhine, Ocean, Nile, all objects of *obstupescent*) and Hor. *Carm.* 4.14.46–48 (Nile, Ister, Tigris, Ocean). Although the Ister and Nile had long been connected (Hes. *Theog.* 338–339, but esp. Hdt. 2.26, 34), Seneca's clustering of rivers will have suggested sublimity to his readers.

[108] The historical Caesar did not implement his calendrical reform until 46 BCE, but Lucan may be thinking of Caesar's lost treatise *De astris*, which came first; see Klotz (1927, 211–229) for testimonia and Meyer (2023, 191–193) for discussion. Wolkenhauer (2011, 235) suggests that Caesar's claim here might constitute a joke at the general's expense, given the problems with his calendar that were well-known already in the Augustan period.

up above," however, are redundant, and the word order is anticlimactic. I propose that *supera*, like the well-attested *sublimia*, functions as a calque for *meteōra*, and that the *-que* in *superisque* is epexegetic (*OLD* 6a). Putting this together, we can render Caesar as follows: "In the middle of battle, I have always had time for the starry expanses of heaven—for things that are sublime—and my arrangement of the year will not be outdone by the calendar of Eudoxus." War and celestial inquiry are equally sublime, both lead to and mutually reinforce *imperium*, and Caesar will relinquish neither.

But the Nile appears to be different (10.188–192):

> sed, cum tanta meo uiuat sub pectore uirtus,
> tantus amor ueri, nihil est quod noscere malim
> quam fluuii causas per saecula tanta latentis
> ignotumque caput: spes sit mihi certa uidendi
> Niliacos fontes, bellum ciuile relinquam.

But, since such potency lives beneath my breast, such passion for truth, there is nothing that I would rather come to know than the river's causes, hidden through great ages, and its unknown head. Let me have sure hope of seeing the Nile's source—I will leave civil war behind.

Caesar's innate potency (*uirtus*, *OLD* 5c) is the irrepressible drive that makes him a devastating lightning bolt, and his passion for knowledge (*amor ueri*) is the relentless eros of philosophical inquiry; individually and together, they encode the dual experience of the sublime and help assimilate him to a vatic figure like Lucretius.[109] These dynamic forces, coiled like springs at the core of Caesar's DNA, drive him not just to venture into the unknown but to attempt that which is actually impossible. Crucially, Caesar knows his questions have no answers, for he wishes only for the hope (*spes*) of intellectual resolution, which he represents dramatically as gazing at the river's source. Merely to pretend to be able to master the Nile is altogether thrilling and reward enough. Strikingly, this sublime adventure has Caesar ready to give up power, to abandon civil war (*bellum ciuile*) and, by implication, everything it represents—the chaos, turmoil, and oppression that

[109] Rossi (2005, 241–243) discusses echoes of Lucretius and the *Georgics* in Caesar's self-representation here. The combination of *uirtus* and *amor ueri* also connects Lucan's Caesar to Seneca's Nero, who is *ut aliarum uirtutum ita ueritatis in primis amantissimus* (*NQ* 6.8.3) in the context of investigating the Nile (see Ch. 3.3).

pulse through the poem's every line, as well as the poem itself. For a time, this is precisely what happens; when he stops speaking, Caesar simply disappears for over 200 lines. The impossible journey vanishes Caesar from view and raises the possibility that it elevates him above and beyond the disorderly mess of a world that he himself has upended; it is a potentially sublime payoff for an act of Phaethon-style daring.

Some have seen Caesar's promise as cynical or disingenuous, and a strong tendency pervades the secondary literature to read Caesar negatively here; his request is megalomaniacal, his actions cross all boundaries, and his desire belies the same moral rot as Cleopatra's perversely excessive banquet.[110] Yes, one voice in the poem may condemn Caesar's bombastic and imperialistic aspirations. At the same time, his yearning for the impossible, his grasping for the sublime, are familiar philosophical values whose overwhelming appeal Lucan recognizes and, ultimately, disavows. Given this leitmotif, we can read Caesar's request and Acoreus's response as a dramatization of what happens when we undertake a Senecan study of nature in the maddening heat of civil war. More specifically, we have seen that Seneca prefaces his own account of the Nile by framing the sublimity of natural inquiry as an antidote to the intoxicating awesomeness of Roman *imperium* (Ch. 3.6, "Lucilius, Sicily, and Roman *imperium*"). Caesar's request for knowledge about the Nile, formulated *media inter proelia*, puts to the test Seneca's desperate hope that one modality of the sublime can successfully counteract the other.

At a basic level, it is important to recognize that Acoreus does not answer Caesar's questions, even though he fills the general with wonder (cf. Tracy 2014, 145). Some conclude that Acoreus's opening discussion of the planets (10.199–218) explains the cause of the Nile's summer flood on astrometeorological grounds, but Acroeus does not make the causal connection explicit, and temporal adverbs and conjunctions structure the key clause ("then the Nile is present... and it does not check its growth until night has reclaimed its hours from the sun," *tunc Nilus... adest, auctusque suos non ante coartat / quam nox aestiuas a sole receperit horas*, 10.215–218). Given what Acoreus subsequently says about the flood's *causae*, it makes more sense to read his opening remarks as an elaborate identification of the date

[110] Ozanam (1990, 284–285), Romm (1992, 154–156), Berti (2000, 165, 212–214), Barrenechea (2010, 260–261), Hardie (2013b, 237), Day (2013, 164), and Tracy (2014, 116–117, 132–133). For a recent assessment of Lucan's Cleopatra, see Feldherr (2021).

at which the flood starts and stops.[111] Nor does the subsequent discussion provide any mechanical explanation for the flood. Although Acoreus does attribute the flood to the providential outlook of nature and the requirements of the cosmos (*sic iussit natura...sic opus est mundo*, 10.238–239), he qualifies this assertion, rhetorically wondering who could account for its causes (*quis causas reddere possit? / sic iussit natura*, 10.237–238). By Acoreus's own admission, a theory based on divine machinery, or on vague references to *natura* and the *mundus*, does not count as a rigorous analysis of philosophical *causae*. At the end of his doxography, the priest suggests that he will solve the matter once and for all (*si tantam ius est mihi soluere litem*, 10.262), but his proposed solution is far from clear: he says only that some rivers arose in the course of time without divine action (*non id agente deo*, 10.265), while others originated with the cosmos itself and follow fixed laws under the rule of god (*quas* [sc. *aquas*] *ille creator / atque opifex rerum certo sub iure coercet*, 10.266–267). Acoreus implies that the Nile belongs to the second category, again appealing to the divine rather than physical or material causes. While the very idea might trigger the sublime, by turning Caesar's mind toward god, the priest's account of the Nile's summer flood conspicuously fails, by his own definition, to *causas reddere*.[112]

But something even more unsettling is unfolding here. Diels (1886, 17–18) identifies in this passage the language and schematic opposition of *NQ* 3.22, where Seneca contrasts eternal Ur-rivers with those that came into being in the course of time. Strikingly, Acoreus transforms this two-fold classification of rivers into a contrast between the random world of Epicureanism, in which the gods do not participate, and the providential universe of the Stoics, which is overseen by a benevolent *deus*. Acoreus may take for granted that we dwell in the latter, but the poem has already voiced, on two occasions in quick succession, its anxious uncertainty about which of these systems we actually inhabit.[113] As Seneca's hierarchical scheme crumbles into an open-ended existential question, the collective weight of the poem's

[111] Thus Tracy (2014, 148–149); cf. Manolaraki (2013, 86–87). For other perspectives, see Bakhouche (2002, 38–39), Schrijvers (2005, 29), Domenicucci (2013, 65–66), Meyer (2023, 197–202, 213–214). None of this is to deny that the passage works on other levels simultaneously (e.g., as a challenge to Caesar's claims to astronomical expertise).

[112] Tracy (2014, 150–152) and Gauly (2018, 92–93) rightly identify Acoreus's failure to engage in natural philosophy proper.

[113] Nigidius Figulus at 1.642–645 and Lucan himself at 2.7–15; cf. Lucan's flirtation with the Epicurean position at 7.445–447. The notorious interchangeability of "fortune" and "fate" effects a similar uncertainty about the nature of the *mundus*. For both points, see Feeney (1991, 279–281).

paralyzing aporia completely undermines any resolution that Acoreus's words may suggest. The mighty rush of the Nile's *inenarrabilis natura* sweeps away all certainty and leaves us in a state of disbelief about the most fundamental question of all. Or, as Diels (1886, 19) dryly concludes, "Achoreus hat die Neugier seiner Zuhörer schlecht befriedigt durch seine mystische Auskunft."

The same is true of Acoreus's treatment of the Nile's source (10.268–331). Here, Acoreus makes only imprecise statements. Seemingly ignoring Caesar by addressing the river itself, he declares "you rise from the middle zone" (*medio consurgis ab axe*, 10.287) and "to you alone is granted to wander through both hemispheres" (*solique uagari / concessum per utrosque polos*, 10.300–301). The idea that the headwaters of the Nile are to be located in the Southern Hemisphere was widespread by the first century CE; such an assertion in no way answers Caesar's question.[114] The narrative that follows traces the course of the river from Meroe to Memphis, from the furthest margins of what is geographically known into the very heart of the familiar, precisely the wrong direction to travel if one hopes to find the Nile's headwaters. As Acoreus closes out his speech, we hear that Memphis forbids the river's banks to establish a limit to its growth (*modumque uetat crescendi ponere ripas*, 10.331). The final image of the river's course is one of boundlessness and infinite expansion, a geographical asymptote the ends of which Caesar can never hope to apprehend.[115]

This is not to say that Caesar learns nothing—Acoreus's doxography of the summer flood is detailed (10.219–261), and he outlines the Nile's geography and has potentially exhilarating things to say about god and *natura*. At the same time, the priest packs his discourse with irrelevant and sometimes peculiar information. Acoreus, for instance, informs Caesar that the sun's rays cause the stationary position or retrograde motion of the other planets (10.202–203); the so-called radio-solar theory of planetary motion is irrelevant to the problem of the Nile.[116] Acoreus's overabundance of

[114] See Romm (1992, 149–151). One suspects that Acoreus is having fun at Caesar's expense, since the first person to propose this theory was Eudoxus (frr. 287–289 Lasserre), the very figure to whom Caesar views himself as superior at 10.187 (*nec meus Eudoxi uincetur fastibus annus*).

[115] For the Nile's geographical and temporal expansiveness in the poem, see Manolaraki (2013, 51–52, 88, 90–93).

[116] For the theory, see Beaujeu (1979, 215 with n.5) and Volk (2009, 51–52). Hannah (1996, 183) observes that "to stand still, or worse, to go back in retrogradation, was a poor sign for a planet." By conspicuously mentioning such negative states of rest and backward movement, Acoreus points to Caesar's current lack of forward progress.

information, especially astronomical and astrological information, reads as sublime. He is, in a sense, a Manilian figure who flies through the heavens, accumulating infinite amounts of data, and he is not particularly interested in the relevance or factual accuracy of that data.[117] This becomes clear when we read the whole of Acoreus's speech through the lens of a statement that he makes about the impossibility of locating the Nile's source (10.295–298):

> arcanum natura caput non prodidit ulli,
> nec licuit populis paruum te, Nile, uidere,
> amouitque sinus et gentes maluit ortus
> mirari quam nosse tuos.

Nature has not revealed your hidden head to anyone, Nile, nor have peoples been allowed to see you when you are small; she removed your curves and preferred races to wonder at your origins rather than to know them.

Acoreus opens up a vast chasm between knowledge (*nosse*) and wonder (*mirari*). His astronomical, meteorological, and geographical lessons have an immobilizing effect that disarms through awe and stupefaction; like Phemonoe and those who hear the spells of the Thessalian witches, both Caesar and reader are dumbstruck. But in the present case, philosophically rigorous theories are dismissed, inaccurate and irrelevant information abounds, and fundamental questions receive no satisfactory answers. In contrast to the philosophical motto *nil admirari* (Ch. 1.6, "Stoics, Epicureans, and the *nil admirari* Tradition"), Acoreus seems to offer *omnia mirari*. His speech is escapist paradoxography that masquerades as hard-hitting natural philosophy and natural history, a radical inversion of Seneca's basic position and modus operandi.[118]

In book 10, then, Lucan doubles down on his critique of the scientific sublime, pushing the Senecan paradigm to its breaking point. The inspiring thrill of aporia and philosophical collaboration, which promise one day to expand the mind and bring the reader into fleeting contact with god, is revealed to be grounded in confusion and misinformation, and when all is said and done, Caesar simply returns to his one true passion, civil war; the sublimity of *imperium* carries the day.

[117] For a prominent instance of factual confusion, see Housman (1926, 334–337).
[118] For the relationship between paradoxography and the scientific sublime, see Ch. 5.5 and the Epilogue.

But Lucan's verdict is harsher still if Acoreus knowingly misleads Caesar.[119] Egyptian priests are traditionally unreliable sources of information about the Nile. Herodotus, for one, emphatically declares that he was unable to learn anything about the Nile's summer flood from the priests he interviewed (2.19), and later states that nobody was able to inform him about the Nile's origins except for one priest who appeared to him to be playing a game (οὗτος δ' ἔμοιγε παίζειν ἐδόκεε, φάμενος εἰδέναι ἀτρεκέως, 2.28). Diodorus Siculus also singles out the Egyptian priests for censure, ridiculing their theory that Ocean constitutes the river's ultimate origin (1.37.7), and later criticizes the theories of the philosophers from Memphis (1.40), the very city from which Acoreus hails (*hunc* [sc. *Acorea*] *genuit... Memphis*, 8.477–478). On this reading, Caesar poses two paradigmatically unanswerable questions to an individual from whom he should not expect a reliable answer, and Acoreus gives a performance that constitutes both a paradox and a game. His long, complicated discourse ironically purports to answer the unanswerable and, in the end, he himself adopts the role of Herodotus's playful priest. One plausible conclusion is that the Senecan investigation of nature is nothing more than a con and the *proficiens* an easy mark. More devilish still is what this means for Acoreus—he cunningly exploits Caesar's naïve predilection for the scientific sublime and essentially holds him captive. When Acoreus disdainfully addresses Caesar as "Roman" and comments on his lust to know the Nile (*quae tibi noscendi Nilum, Romane, cupido est*, 10.268), he derisively brands all Romans as hopelessly enchanted with the scientific sublime. Caesar's delusional deipnosophistic conversation with Acoreus parallels the absurd celestial flights of Nero and Pompey.[120]

As civil war rages, the sublime inquiry into nature becomes a sad, frustrating, and irrelevant joke, but the last laugh is ultimately on Acoreus, for Caesar will not be stopped. To read Acoreus's performance as intentionally duplicitous is extremely attractive. More certain is that the episode

[119] This is the argument of Berrenechea (2010).

[120] In the *Trojan Discourse* (*Or.* 11), Dio Chrysostom manipulates Herodotus's playful priest in a similar way. From 11.37–70, Dio reports the speech of an aged Egyptian priest who allegedly informed him of the "true" causes of the Trojan War. The priest, who disdains Greek intellectual history and literature, gives an account of the war that repeatedly and ostentatiously contradicts the Homeric tradition and so tests the reader's knowledge and acuity, just like the *Trojan Discourse* itself. The nature of the game is clear from the outset when the priest ridicules the Greeks *en masse* for having been tricked by merely one man (11.37), ostensibly Homer but, by implication, the priest/Dio himself.

dramatizes the extent to which natural inquiry constitutes a dangerous distraction in a time of civil war, a reading that still constitutes a brutal critique of Seneca. While Acoreus serenades his audience, his song allows Pothinus time to plot an attack (10.332–335):

> sic uelut in tuta securi pace trahebant
> noctis iter mediae. sed non uaesana Pothini
> mens inbuta semel sacra iam caede uacabat
> a scelerum motu.

This is how they dragged out the course of middle night, as if free from care in the safety of peacetime. But the mad mind of Pothinus, now that it had once been intoxicated with sacred slaughter, was not free from the instigation of crimes.

Caesar behaves as if in a state of *securitas*, which now looks as preposterous as the celestial journey of Pompey's soul (Ch. 4.2, "Pompey's Soul"). The general thinks he has time to unearth hidden causes (*causas... latentis*, 10.190), when he should have followed Lucan's lead and bid them remain concealed (Ch. 4.2, "The Gallic Tides"). Indeed, he has allowed an avalanche of data and (mis)information to pin him down (cf. Ch. 4.3). When he awakens from his daydream, he immediately springs into action and leaves the *mundi labor* (1.417) to others, but he finds himself trapped within the palace: "One for whom the extent of the Roman world was not enough... like a boy unfit for war or a woman within captured walls, seeks the safety of the house" (*hic, cui Romani spatium non sufficit orbis... ceu puer imbellis uel captis femina muris, / quaerit tuta domus*, 10.456–459). Lucretius's Epicurus seeks truth beyond the boundaries of the universe (1.72–74), and Longinus declares the whole of the cosmos not to be enough for the intellectual faculty and investigative powers of humankind (οὐδ' ὁ σύμπας κόσμος ἀρκεῖ, *Subl.* 35.3); Caesar is a travesty of the sublime investigator of nature. In a time of civil war, those who possess expert knowledge about the natural world are laughably irrelevant (cf. Ch. 4.4, 4.5) or downright deadly, and those who follow their example wind up hoist by their own petard. Such are the dangers of studying nature's *miracula* (10.196), and it is a lesson that the historical Seneca ultimately learned quite well.

Lucan's indictment of the scientific sublime is scathing. If future generations are to have any hope of rehabilitating the experience, the program of study is going to need to look radically different than what Manilius, Seneca, and Acoreus have to offer. Instead of far-flung flights of fancy and a pointless or even dangerous obsession with the stars, something fundamentally more concrete and down to earth might be in order. This is precisely what the poet of the *Aetna* has to offer, and it is to this text that we now turn.

The Scientific Sublime in Imperial Rome: Manilius, Seneca, Lucan, and the Aetna. Patrick Glauthier, Oxford University Press. DOI: 10.1093/9780197787588.003.0005

5
Aetna and the Wonders of the Earth

5.1. Introduction

The scientific sublime emerges from the wreckage of Neronian Rome beleaguered and badly bruised. If Seneca is anxious about the potential of meteorological inquiry to fortify and elevate the soul, Lucan mercilessly exposes the naïveté and, at times, aggressive duplicity of those who study nature's causes. By the middle of the 60s CE, a pervasive air of uncertainty and doublespeak has hopelessly undermined the legitimacy of the tradition, and it becomes desirable, if not necessary, to find a new direction. Enter the now-anonymous *Aetna*, most likely written in the late 60s or 70s.[1] The author explicitly rejects the exploration of that which is up above and articulates a dual imperative, to study the physics of the earth and to celebrate its wonders. Mount Aetna affords the poet ample scope for both pursuits; to read the poem, to study the mountain and to contemplate its marvels, is to undertake a harrowing journey through a volatile landscape that teeters on the verge of collapse. By comparison with earlier texts, however, the poem's technical focus and the scope of its journey are significantly restricted. This narrowing of horizons becomes particularly apparent in those passages that frame the mountain as a piece of machinery that might appear in an imperial spectacle; it's a dazzling contraption that is wildly entertaining but poses no vital threat. More striking still, even if some of the mountain's causes can be known, the author, following Lucan's lead, abandons any pretense that the study of nature might lead to moral progress, and he ultimately subordinates the analytic framework of philosophical inquiry to the stupefied observation

[1] The poem was attributed to Virgil at an early date, although not without trepidation; the *Vita Donatiana* expresses uncertainty (*scripsit* [sc. *Vergilius*] *etiam de qua ambigitur Aetnam*, 19). Goodyear (1984, 348–353) and De Vivo (1989) compellingly argue for Senecan influence on the poem, esp. that of the *Natural Questions*, which establishes a terminus post quem c. 63. The author's characterization of the area between Cumae and Naples as volcanically dormant for many years (*locus*... *multis iam frigidus annis*, 431) strongly suggests a date prior to Vesuvius's eruption in 79 CE. While a date in the mid to late 60s or 70s seems most likely, my arguments do not depend on it. For more, see Goodyear (1965, 56–59), Iodice (2002, 94–95), and Volk (2005a, 69–72).

of nature's marvels and of a personified nature. At the beginning of the Flavian period, some aspects of the natural world are simply not susceptible to causal analysis, at least not entirely, and that is a reason to celebrate. In elevating the thrill of the unknowable and isolating it from the specious allure of moral advancement, the *Aetna* poet tacitly acknowledges the compromised status of the scientific sublime. After the excesses of the preceding decade, it's best to keep one's head down and not to overreach.

From the fifth century BCE into late antiquity, ancient writers regularly represent the sights and sounds of Mount Aetna through the dual logic of the sublime—its ominous rumblings and explosive fireworks are exemplars of nature's grandeur and frequently suggest the existence of titanic forces precariously confined below the surface of the earth.[2] Recent assessments of the *Aetna* and the sublime, especially those of James Porter and Gareth Williams, have focused on the poem's indebtedness to the language and enthusiastic mindset of Lucretius and Manilius. Over the course of this chapter, new dimensions of this complex relationship will become apparent.[3]

The idea that a basic tension permeates the poem or that the author is pulled in two intellectually irreconcilable directions at once is widespread in the secondary literature.[4] A major contributing factor to such an assessment is the poem's finale: the author abandons philosophical analysis to tell the quasi-historical, quasi-mythological tale of the pious brothers of Catania, which endows the mountain and its lava flows with moral agency. Like the story of the Athenian plague that ends *De rerum natura* and the myth of Orpheus that ends the *Georgics*, the tale of the pious brothers poses a challenge to readers looking for philosophical continuity. From our perspective, however, the poet simply negotiates two different modes of sublimity, one grounded in the causal analysis of natural phenomena, the other in the recognition that nature herself sometimes exceeds the categories of philosophical inquiry. I read the negotiation of this tension as constituting

[2] For Aetna in ancient texts, see Leroux (2004), Muecke (2006), Buxton (2016), and esp. Williams (2017b, 23–71).

[3] Williams (2020) views the poem as an "answer" (121) or "complement of sorts" (122) to the *Astronomica*. Porter (2016, 508–517) stresses that the poem inhabits the same psychological plane as Lucretius and Manilius, creating a world that is full of chaotic energy, all while summoning the reader to "a higher vocation" (153). For the *Aetna* and Lucretius, see Hunink (1989) and Le Blay (2006, 336–341).

[4] Particularly helpful on the poem's tensions are Volk (2005a, 77–82) and Porter (2016, 516–517). On the tendency for a plurality of discourses to interact in Aetna narratives, Williams (2017b, 23–71) is fundamental; cf. Buxton (2016, 38, 43) and Glauthier (2018, 63–65, 75–76).

the poem's volcanic aesthetics, a worldview and modus operandi that takes the place of the Phaethon paradigm after Nero's collapse rendered it an unappealing model. The mountain's topographical chaos, its deep fissures and broken crags, and its seismic volatility, which is always on display, figure cognitive dissonance and thematic disjunction. As the author careens over Sicily's rugged landscape, now diving below its surface, now rocketing to the mountain's peak, he propels the reader in multiple directions at once, challenging us to hold seemingly incompatible positions, filling us with the mountain's dynamic energy, which is ultimately revealed to be that of *natura artifex* ("nature the demiurge"), whose mysteries cannot be explained. In the process of articulating this vision, the author rejects the study of celestial and meteorological phenomena, combines the subjectivity of the student of nature with that of the paradoxographer, and thereby stakes out a position in the ongoing debate about how and why one studies the natural world.

Since I am not primarily concerned with the details of the author's theoretical account of Aetna, it will be helpful to outline them here. It is well-known that the Greeks and Romans had no word for "volcano." Beginning in the fifth century BCE, philosophically grounded accounts of nature began to explain seismic and volcanic activity through the violent movement of subterranean air; I call this the subterranean wind theory.[5] Various studies have attempted to extrapolate a coherent philosophical orientation from the *Aetna*'s analyses and/or to identify its historical source(s); such lines of inquiry are not particularly useful in the present context.[6] According to the version of the subterranean wind theory that our poet adopts, the earth contains hollow cavities and passageways. Air regularly enters these subterranean spaces and occasionally becomes trapped and compressed within them. As this compressed air struggles break to free, it can cause earthquakes

[5] In Glauthier (2018), I argue that the subterranean wind theory was first formulated in the mid-to-late fifth century. For an overview of ancient approaches to seismic and volcanic phenomena, see Hine (2002). For the *Aetna* in particular, note Volk (2005a, 72–77). The account of Paisley and Oldroyd (1979, 11–14) should be treated with caution.

[6] Stoic: Sudhaus (1898) and Effe (1977, 206–214), with an emphasis on the promotion of "true religion"; cf. Lapidge (1989, 1409–1411). Epicurean, with various admixtures: Rostagni (1933, 281–339), De Lacy (1943), Lassandro (1993), and Verde (2020, 97–99). Nature religion: Volk (2005a, 80–82) and Holzberg (2020, 28). "Science for science's sake": Taub (2008, 48; 2009, 131). For skeptical assessments of such endeavors, see Goodyear (1984, 355), De Vivo (1989, 81–83), and Payne (2016, 96). Theophrastus, Posidonius, and Seneca's *NQ* (or one of Seneca's own sources) are regularly canvassed as possible sources; see Goodyear (1965, 54–56; 1984, 346–356), De Vivo (1989), Garani (2009, 103 n.3, 120–121), and Verde (2020, 92–99).

or ignite underground fires, which eventually erupt in spectacular explosions at the earth's surface.[7] Although the fire itself draws nourishment from and liquefies a variety of combustible fuels, such as sulphur and alum, the *lapis molaris* ("mill stone") constitutes Aetna's primary fuel, which it alone possesses.[8]

Despite the relatively simplicity of the argument, the poem's structure is not always self-evident, which is partially a result of the problematic state of the text. I adopt the following organizational scheme:

A. Proem (1–93). Through a series of indirect questions, the author indicates that his theme is Aetna and its causes. After invoking Apollo and the Muses, he establishes a poetics of novelty, in the tradition of Callimachus, and contrasts the tired topics of mythological poetry with the powerful newness of his own song (*fortius ignotas molimur pectore curas*, 24).[9] From lines 29 to 93, he criticizes those poets who falsely explain Aetna's behavior through mythological narratives and misrepresent the nature of the earth and the heavens. The author does not deny the existence of the Olympian gods, only that Vulcan's forge causes Aetna to erupt or that Jupiter threw the mountain on top of the defeated Giant Enceladus.
B. The existence of hollow spaces within the earth (94–176).
C. The subterranean wind theory (177–384). The exposition of the theory contains the poet's most explicit statements about how and why one should study the natural world (see Ch. 5.2).
D. The nature and fuel of Aetna's fire, with emphasis on the *lapis molaris* (385–567).

[7] The poet does not explicitly state how this air (*spiritus*) ignites a fire. Volk (2005a, 76–77) suggests either that *spiritus* causes friction, which then generates a spark, or that *spiritus* spontaneously turns into fire. The spontaneous transformation of the elements from one to another is a relatively widespread idea; see Vottero (1989, 320 n.2) on Sen. *NQ* 2.26.2.

[8] *Lapis molaris* is traditionally rendered "lava stone" but more accurately means "mill stone"; it is "the rock that, once cooled after liquefaction, becomes solid and is used as a stone mill" (Verde 2020, 93). Its nature was discussed by Posidonius (Strab. 6.2.3 [269] = fr. 234 Edelstein-Kidd).

[9] In the Callimachean tradition, such claims to newness are regularly made in the language of earlier poets. Hence, the *Aetna* poet's prayer that the Muses hasten "to new wishes" (*in noua... uota*, 7–8) echoes Ov. *Met.* 1–2 (*in noua...corpora*) and Man. 3.1–3 (*in noua*); see Wheeler (2009, 155–157). The allusion to Ovid gains point when we consider the contrasting scales and approaches toward mythological narrative; for all the Callimachean elements of the *Met.*, the *Aetna* will be something very different. Poetic claims to newness are also a common feature of Latin didactic; for the *Aetna* poet's familiarity with that tradition, see Volk (2005a).

E. Finale (568–645). The poem's concluding movement (603–645) tells the story of two brothers, often referred to as the *pii fratres* ("the pious brothers"), who save their aged parents from an eruption when Aetna's fires miraculously part in deference to the youths' piety.

The boundaries of some of the sections of the poem are up for debate, but the precise lines of demarcation do not matter for my position.

5.2. A Turn toward the Earth

In an extended discussion of his own intellectual commitments, which occupies lines 219–281, the poet proclaims that the search for celestial knowledge is a thrilling endeavor but that the ground beneath our feet takes precedence. The prioritization of terrestrial study is part of a dual imperative to study the earth and to draw attention to its wonders, and the contemplation of Aetna is the paradigmatic example of this pursuit. The poet's stance is both conservative and daring—in virtually the same breath, he refuses to embark on an upward adventure, reveals that he is already plumbing new depths in pursuit of the divine, and announces an intellectual program that combines causal analysis with a celebration of that which is fundamentally unexplainable.

In an extraordinary sentence that runs from lines 224 to 250, the poet lists a variety of wondrous topics (*miranda*, 224) for systematic study. The topics are primarily astronomical, such as the movements of the fixed stars, but the catalog begins with some basic items of natural philosophy, such as the elements and whether the *mundus* endures eternally. At the beginning and end of the sentence, the poet spells out what is at stake (226–227, 247–250):

> nosse fidem rerum dubiasque exquirere causas,
> ingenium sacrare caputque attollere caelo...
> et quaecumque iacent tanto miracula mundo
> non congesta pati nec aceruo condita rerum,
> sed manifesta notis certa disponere sede
> singula, diuina est animi ac iucunda uoluptas.

To know the proof of things and to investigate uncertain causes, to deify the intellect and to raise one's head toward heaven...and not to suffer whatever wonders exist in so great a world to be heaped up or buried in a pile of things, but to arrange individual items with marks in a sure location so that they will be manifest—this is the divine and delightful pleasure of the mind.

The study of celestial phenomena is an act of devotion that elevates the mind, rendering it divine.[10] At the end of the passage, this activity becomes a form of cosmogony. In contrast to those who would simply allow the wonders of the heavens to lie idle in undifferentiated heaps, the student arranges each item in its proper place and labels it, thereby making it identifiable (*manifesta notis... disponere... singula*). Such "arrangement" (*disponere*) is a generative act that mirrors the creation of an ordered cosmos in Lucretius, Ovid, and Seneca.[11] The study of the heavens allows one to play the role of god; as such, it is a form of *homoiōsis theōi* (see Ch. 1.6, "Plato and Aristotle"). To drive home the sublimity of the endeavor, the poet adapts Lucretius's powerful tag from the opening of *De rerum natura* 3. When Lucretius reads the work of Epicurus, *diuina uoluptas... atque horror* seize him (3.28–29; see Ch. 1.3). In the present context, however, there is no *horror*, only *diuina uoluptas*, a mode of sublimity that registers the expansion of the reader's consciousness without the accompanying disorientation of the Lucretian model.[12] In further contrast to Lucretius, our author does not encode the experience through a first-person narrative. The transcendent pleasure of celestial study is universal, a fundamental fact of the human mind, an experience to which we are summoned by our very nature.

To the reader fresh from Seneca and Lucan, such optimistic enthusiasm constitutes a striking change of outlook. Indeed, it reminds one of Manilius's ardent insistence that through mathematical astrology one reprises the role of the cosmic "builder" (*conditor*, 2.701; see Ch. 2.4). The *Aetna* poet, however, immediately strikes a rather different tone (251–256):

[10] For *sacro* = "deify," see *OLD* 4b, where this passage is cited. Goodyear (1984, 351) compares *ingenium sacrare* with Sen. *NQ* 4a praef. 10, *ingenium suspicere coepisti... quod consecrari malles quam conteri*. De Vivo (1989, 66–78) collects numerous Senecan parallels for lines 219–281.

[11] Lucr. 5.445: *magnas disponere partis*; Ov. *Met.* 1.32–33: *sic ubi dispositam quisquis fuit ille deorum / congeriem secuit*; Sen. *Ep.* 16.5: *arbiter deus uniuersi cuncta disposuit*. Bömer (1969, 29) glosses *disponere* as διατιθέναι, διακοσμεῖν, and cites *SVF* 1.160, where Zeno is said to have called *logos* the *rerum naturae dispositorem atque artificem uniuersitatis*. The *Aetna* poet's *congesta* perhaps recalls Ovid's jumble of atoms that are *congestaque eodem* (*Met.* 1.8).

[12] The move resembles Cic. *Fin.* 4.11–12 (see Ch. 1.6, "Stoics, Epicureans, and the *nil admirari* Tradition"). Giuliese (2001, 115) reads the positive assessment of astronomical inquiry as a response to Lucr. 5.1204–1210, where the sight of the heavens prompts philosophical backsliding.

> sed prior haec homini cura est, cognoscere terram
> quaeque in ea miranda tulit natura notare:
> haec nobis magis adfinis caelestibus astris.
> nam quae mortali spes est, quae amentia maior,
> in Iouis errantem regno perquirere diuos,
> tantum opus ante pedes transire ac perdere segnem?

But the following concern takes precedence for humankind, to come to know the earth and to call attention to the wonders that nature has brought forth in it—it is more closely related to us than heaven's stars. For what hope is there for a mortal, what greater madness than[13] to seek out gods while wandering in Jupiter's realm, and to pass by so great a work before our feet and to destroy it in idleness?

The poet issues a dual imperative, to come to know the earth (*cognoscere*) and to call attention to its wonders (*miranda... notare*).[14] This dual imperative resembles the steps of celestial inquiry, whereby the student arranges and labels wonders, thereby making them manifest (*miracula... manifesta notis... disponere*). Significantly, however, the study of the earth is not a generative or cosmogonic act of "arrangement."[15] While *cognoscere* and *notare* can and do overlap—to explain Aetna's causes is to draw attention to one of the earth's wonders—they can and do diverge. At the end of the poem, the celebration of earth's wonders crowds out the systematic investigation of its causes, leaving the reader in a state of aporia. This is not the mindset of a god or of one who assimilates himself to god as far as possible, to adopt the words of Plato's Socrates (*Theaet.* 176b). Indeed, our poem champions a less lofty intellectual pursuit precisely because human beings are not divine. We are earth-born, and it is implicitly easier and more appropriate for us to investigate that to which we are physically and qualitatively close (*haec nobis magis adfinis*) than that from which we are physically and qualitatively removed (*caelestibus astris*).[16]

[13] For the omission of *quam* in the Latin, see Goodyear (1965, 154).

[14] Lassandro (1993, 320) and Taub (2008, 47; 2009, 130) briefly distinguish these two activities.

[15] At lines 510–511, the poet frames his inquiry as an attempt "to arrange individual items with fixed causes" (*certis disponere singula causis / temptamus*); this comes close to the cosmogonic use of *disponere*.

[16] The basic idea resembles the argument that Aristotle articulates at *Parts of Animals* 1.5.644b–645a: the sublimity of celestial inquiry is a given, but an infinity of inconceivable pleasures awaits those who study animal life precisely because animal life is more accessible and closely related to us; see Ch. 1.6 ("Plato and Aristotle").

From this perspective, to investigate the heavens is an act of hybris, a hollow or even comedically misinformed assertion of greatness of soul. The image of the student of the stars wandering through the cosmos studying the divine (*in Iouis errantem regno perquirere diuos*, 255) recalls the opening of the *Astronomica*, where it is thrilling to stroll and live in the boundless heavens (*iuuat... immenso spatiantem uiuere caelo*, 1.13–14), and the preface to *Natural Questions* (*NQ*) 1, where it is similarly thrilling to wander among the stars and laugh at the misguided priorities of those on earth (*tunc iuuat inter ipsa sidera uagantem diuitum pauimenta ridere*, 1 praef. 7). Although the verbal parallels are not exact, the presence in all three texts of an accusative singular participle indicating leisurely movement through the celestial region (*errantem, spatiantem, uagantem*) strongly suggests that rather than composing a vague *recusatio* to celestial study, the *Aetna* poet is thinking specifically about Manilius and Seneca.[17] In pointed contrast to the grandeur of the study of the stars in the *Astronomica* and *Natural Questions*, "to pass by so great a work before our feet" (*tantum opus ante pedes transire*) draws on a well-established tradition for criticizing those who study that which is up above. The phrase may point specifically to Thales, who, according to legend, was absorbed in astronomical observation, ignored what was before his feet, and fell into a well (see Ch. 1.5, "The Leap"). Or, perhaps more likely, it looks to a well-known line from Ennius's *Iphigenia* that criticizes astrologers: "What is before one's feet no one watches, they probe the expanses of heaven" (*quod est ante pedes nemo spectat, caeli scrutantur plagas*, fr. 187 Jocelyn, 82 Goldberg-Manuwald).[18] In the *Aetna*, such inquiry constitutes a grave error (*errantem*), dishonoring the earth and spoiling the investigator's own claims to sublime subjectivity; greatness of soul suddenly becomes the greatest *amentia* (254). The poet's stance, however, is not that one should study the earth before studying the heavens. Lines 257–269 castigate those who physically assault or even torture the earth in search of precious stones or through agricultural work; the criticism

[17] For the connections between the Manilian and Senecan passages, see Ch. 3.4. In *Aetna* 225, Lühr (1971, 145–147) locates a direct criticism of Manilius; his arguments have gained wide acceptance, but Porter (2016, 514) remains skeptical. For evidence of the *Aetna* poet's engagement with Manilius elsewhere, see Effe (1977, 214–220), Goodyear (1984, 357–358), De Vivo (1989, 71), Wheeler (2009, 155–57), Welsh (2014, 102–103), and Williams (2020). For the *recusatio* to celestial study, see Volk (2005a, 86–87; 2005b, 169–171).

[18] For the proverbial nature of Ennius's line and its connections with stargazing, see Glauthier (2011, 190 n.9). For the line's reappearance at Plin. *HN* 18.252–253, see the Epilogue.

belongs to the diatribe tradition.[19] When set against line 256, the implication is that to fail to investigate the earth and to call attention to its wonders constitutes an equally reprehensible violation. To ignore the earth is to act immorally, regardless of one's interest in *caelestia*; Manilius and even Seneca have their priorities all wrong.

The exclusivity of the poet's turn toward the earth becomes clear at the end of this section. After criticizing those who mistreat the earth, the poet suggests that terrestrial study is mankind's highest calling (271–275, 279–281):[20]

> haec rerum maxima merces,
> scire quid occulto terrae natura coercet,
> nullum fallere opus, non mutum cernere sacros
> Aetnaei montis fremitus animosque furentis,
> non subito pallere sono, non credere subter
> caelestis migrasse minas aut Tartara rumpi,
> nosse quid intendat uentos, quid nutriat ignes,
> unde repente quies et nullo[21] foedere pax sit.

This is the most profitable thing in the world, to know what nature keeps hidden in the depths of the earth, not to speak falsely about her work—not to gaze in silence at the holy roar and raging anger of Mount Aetna, not to grow pale at the sudden sound, not to believe that heaven's threats have moved underground or that Tartarus is being broken open, to know what strengthens the winds, what feeds the fires, what suddenly causes silence and peace without a treaty. (Trans. based on Hine 2012)

Once again, humankind has a twofold calling, to know the inner workings of the earth (*scire quid occulto terrae natura coercet*) and not to misrepresent its mechanics (*nullum fallere opus*).[22] Now, however, celestial inquiry has fallen by the wayside. After formulating the imperative in general terms, the poet adduces a particular example—the study of Mount Aetna. In what

[19] Note Rossetti (2017, 88–89). Di Giovine (1981) reads the passage as a polemical engagement with the *Georgics*; cf. De Vivo (1985, 271–273; 1989, 71–73), Volk (2009, 197), and Welsh (2014, 103–105). Giuliese (2001, 119) compares several passages from Sen. *NQ*.

[20] The manuscripts are not consistent in the placement of lines 276–278, and editors since Scaliger regularly print them after 257.

[21] Following Hine (2012), I read *nullo* for the manuscripts' *multo*, which Goodyear (1966) obelizes.

[22] For *fallo* = "conceal the nature of, disguise," see *OLD* 8.

follows, the order of the elements is reversed. The injunction not to speak falsely about any aspect of the earth becomes a command not to watch the mountain in silent fear or to explain its operations in mythological terms, while the exhortation to know the earth is answered by a call to understand the physics of the mountain.

The study of the mountain, which the poem itself embodies, is the paradigmatic example of a turn toward the earth that the poet upholds as an intellectual ideal. It is a deep dive into a hidden world that overloads the senses with its imposing and unexpected sounds and movements, which, the poet insists, are divine (*sacros...fremitus animosque furentis...subito...sono*). The sacred status of the mountain is a core tenet of the text that we will explore further in Ch. 5.3. For now, we can observe that to study the mountain is to approach the divine, in all its electrifying awesomeness. If the confrontation threatens to reduce the student to passive silence (*mutum*), the implicit challenge is to transform the mountain's sacred energy into action, to come to know that which is more than human and to publicize that knowledge. It is a task structured according to the dual logic of the sublime, and the goal is both to honor the mountain and to realize a form of high-mindedness. By comparison, the study of the stars appears unexpectedly mundane, and its *diuina...ac iucunda uoluptas* (250) looks woefully one-dimensional. The study of the earth adds *horror* back into the equation, and we are left with the impression that true elevation only comes from a bold plunge to new depths.[23]

The *Aetna* poet's turn toward the earth participates in a long-standing debate about what area or areas of the natural world call for investigation and are worthy of wonder and awe—a debate that has gradually revealed itself throughout this study. For Manilius, the study of the heavens subsumes all other forms of knowledge and renders them pointless. Since celestial knowledge constitutes a mystical experience that collapses the distinction between god and man, Manilius confidently urges us to dismiss all other arts (*mitte alias artes*, 2.109; see Ch. 2.6). The *Aetna* poet rejects the premise and the conclusion. In the *NQ*, Seneca treats meteorology as a propaedeutic to the study of *caelestia*—through the investigation of atmospheric and certain terrestrial and subterranean phenomena, we prepare our minds for the contemplation of higher things and the expansiveness of vision that

[23] For very different readings of 271–281, see De Lacy (1943, 173–175), Effe (1977, 208), and Toohey (1996, 189).

momentarily comes with such a pursuit (see Ch. 3.2). This position allows Seneca to spend considerable time underground, but the *Aetna* poet construes the study of the earth as an autonomous area of inquiry that has intrinsic value. It is its own *telos*, and it is uniquely suited for the expansion and ennoblement of human thought. Lucan, by contrast, reveals the ways in which the study of nature fails to achieve the lofty goals to which it aspires, an analysis that exposes the preposterous vanity of such debates. Amid all-consuming crisis, when the gods have seemingly withdrawn, the study of nature is subordinated to political engagement (see Ch. 4). While the *Aetna* poet momentarily shines a light on the absurdity of those who claim to soar through the heavens, he unequivocally embraces the study of the earth. In the wake of Lucan's devastating critique of natural inquiry, the *Aetna* poet unflinchingly asserts that a pathway to the divine lies open through the investigation of the terrestrial world. Lucan's Acoreus may treacherously distinguish knowledge from wonder (see Ch. 4.6), but the *Aetna* poet embraces both.

5.3. An Epiphany in the Amphitheater

A significant part of the literary tradition against which our poet is working (see lines 41–73) configures Aetna's sights and sounds as extensions of the bodies of conquered monsters. In *Pyth.* 1, Pindar describes springs of fire that "belch" from the mountain (ἐρεύγονται, 1.21–22), below which lies the hundred-headed monster Typhos. In *Prometheus Bound*, the same monster sends forth rivers of fire that "devour" Sicily's smooth fields "with savage jaws" (ποταμοὶ πυρὸς δάπτοντες ἀγρίαις γνάθοις / τῆς καλλικάρπου Σικελίας λευροὺς γύας, *PV* 368–369). In *Aeneid* 3, Aetna "lifts up balls of flame and licks the stars; sometimes, with a belch, it raises aloft boulders and the mountain's innards, torn free" (*attollitque globos flammarum et sidera lambit; / interdum scopulos auulsaque uiscera montis / erigit eructans*, 3.574–576). This monstrous Aetna resembles the Cyclops Polyphemus who strikes the stars (*altaque pulsat / sidera*, 3.619–620), feeds on the bloody innards of his wretched victims (*uisceribus miserorum et sanguine uescitur atro*, 3.622), and belches them back up (*saniem eructans*, 3.632). In such contexts, an eruption becomes a kind of epiphany, a moment when a superhuman power bursts into the world of human experience, staging a sublime spectacle.

Although the *Aetna* poet rejects mythological aitia, he regularly blurs the distinction between the mountain and the monsters of myth.[24] While the poet dismisses the idea that the monster Enceladus lies beneath the mountain and breathes fire from his "jaws" (*Enceladon... qui... petulans expirat faucibus ignem*, 72–73), he writes repeatedly of Aetna's *fauces*.[25] Similarly, no monster belches through the mountain, but fiery wind "vomits" furiously all over Aetna (*furens tota uomit igneus Aetna*, 328).[26] Shockingly, the rocks that the mountain ejects resemble human beings ("the uncertain appearances and semblances of men," *incertae facies hominumque figurae*, 469); it is as if the monster, Polyphemus-like, has disgorged its gruesome feast. Such language fleetingly conflates mountain and monster and gestures toward an epiphany.

Even more striking, however, are the ways in which the poet replaces the monster of myth with a personified *natura artifex*. Like Typhos and Enceladus, this figure cannot be constrained. The jagged topography that the poem repeatedly celebrates attests to the vital energy that courses below the surface, and when the mountain erupts, *natura artifex* reveals herself in an impressively staged spectacle worthy of the soon-to-be-completed Flavian amphitheater.

The chaotic and intensely discontinuous nature of the earth is a driving force of the poem. After promising to give "you," the reader, and "your eyes" (*tibi... oculis... tuis*, 135–136) sure signs of the existence of subsurface cavities, the poet forces us to stare into the primordial abyss (137–139):

> inmensos plerumque sinus et iugera pessum
> intercepta licet densaeque abscondita nocti
> prospectare: procul chaos ac sine fine ruina est.

Very often, you may gaze out at boundless chasms and stretches of land that are cut off and plunge downward and are buried in dense night: far below there is chaos and ruin without end.

The fractured landscape opens a wormhole through which we travel back in time to stare at an almost Hesiodic Chaos (*Th.* 116). Indeed, a consideration

[24] Cf. Buxton (2016, 34–35). Goodyear (1984, 360–362) and Garani (2009, 108–111) discuss the poem's pervasive use of personifying metaphors, which collectively make the mountain pulse with life.

[25] Lines 168, 319, 330, 374, 378, 561; cf. 30. Enceladus's position under Aetna appears to be at least as old as Euripides (see *HF* 908 with Apoll. *Bibl.* 1.6.2). Subsequently, he appears at, e.g., Callim. fr. 1.36 Pfeiffer, Virg. *Aen.* 3.578.

[26] Cf. line 394, where waters tainted by sulphur and alum "belch forth" (*eructantur*).

of earthquakes demonstrates the "fury" of subterranean wind (*uenti rabies*, 171), and the destruction that such fury unleashes is itself a sure sign that Chaos is about to return (*neque est aliud, si fas est credere, mundo / uenturam antiqui faciem, ueracius omen*, 173–174). In both passages, the poem brings to the surface infinite dimensions and infinitely unruly forces that are normally concealed from view or relegated to the distant past, like defeated Titans. The study of the earth and the consideration of its marvels threaten to bring our world crashing down.[27]

But that is only half the story. If it were entirely solid, "the earth would produce no wondrous spectacles of itself and would be inert, compressed into dead weight, immobile" (*nulla daret miranda sui spectacula tellus / pigraque et in pondus conferta immobilis esset*, 156–157). The boundless dimensions, infinite hollows, and irresistible energy of the earth transform volcanic eruptions and seismic disturbances into thrilling shows that are evidence of vitality and life.[28] The same holds true when the poet pivots directly to Aetna. At the outset, he insists that Aetna will spontaneously reveal itself (177–180):

> Aetna sui manifesta fides et proxima uero est.
> non illic duce me occultas scrutabere causas;
> occurrent oculis ipsae cogentque fateri.
> plurima namque patent illi miracula monti.

Aetna is clear proof of itself and closest to the truth. You will not probe hidden causes there with me as your guide—they will meet your eyes of their own accord and compel agreement, for very many wonders are open to view on that mountain.

The mountain promises to turn itself inside out; its subterranean secrets will appear, like a god, and the reader will be powerless to resist their awesome truth.[29] As proof that Aetna will behave accordingly, the poet declares that the mountain itself is a landscape that teems with wonderous sights, all easily accessible to the senses. An energetic account of Aetna's fractured topography follows: terrifying chasms, broken crags, unfathomable disorder

[27] I have discussed the experience of the sublime as time travel at Glauthier (2022, 87–89, 92–96). For the sublimity of air and wind, see Ch. 1.4 and 3.5.

[28] Cf. lines 366–384, where the mountain's periods of dormancy, the cause of which cannot be explained, lead to fiery explosions that are "the spectacles of wind" (*spectacula uenti*, 384).

[29] Aetna thus resembles Manilius's configuration of the Milky Way at 1.701–702 (Ch. 2.2).

(*discordia... ingens*, 183), and everywhere evidence of the overpowering fire that seethes within.[30] The geological survey concludes by emphasizing the power and divinity of the mountain: "This is the appearance and home of the sacred work, which must be seen, this is the location and site of its tremendous power" (*haec operis uisenda sacri faciesque domusque, / haec illi sedes tantarumque area rerum est*, 187b–187).[31] Aetna's chaotic landscape is evidence of a powerful divine presence that dwells within the mountain and is eager to break free. As the mountain opens up, through the verses of the poem, the monster of the mythological tradition becomes something both more intangible and simultaneously entertaining.

A few lines later, the poet adds a twist. He begins by stating that the argument calls for identifying the "craftsman" and "cause" of the volcanic fire (*nunc opus artificem incendi causamque reposcit*, 188). Once again, the reality of the mountain will confront the reader's eyes and compel assent (191). Unfortunately, we cannot approach Aetna and use the sense of touch to examine the cause of its fires (192–198):

> quin etiam tactu moneant [*sc.* res], contingere tuto
> si liceat; prohibent flammae custodiaque ignis
> illi operi est, arcent aditus diuinaque rerum,
> ut maior species et ne succurrat inanis,[32]
> cura sine arbitrio est: eadem procul omnia cernes.
> nec tamen est dubium penitus quid torqueat Aetnam,
> aut quis mirandus tantae faber imperet arti.

Indeed, [the facts] would inform you [about this topic] by touch, too, if it were possible to touch safely. The flames prevent this: the fire guards the machine,[33] they stop you from approaching. The divine supervision of the world is without witness, so that the show may be greater and not occur [to the mind] to no purpose; you will behold the same things from a distance. Nevertheless, there is no doubt about what tortures Aetna deep within or the identity of the wondrous artisan who gives orders to so great a work of art.

[30] "Disorder" (*discordia*) points to Empedoclean Strife, regularly translated as *discordia* (e.g., Cic. *Amic.* 24). For *discordia* and the sublime, see Hardie (2009, 99–109, 144–146).

[31] I borrow "tremendous power" from Hine (2012). Ellis (2008) renders *tantarum rerum* "marvelous operations."

[32] Goodyear (1966) and Iodice (2002) bracket line 195. I follow Ellis (2008) in retaining it.

[33] For *opus* in this sense, see Ch. 5.4.

Aetna's fires are an instrument of a divine force that oversees the world (*diuina...rerum...cura*). Through them, this force keeps would-be spectators at a distance, intending thereby to enhance the sublimity of the spectacle (*ut maior species*) and to give it purpose (*ne succurrat inanis*). In other words, this divine force is an obstacle to the systematic study of nature—whether one contemplates an eruption in real time, through recollection, or via the mediation of the written word.[34] The poet insists, however, that the student can overcome this obstacle (*nec tamen est dubium*) and learn the identity of the wondrous artisan (*mirandus...faber*) who operates the mountain. This ability implies the superiority of the investigator of nature over the divine force. Through the study of the earth, which is configured as a kind of amphitheatrical performance, the student achieves true greatness.[35]

The lines that follow support this reading. As the poet describes an eruption, he imagines that Jupiter himself wonders at the fires from a distance (*ipse procul tantos miratur Iuppiter ignes*, 203) and grows fearful that the Giants may be launching a new assault on Olympus or that Dis may be attempting to exchange heaven for hell. The contrast with the more conventionally heroic Jupiter who marshals the other Olympians to fight the Giants and then triumphs over Enceladus at lines 54–73 is striking. Equally striking is the similarity between Jupiter's gaze and that of the reader (*eadem procul omnia cernes*, 196). *Miratur* confirms that the poet is thinking in terms of the sublime. Jupiter's fear, however, falls far short of the *megalopsuchia* and *megalophrosunē* to which a confrontation with Aetna ought to lead. In pointed contrast to the poet's own exhortation "not to gaze in silence at the holy roar and raging anger of mount Aetna, not to grow pale at the sudden sound, not to believe that heaven's threats have moved underground or that Tartarus is being broken open" (273–275, 279), this Jupiter is stuck in a superstitious mode of mythological thought that can find only terror in seismic and volcanic activity. The student of the earth recognizes such phenomena for the divinely orchestrated spectacles that they are,

[34] Ellis (2008, 113–114) construes *succurrat* differently on the grounds that it means "to occur, as a thought, to the *mind*, not the *eyes*" (emphasis original). The poet, however, primarily conceives of the mountain as a textual artifact, precisely the kind of thing that is more likely to occur to the mind than the eyes.

[35] To illustrate secrecy's ability to enhance grandeur, Ellis (2008, 113 n.2) compares Strab. 10.3.9 (467), where the secrecy of the mysteries enhances the grandeur of the divine (ἡ κρύψις ἡ μυστικὴ τῶν ἱερῶν σεμνοποιεῖ τὸ θεῖον). At Strab. 6.2.8 (274), Aetna's fires are never dormant long enough for one to be able to approach so great a force (ὥστ᾽ ἀντὶ τῆς τοσαύτης βίας ἐφικτὸν ἀνθρώπῳ γενέσθαι τὸν πλησιασμόν); the *Aetna* poet gives this reality purpose.

and "you" can cheer, applaud, or write poetry about them as you wish. For an instant, at least, this student becomes superior to Jupiter himself.[36]

The passages examined above do not name *natura artifex* as the force within the earth or the agent who manages the *mundus*, but comparison with other passages leads to this conclusion. As we have seen, one of the tasks of the student of the earth is to call attention the wonders that *natura* has brought forth in it (*quaeque in ea* [sc. *terra*] *miranda tulit natura notare*, 252). Similarly, the greatest prize for a human being is to know what *natura* conceals in the depths of the earth (*scire quid occulto terrae natura coercet*, 272). Both passages personify *natura* and assign her a creative role in world. In the poem's final movement, the author exhorts the reader to behold the massive work of "nature the demiurge" (*artificis naturae ingens opus aspice*, 600).[37] The phrase *natura artifex* recalls the poet's statement that he must identify the *artificem incendi* (188), which leads into his remarks about the *diuina rerum cura* (194–196) and the assertion that a marvelous craftsman controls the *ars* that is the mountain (*mirandus tantae faber imperet arti*, 198). From the perspective of the whole poem, the *artifex incendi* and the *faber* are one and the same—namely, *natura artifex*. Nature the demiurge is a bold personification. Her presence within the mountain explains why the poet repeatedly refers to Aetna and its fires as sacred (187b, 273, 464; cf. 351, 559–560). For the angry protests of conquered monsters, the poet substitutes an infinitely more elusive divinity, and we must stay in our seats and watch the show from a distance.

But who or what is *natura artifex*? The phrase first occurs in Cicero, who attributes it to Zeno,[38] and it seems particularly well suited to express the basic Stoic position that god is the maker of the world and simultaneously immanent in or even identical with the world.[39] At *NQ* 2.45, for instance, Seneca describes Jupiter as the *animus* and *spiritus* of the world, "the master and demiurge of this work" (*operis huius dominum et artificem*), and then declares that we may call him, among other things, *fatum*, *prouidentia*, *natura*, or *mundus*. Although none of Seneca's extant works contains the exact

[36] Welsh (2014, 121 with n.63) makes the connection between Jupiter's gaze and that of the student and compares 464–465 and 505; cf. Stoneman (2020, 200–201). For a similarly amazed Jupiter, see Ch. 4.3 on Luc. 6.463–465.

[37] Cic. *Ti.* 6 translates δημιουργός at Pl. *Ti.* 29a with *artifex*; cf. the uses of *artifex* cited under the heading "de deo siue natura mundum fabricantibus" at *Thesaurus Linguae Latinae* 2.700.58–701.11.

[38] Cic. *Nat. D.* 2.58 (*SVF* 1.172, cf. 1.160, 2.599).

[39] See D. Furley's succinct account at *CHHP* 448–51. L-S 43 and 54 collect some of the relevant texts. Kruschwitz (2015, 93) understands *natura artifex* at *Aetna* 600 along these lines.

phrase *natura artifex*, this passage shows that the concept is at home in his thought (cf. *Dial.* 12.8.3). It makes perfect sense, then, that the author of the *Octauia*, which likely dates to the late 60s or 70s CE, puts the expression into Seneca's mouth.[40] In the second act, a frustrated Seneca, reflecting on the bygone "thrill" of celestial inquiry, describes the heavens as the most magnificent creation of our parent *natura*, "the demiurge of [this] boundless work" (*o quam iuuabat, quo nihil maius parens / natura genuit, operis immensi artifex, / caelum intueri*, 385–387). The impersonal *iuuabat* encodes the sublimity of studying the stars, just as it does in the *NQ*.[41] Writing at precisely the same time and in a context that is saturated with the ethos of the scientific sublime, the *Aetna* poet employs strikingly similar language (*artificis naturae ingens opus* vs. *natura... operis immensi artifex*). If we suppose that the *Octauia* predates the *Aetna*, a dialogue emerges. When Seneca's character laments the impossibility of realizing the thrill of celestial inquiry, the *Aetna* poet redirects his gaze toward the earth and insists that it is still possible to experience the power of nature the demiurge. This is the same basic back-and-forth that we detected when we read the *Aetna*'s exhortation to study the earth alongside Lucan's critique of natural inquiry; the parallelism is unlikely to be accidental. Still, whichever text came first, *Aetna* or *Octauia*, together they constitute further evidence of the ongoing argument about intellectual priorities that we considered at the end of Ch. 5.2.

We should not assume, however, that *natura artifex* points only to Stoic influence. In Greek, similar expressions predate Stoicism and occur in authors who do not align themselves with the Stoa.[42] Moreover, a belief in a divine *dēmiourgos* is widespread, from one end of antiquity to the other (see Theiler 1957), and *natura artifex* has a long post-classical afterlife,

[40] For the date, see Boyle (2008, xiii–xvi), who prefers the early to mid-70s CE.

[41] See Ch. 3.4 on *NQ* 1 praef. 7. For *iuuat* and the affective intensity of the scientific sublime, see Ch. 4.2 ("Nero-Phaethon").

[42] I cite a few examples from Aristotle and Galen. Aristotle: "nature the demiurge" (ἡ δημιουργήσασα φύσις) provides inconceivable pleasures to those who study causation (*Part. an.* 1.5.645a; see Ch. 1.6, "Plato and Aristotle"); "nature crafts nothing haphazardly" (ἡ φύσις οὐθὲν δημιουργεῖ μάτην, *IA* 12.711a), a variation on the refrain that nature "does/makes" nothing haphazardly (e.g., *IA* 11.711a). Galen: nature is "artistic" (τεχνική) and operates "artistically" (τεχνικῶς, *Nat. Fac.* 1.12/2.26–30 Kühn); nature resembles a demiurge and craftsman (ὁ δ' οἷον δημιουργός τε καὶ τεχνίτης ἡ φύσις, *MM* 3.3/10.174 Kühn); "nature is the demiurge, the physician her servant" (ἡ μὲν φύσις ἐστὶ δημιουργός, ὁ δ' ἰατρὸς ὑπηρέτης, *Ars med.* 26/1.378 Kühn); cf. the hymn to nature's art and the (male) demiurge at *UP* 3.10/3.234–242 Kühn. Galen also provides evidence that the Hellenistic physician Erasistratus of Ceos used the expression ἡ τεχνικὴ φύσις (frr. 79–80 Garofalo; cf. *Nat. Fac.* 2.3/2.87 Kühn, 2.9/2.131–132 Kühn).

extending through late antique Neoplatonism well into the Middle Ages.[43] As we will see (Ch. 5.5, "*Natura artifex*"), the expression often implies a general enthusiasm for the grandeur and magnificence of the natural world precisely at those moments when the natural world's logic baffles human understanding.

Even at this point, though, we can apprehend how the concept introduces friction into the text since the poet does not explain the relationship between *natura artifex* and the Olympian gods. Does his apparently pantheistic worldview accommodate the Olympians through allegory? Are nature the demiurge and Jupiter interchangeable constructs, two sides of the same cosmic coin? Unlike Seneca or Manilius, our poet has no interest in presenting an explicitly teleological system in which *ratio* reigns supreme and an unnamed *deus* is immanent in the cosmos. Instead, we get a parade of Olympians, two irreconcilable Jupiters, and a mysteriously personified nature. These characters are intellectual fragments uprooted from their respective philosophical, poetic, and religious systems, and they are not in dialogue.[44] For now, then, we can note that the idea *natura artifex* cuts two ways. At one level, the poet signals, quite vaguely and much to our surprise, that nature the demiurge is the organizing principle of his *mundus*, whatever that means. Through the study of the earth and through the poet's own verses, she reveals herself and the orderliness of her seemingly chaotic creation, her carefully orchestrated amphitheatrical show. At another level, the reader understands that some aspects of the world lie beyond our comprehension. The wonders of the mountain are no longer the monsters of myth, but when we call attention to them, we are venturing into territory where we should not expect logical answers or perhaps any answers at all—it is all chaos, fragmentation, and kinetic energy.

5.4. The Mechanical Sublime

The substitution of nature the demiurge for the monsters of myth makes Aetna an example of divine craftsmanship. As part of the maker's plan,

[43] Truitt (2015, 42–47) shows how *natura artifex* passed from Chalcidius, Macrobius, and Boethius into the twelfth and thirteenth centuries; cf. Economou (2002, 19–20, 26, 153).

[44] As I point out in Ch. 1.6 ("Stoics, Epicureans, and the *nil admirari* Tradition"), the second volume of Festugière (1944–1954) remains the classic survey of "le dieu cosmique." On the two Jupiters, note Payne (2016, 101).

its chaotic topography and unpredictable explosions become exciting yet harmless forms of entertainment (*miranda...spectacula* at 156, *spectacula uenti* at 384). But we can say more. One of the major ways in which the author seeks to know Aetna and to call attention to its marvels is by assimilating it to a breathtaking machine, a manmade contraption that one might encounter at a historical spectacle in the imperial period. By reconfiguring the mountain's geological fragmentation and seismic wildness as smoothly functioning components of a mechanical instrument, such analogies downplay the danger that the mountain poses without sacrificing the thrill that it inspires, and they allow the reader to comprehend in miniature the exquisite handiwork of the engineer who made the cosmic machine (*machina*, 230). Through this "mechanical sublime," human ingenuity figures divine creation and becomes a conduit to divinity itself. At the same time, the study of the earth becomes a form of imperial spectacle, an emblem of imperial power and largess that is altogether captivating and seemingly inescapable.[45]

While listing several theories that could account for the development of winds beneath the mountain (282–305), the poet draws a comparison with an unidentified musical instrument and then with the water organ (293–300):[46]

> nam, ueluti sonat †ora duc† Tritone canoro
> (pellit opus collectus aquae uictusque moueri
> spiritus et longas emugit bucina uoces)
> carmineque irriguo magnis cortina theatris
> imparibus numerosa modis canit arte regentis,
> quae tenuem impellens animam subremigat unda,
> haud aliter summota furens torrentibus aura
> pugnat in angusto et magnum commurmurat Aetna.

For, just as...sounds with the sonorous Triton (the pool of water and the air that is compelled to move strike the machinery and the trumpet bellows forth its long-drawn notes) and in great theaters the water organ with its watery

[45] Both "technological" and "mechanical sublime" have gained wide currency in the twenty-first century. On the sublimity of technology, see Miller (1965) and Nye (1994). On the continuum between "science" and "technology" in the ancient world, see Rihll (1999, 13–15) and Keyser (2013, 29–30). In a sense, the *Aetna* poet reconfigures the mountain as one of the seven (manmade) wonders of the world (e.g., Antipater of Thessalonica 91 Gow-Page) while still exploiting its sublimity as a natural phenomenon.

[46] Garani (2009) is a comprehensive treatment of the poet's analogies. At 112–115, she offers a complementary reading of this passage, stressing the originality of its theoretical implications.

song, melodious through pipes of unequal length, sings due to the controller's art, which strikes slender currents of air and causes a rowing motion in the waves below—in no way differently does the raging breeze, set in motion by the streams, fight in the confined space, and Aetna makes a great murmur.

By comparing the mountain with large-scale musical instruments, the poet momentarily turns Aetna into a complex piece of machinery; it is something that can be assembled, disassembled, and understood. The precise nature of the Triton contraption is not immediately clear. If, however, we read *hora duci* for *ora duc* and translate "just as the hour of battle (*hora*) is sounded to the general (*duci*) by the trumpeting tones of Triton," then a very close parallel may be found in Suetonius's biography of Claudius.[47] There, Suetonius describes an astounding *naumachia* staged by the emperor at Lake Fucinus in which a musical Triton played a key role: "At this spectacle, a Sicilian and Rhodian fleet clashed, each numbering twelve triremes, when they were set in motion by a silver Triton and its trumpet, which had risen from the middle of the lake by means of a mechanical device" (*hoc spectaculo classis Sicula et Rhodia concurrerunt, duodenarum triremium singulae, exciente bucina Tritone argenteo, qui e medio lacu per machinam emerserat, Claud.* 21.6). If the *Aetna* poet has this kind of Triton in mind, with its melodious *bucina*, the mountain suddenly becomes an astonishing piece of machinery that could figure prominently in an imperial spectacle—it will enthrall and entertain the reader.[48]

This seems all the more likely when we consider that the second half of the simile compares the mountain to a water organ or *hydraulus*. Although *cortina* does not elsewhere refer to water organs, Robinson Ellis has shown quite clearly that this is its meaning here.[49] Notably, the poet connects the instrument closely with public performances in grand theaters (*magnis cortina theatris*). We know from Petronius that a water organist could perform during gladiatorial shows ("the carver, dancing to the music, mutilated the meal in such a way that you might have thought a chariot-fighter was fighting to the accompaniment of a water organist," *scissor... ad symphoniam gesticulatus ita lacerauit obsonium ut putares essedarium hydraule cantante*

[47] The emendation is Munro's; see Ellis (2008, 140), whose translation I give here. Although he obelizes the phrase, Goodyear (1965, 157) dubs *hora* the "obvious correction."

[48] Coleman calls Claudius's event "the most spectacular" (1990, 71) and "the most ambitious *naumachia*" on record (1993, 56). Mart. *Spect.* 28 provides evidence of another magnificent Triton; see Coleman (1993, 65–67).

[49] Ellis (2008, 141–145); the *OLD* s.v. 2 follows his lead.

pugnare, *Sat.* 36.6).[50] Similarly, Suetonius's biography of Nero contains a story in which the emperor exhibited a collection of previously unknown types of water organ: "Pointing out their individual features and lecturing on the theory and complexity of each one, he even declared that he would produce them all in the theater" (*ostendensque singula, de ratione ac difficultate cuiusque disserens, iam se etiam prolaturum omnia in theatrum affirmauit*, *Ner.* 41.2). Later, we learn that Nero had publicly vowed that if he retained power, he would play the water organ, along with other instruments, at the games in celebration of his victory (*Ner.* 54). Against this backdrop, the water organ emerges as a familiar kind of musical instrument that has close connections with imperial spectacles.

When we consider technical accounts of the *hydraulus*, the device appears even more wondrous and amazing. Vitruvius devotes one chapter of *De architectura* to water organs (10.8). He describes the workings of the device as obscure and difficult to understand but emphasizes that anyone who acquires firsthand knowledge of the instrument will see that it has been constructed with great ingenuity and precision (*profecto inueniet curiose et subtiliter omnia ordinata*, 10.8.6). Another detailed account appears in Heron of Alexandria's *Pneumatica* (pp. 192–202 Schmidt).[51] In his introduction, Heron tells the reader that all the devices he describes fall into two categories, those that fulfill some sort of necessary function and those that make an epideictic display of the astounding and the marvelous (ἐκπληκτικόν τινα θαυμασμὸν ἐπιδεικνύμεναι, p. 2 Schmidt). The water organ falls into the latter category. At the end of the preface, Heron explicitly characterizes the contents of his work as spectacles produced through the mechanical manipulation of the elements (τὰ διὰ τῆς συμπλοκῆς τῶν εἰρημένων στοιχείων ἐπιτελούμενα θεωρήματα γράψομεν, p. 28 Schmidt), and proclaims that they make it possible to discover especially intricate and wondrous kinds of motions (ἔστι γὰρ δι' αὐτῶν εὑρίσκειν πάνυ ποικίλας καὶ θαυμασίας κινήσεις, p. 28 Schmidt). Through such language, Heron comes close to framing the *Pneumatica* a work of "wonder-making" (ἡ θαυματοποιϊκή).[52]

[50] The popularity of the water organ is evident already at Cic. *Tusc.* 3.43.

[51] Heron was a mathematician and inventor. His dates are unclear; he cites Archimedes (d. 212/211 BCE), is cited by Pappus (fl. first half of the fourth century CE), and is generally placed in the first century CE. The two-book *Pneumatica* explains the construction of devices that operate by compressed air, steam, and water; for the Greek text, with German translation, see Schmidt (1899). For the importance of wonder in his works, see Lightfoot (2021, 208–214).

[52] A fragment of Geminus, filtered through Proclus, cites two preeminent practitioners in mechanical θαυματοποιϊκή: Ctesibius and Heron (Aujac 1975, 116). Heron uses similar language in the preface to his one-book *On Automata-making* (Περὶ αὐτοματοποιητικῆς) (Schmidt 1899,

A comparison with the water organ makes Aetna just the type of breath-taking contraption to figure in Heron's mechanical notebooks.

As a musical Triton or sonorous *cortina*, Aetna is both spectacular and familiar, both infinitely powerful and easily controlled by the skilled practitioner. This is the point when the poet draws attention to the *ars* of the musician who operates the water organ (*cortina... canit arte regentis*, 296–297); he is an image for nature as *artifex*. The mountain, then, constitutes a marvelous wonder that operates according to readily accessible mechanical principles. Through these similes, Aetna dramatizes the mechanical sublime.[53]

While describing the constrained and precipitous movements of the wind that rushes toward the mountain's crater and thereby unleashes an eruption, the poet employs another suggestive simile (326–328):

> et, quacumque iter est, properat [sc. spiritus] transitque moramen,
> donec confluuio, ueluti siponibus actus,
> exilit atque furens tota uomit igneus Aetna.

And wherever the path lies, [the air] hastens and passes through [all] impediments, until, at the meeting-place [of other currents of air], as if compelled by pumps, it leaps forth and vomits furiously and with fire all over Aetna.

The point of the comparison is that erupting *spiritus* climbs to such heights and moves with such power that it resembles jets of water forced out through pumps. The pumps in question are those used for putting out fires, which could send water to great heights.[54] Much like the Triton contraption and the water organ, these devices constituted wondrous devices. Once again, Heron describes them at length among his mechanical spectacles (*Pneum.* pp. 130–136 Schmidt). At the conclusion of Vitruvius's account of the *sipo*, the architect characterizes such pumps as just one of the many

338–342). For automata and the Alexandrian interest in *paradoxa*, see Schepens and Delcroix (1996, 407). For Ctesibius, whom some ancient authorities cite as the inventor of the water organ (Plin. *HN* 7.125, Athen. *Deip.* 4.174b–f), see below. Medieval automata were similarly "wondrous"; see Truitt (2015, 41–42).

[53] Williams (2017b, 65; 2020, 115) remarks, in passing, on the mountain's transformation into "a mechanistic marvel." For examples of musical technology and the sublime in other cultural contexts, see Richards (1999, on "Mozart and the Mechanical Sublime") and O'Gorman (2021, on "The Pipe Organ and Immortality").

[54] See Ellis (2008, 151). The *sipo* regularly figures in analogies in technical writers; Vottero (1989, 312), on Sen. *NQ* 2.16, cites numerous examples. Cf. Plin. *Ep.* 5.6.36–37 for an account of impressive miniature pumps (*sipunculi*).

remarkable inventions of Ctesibius, the founder of pneumatics, that charm and delight the senses (*delectationibus oculorum et aurium usu sensus eblandiantur*, 10.7.4).[55] Once again, a striking simile turns Aetna into a stunning piece of technical craftsmanship that utilizes mechanical expertise. From an ancient perspective, the passage stages the mechanical sublime.[56]

In the Triton simile, the author notes that sound is produced when water and air strike the "machinery" (*opus*, 294). Numerous passages throughout the poem associate Aetna with the noun *opus*, and even though it is tempting to translate *opus* abstractly as the workings or activities of the mountain, many of these passages suggest a more concrete meaning and so characterize Aetna as an actual piece of workmanship that has been carefully or methodically built—it is a machine, a device, a contraption.[57] In the proem, the poet declares that he will sing about "the movements that so great an *opus* has" (*qui tanto motus operi*, 25). Later, while noting that the volcano's flames make it difficult to approach the mountain, he states that "that *opus* has fire as its guard" (*custodiaque ignis / illi operi est*, 193–194). In his proclamation of the value of terrestrial study, the poet does not wish "to pass over so great an *opus* at our feet" (*tantum opus ante pedes transire*, 256). More explicit is the following passage (334, 336, 335):[58]

> illinc obscura semper caligine nubes
> prospectat sublimis opus uastosque recessus
> pigraque defixo circum stupet undique uultu.

From there [= the summit], a lofty cloud, dark and dense, always looks out on the *opus* and its vast recesses and, inert, gazes all around in stupefaction, with its visage transfixed.

The elevated (*sublimis*) gaze and immobilized stupefaction of the personified cloud encode the sublimity of the *opus*. Moreover, the combination of

[55] Ctesibius was an inventor who worked in Alexandria and was employed by Ptolemy II (fl. 270 BCE). For his status as a practitioner of θαυματοποιϊκή, see above.

[56] Ironically, the poet likens the explosion of *spiritus*, which ignites the fiery lava flow, to powerful jets of water that extinguish fires. In both Greek and Latin, however, volcanic lava is said to "flow" and to behave like water; cf. Pind. *Pyth.* 1.21–23, [Aesch.] *PV* 367–368, and Buxton (2016).

[57] Cf. Wolff (2004, 83), Welsh (2014, 119–120), and Williams (2017b, 63; 2020, 114–117).

[58] For the transposition of lines, see Goodyear (1965, 164). In his note to 335, where the manuscripts read *def(f)uso/decusso*, Goodyear observes that the emendation *defixo* "goes well with *stupet*," although he prints *diffuso*.

stupet and *defixus* in 335 echoes Aeneas's experience of the images on Dido's temple in the first book of the *Aeneid* (*stupet obtutuque haeret defixus in uno*, 1.495), a passage that itself encodes the sublime. The reminiscence, in other words, suggestively likens the mountain to a work of art that induces an experience of sublimity.[59] Even when it is not in a state of active eruption, the mountain's machinery activates the mechanical sublime, which, in this case, is superimposed over a more familiar mode of sublimity.

In one instance, the poet rejects an analogy that does not elevate the mountain in this way. After explaining the nature and activity of the *lapis molaris*, the poet complains that certain readers may still cling to a *fabula mendax* (511) according to which some other substance or combination of substances are liquefied by the fire and so make up the lava stream (515–519):

> nam posse exusto cretam quoque robore fundi,
> et figulos huic esse fidem, dein frigoris usu
> duritiem reuocare suam et constringere uenas.
> sed signum commune leue est atque irrita causa
> quae trepidat: certo uerum tibi pignore constat.

For clay, too, can be melted after its substance has been burned up—potters are the proof of this—and then, through contact with the cold, it summons back its hardness and contracts its veins. But [this] common piece of evidence is insubstantial and [this] explanation is invalid and uncertain: the truth is established for you by a sure token.

The basic premise of the comparison is that clay is a compound substance such that one part of it can be burned up while the other persists; perhaps the *lapis molaris* is similarly constituted.[60] The poet rejects a line of argument that likens the workings of the mountain to something as unimpressive and ordinary as the production of ceramics. He emphatically calls the comparison a *signum commune*, which F.R.D. Goodyear glosses as an "analogy from everyday life," and promises to produce more compelling evidence.[61] The *lapis molaris*, by contrast, is extraordinary in the highest

[59] Note that Aeneas experiences the scenes depicted on the temple as *miranda* (*Aen.* 1.494), too.
[60] Ellis (2008, 200) paraphrases the comparison thus: "just as when potter's clay is submitted to the action of a furnace, it has an inner substance which fuses, distinct from the rest."
[61] Goodyear (1965, 193). Ellis (2008, 201), following Sudhaus (1898, 195) and later expanded upon by De Lacy (1943, 175–177), suggests that a *signum commune* is the Latin equivalent of a κοινὸν σημεῖον, "a sign which has more than one application."

degree. While other flammable objects perish (*moritur*, 419) and are reduced to ash, the *lapis molaris* burns again and again, repeatedly renewing itself (421–424). Its longevity defies logic, making it a kind of phoenix or perhaps a perpetual motion machine, a wondrous power that is full of energy and life (*miranda est lapidis uiuax animosaque uirtus*, 417). The *lapis molaris* grounds the spectacular nature of the mechanical mountain, ensuring that it does not run the risk of becoming a commonplace phenomenon familiar from the operations of everyday artisans; it is evidence of the work of *natura artifex*.

Imperial spectacles regularly presented audiences with stunning examples of technological creativity. As Kathleen Coleman (1990, 52) writes:

> With the construction of amphitheatres came the installation of machinery that increased the scope and ingenuity of the displays that could be mounted. Calpurnius Siculus [*Ecl.* 7.69–72] marvels at an arena (probably Nero's wooden amphitheatre constructed in A.D. 57) in which a "forest" rose into view out of the *hypogeum*. With technological skill came realistic effects: for the pantomime of the Judgement of Paris that occupied part of a composite entertainment at Corinth, Apuleius [*Met.* 10.30.1–34] describes a wooden mountain erected through the floor to simulate Mt. Ida, complete with real plants and incorporating plumbing to facilitate a stream flowing from the top; live goats added verisimilitude. At the end of the performance it sank out of sight.

Coleman also cites Strabo's eye-witness account of an execution during which a Sicilian rabble rouser associated with the area around Aetna was placed on top of a lofty—or, quite simply, sublime—contraption built to resemble the mountain (ἐπὶ πήγματος γάρ τινος ὑψηλοῦ τεθεὶς ὡς ἂν ἐπὶ τῆς Αἴτνης). At the appropriate time, the contraption collapsed, and both it and the prisoner fell into the fragile cages full of wild beasts that had been strategically placed below.[62] Readers of *Aetna* will have been familiar with a variety of technically complex devices designed for just such shows, and our poet's conception of the mountain will have seemed quite plausible.[63]

[62] Coleman (1990, 53) on Strab. 6.2.6 (273). Seneca counts the inventors of *pēgmata* as practitioners of the *ludicrae artes*, those that attempt to please the eyes and ears (*Ep.* 88.22 = Posidonius fr. 90 Edelstein-Kidd), not unlike the inventions described at length by Heron.

[63] Connors (2015, 133–134) notes that Cassius Dio describes the Flavian Amphitheater and the crater of Vesuvius as a hunting theater (*kunēgetikon theatron*); vulcanism and imperial spectacle go hand in hand.

There is, however, another side to this astounding *opus* (1–4):

> Aetna mihi ruptique cauis fornacibus ignes
> et quae tam fortes uoluant incendia causae,
> quid fremat imperium, quid raucos torqueat aestus,
> carmen erit.

Aetna, the fires that burst from its hollow furnaces, the causes that are so strong and that roll forth its conflagrations, why it grumbles at being ruled, why it shoots forth its harsh-voiced blazes—this will be my song.

The work's opening sentence equates the mountain, in all its volcanic grandeur, with the poem itself. To be sure, *Aetna mihi... carmen erit* means "the subject of my poem will be Aetna." It also means "Aetna will be my poem" or, conversely, "my poem will be Aetna." On this reading, the poem itself is the volcanic machine—its sounds are those of the "sonorous" Triton (*Tritone canoro*, 293), its flowing verses are the "wet song" of the water organ (*carmineque irriguo*, 296). At the same time, the poet is the skilled worker who runs the Triton, the musician through whose "art" the water organ "sings" (*canit arte regentis*, 297), the technician who operates the *sipo*, setting the machine in motion through the lines of his poem. That ancient poems were regularly referred to by their first word or words only reinforces the point; we call the composition "Aetna," and this is what it is.[64]

The author of the *Aetna* is not the first to suggest a connection between the mountain and sublime poetry. Lucretius, for one, makes Aetna's fires and Empedocles's incendiary verses mirror one another,[65] and Lucan compares the process of divine inspiration to an eruption of Aetna (see Ch. 4.3 on Luc. 5.97–101). In Seneca's *Epistulae*, Aetna is a symbol of the dynamics of poetic imitation and emulation, "a topic common to all poets" (*sollemnem omnibus poetis locum*, 79.5).[66] The very idea of writing about Aetna physically excites Seneca's correspondent Lucilius: "Either I don't know you or Aetna makes your mouth water (*Aetna tibi saliuam mouet*); now you are eager to write something magnificent (*grande aliquid*) and equal to

[64] For complementary readings, see Welsh (2014, 119–120, 128–129), Kruschwitz (2015, 93–94), Payne (2016, 100), and Williams (2017b, 63–65; 2020, 114–117). De Vivo (1985, 1992a) treats the proem in detail.

[65] See Hardie (1986, 211–212; 2009, 88–90, 216–217) and Williams (2017b, 38–45).

[66] For Seneca's Aetna, see Williams (2017b, 45–49); cf. Delarue (2000, 28–29) on this passage and Gowers (2011, 187) on "sublime Mount Etna" at *Ep.* 51.1.

[the works of] previous generations" (79.7). The imagined arousal of Lucilius's appetite(s) figures the affective intensity of the sublime, and the chance to match oneself with the great writers of the past constitutes a kind of transcendence.[67]

Our poet, however, is up to something quite different. By mapping poem and mountain onto one another, he establishes a new paradigm for the study of nature—Aetna as a locus of volcanic energy, an instrument of *natura artifex*, a concrete example of imperial largess. Because both poem and poet are rooted in the earth, Phaethon does not provide a useful intellectual model, and the fall is not a particularly relevant construct—Phaethon, moreover, may have become tainted by his associations with Nero following the great fire in 64 and the emperor's eventual demise (see Ch. 4.2, "Nero-Phaethon"). Despite its relatively restricted ambit, though, the poem shows the same multiplicity of perspectives and voices, the same exuberant disregard for the demands of logical consistency and tidiness, that the Phaethon paradigm has come to represent. In the present case, this is an attitude that Mark Payne (2016, 97) describes as the poem's "noncommittal *poikilia.*" The mountain's fractured landscape, unpredictable explosions, and untamable excess figure the modus operandi of an author who defines the study of nature as an examination of causes and a celebration of wonders, who denounces mythological aitia and repeatedly suggests that the his subject is a living monster, who imagines a cowering Jupiter and extols a personified nature, whose theoretically grounded exposition of the earth will terminate in an intellectual landscape where the premises of such arguments are null and void. As the mountain erupts, over and over, the poet plays the seemingly incompatible roles of nature the demiurge and the monster of myth, the water organist or pump operator whose wet instruments flow with unquenchable fire, the singer whose song is inarticulate noise. The dynamic tension in these oppositions dramatize the vitality of the sublime, which bursts into view through the twofold spectacle that is both mountain and poem. Like the mountain, the poem is fractured and multiple, and deep intellectual chasms make it difficult to traverse—it is unapologetically *chaos ac sine fine ruina* (139), defiantly *discordia ingens* (183). This is the poem's volcanic aesthetics.[68]

[67] At *Dial.* 9.17.10–11, in the context of poetic madness, Seneca deploys the phrases *grande aliquid, aliquid...grandius ore mortali*, and *sublime quicquam*, which all point to the sublime. For discussion, see Schiesaro (2003, 22–23, 52–53), Littlewood (2004, 121), and Staley (2010, 41–42).

[68] For a partial parallel to this concept, see Billault (2004) on "volcanisme et esthétique" in Long. *Subl.* 35.4.

The poem's discontinuities are not an accident of transmission.[69] On the contrary, the work's volcanic aesthetics and poetics of novelty are mutually implicated in one another. After invoking Apollo and the Muses, the poet states that it is safer to make a journey over unfamiliar terrain if the god leads the way (*per insolitum Phoebo duce tutius itur*, 8). The act of writing and implicitly of reading the poem is a dangerous journey across a new landscape. That landscape is the mountain itself, and its combustible instability contrasts starkly with the predictable, and predictably safe, mythological plotlines that the poet immediately denounces (9–23). The poem's novelty, however, consists not just in its choice of topic. Having rejected the stories of myth, the author equates the mountain's movements, sounds, flames, and fires with the "mind" of his poem (*mens carminis haec est*, 28). The poem's very arguments—its foundational principles, analogies, and conclusions—are going to be raucous, explosive, and impossible to pin down. We should locate its novelty in both its subject matter and its discursive *poikilia*. Put differently, the text's intellectual multiplicity and fragmentation vis-à-vis the mountain contribute fundamentally to its poetic newness, and they stage a wild show.

5.5. From Causal Analysis to Paradoxography

After a succinct recap of the mountain's mechanics (565–567), the poet changes direction. From lines 568–598, he lambasts the zeal with which his contemporaries travel throughout the Mediterranean world to see famous cities and works of art. In contrast to such spectacles (*spectacula*, 601), the poet exhorts the reader to behold "the massive work of nature the demiurge" (*artificis naturae ingens opus aspice*, 600), which leads into "the wondrous story" (*miranda...fabula*, 603) of two brothers who rescued their parents from an eruption when Aetna's fiery rivers parted in deference to the siblings' *pietas* and *uirtus* (605–645); this will be the ultimate show. Although the author has previously alluded to the operations of *natura artifex*, the climactic narrative comes as a shock—in a world that operates according to mechanical principles, Aetna is suddenly endowed with moral agency. Some readers insist that the contradiction is only apparent and/or

[69] Similarly Wolff (2004, 84), Buxton (2016, 34–35), Payne (2016, 96–102), and Porter (2016, 511–512, 516–517).

attempt to integrate this new data point into an ostensibly coherent worldview that remains grounded in the analysis of material causes. I side with those who see the intellectual disjuncture as unbridgeable.[70] There is, however, another layer to consider—namely, that the story of the two brothers itself encodes the sublime and operates according to its dual logic. The sublimity of the *pii fratres* narrative is rooted in the values and concerns of the paradoxographical tradition, which catalogs nature's marvels without attempting to explain their causes. As the poem draws to a close, the author pivots from one mode of sublimity to another. In so doing, he blocks access to the divine power that oversees the world and thereby enhances the spectacle that is both the mountain and the poem (cf. Ch. 5.3 on 192–198). The sudden intellectual shift from causal analysis to the promotion of inexplicable wonder should be read from the perspective of the poem's volcanic aesthetics; it is the final burst of imperial fireworks. Before we can consider the details, we need to examine Aetna's longstanding connections to the paradoxographical tradition and the contemporary resonances of *natura artifex*.

Aetna and the Paradoxographical Tradition

The Greek adjective *paradoxos* means "contrary to expectation, extraordinary, marvelous." While keen interest in the marvelous is evident already in the Homeric poems and spreads throughout the classical period, it is not until the third century that we find a systematic attempt to catalog marvelous phenomena that are reliably attested and hence supposedly factual and/or historical; this is the beginning of the paradoxographical tradition.[71]

The first person to produce such a text may have been Callimachus, whose wonder-book was evidently a methodical and well sourced work of scholarship.[72] Similar works began to appear at the same time and in the

[70] Note Payne (2016, 96–97). Others who are comfortable with the discontinuity include Toohey (1996, 190–193), Wolff (2004, 84), Pingoud (2008, 211), Buxton (2016, 34–35), and Porter (2016, 511–512). Contrast the approaches of, e.g., Effe, (1977, 210–213), Goodyear (1984, 357), De Vivo (1989, 78 n.51), Volk (2005a, 82–83, 88–90), Taub (2008, 54–55; 2009, 135–137), Santelia (2012, 44–45), Welsh (2014, 112–118), and Williams (2017b, 61–62l; 2020, 118–119).

[71] For the paradoxographical tradition, see Ziegler (1949), Schepens and Delcroix (1996), Geus and King (2018), and Lightfoot (2021, 42–88). Throughout Ch. 5.5, I cite the titles and fragments of the paradoxographers from Giannini (1966).

[72] The title is variously recorded; one version in the Suda is *Collection of Wonders that Occur over the Whole Earth, Arranged by Location* (Θαυμάτων τῶν εἰς ἅπασαν τὴν γῆν κατὰ τόπους ὄντων συναγωγή). In Pfeiffer's edition, the fragments take up over nine pages (frr. 407–411), most of which are preserved by Antigonus (Giannini 1966, 88–106).

following centuries, with titles such as *On Wondrous Things* (Περὶ θαυμασίων), *Investigations into Wondrous Things* (Ἱστορίαι θαυμάσιαι), and *Collection of Wonders* (Θαυμασίων συναγωγή). Although what constitutes a *paradoxon* is subjective, the paradoxographers have predictable interests, and the compilatory nature of the tradition means that the same subjects often appear in multiple authors. The natural world provides most of these subjects, which tend to involve "extraordinary" animals, plants, rivers, springs, and metals, but ethnographic and even historical matters attract attention, too.

According to Diogenes Laertius, the Stoics insisted that the sage not wonder at apparent marvels such as supposed entrances to the underworld, tides, hot springs, and eruptions of fire (7.123, *SVF* 3.642). These quintessential *paradoxa* are all subjects that appear in philosophical analyses of meteorological phenomena.[73] What, then, is the difference between the two traditions? One crucial distinction is that paradoxographers regularly and deliberately describe wondrous phenomena without analyzing their causes. The goal is to inspire a sense of the marvelous, even to activate the sublime, by focusing squarely on the strangeness or extraordinariness of the sublunary world while setting philosophical speculation to the side. In chapter 60 of the *Collection of Investigations into the Marvelous* (Ἱστοριῶν παραδόξων συναγωγή), which has traditionally been ascribed to Antigonus of Carystus (fl. c. 240 BCE),[74] the author makes this clear: whereas Aristotle, in the *History of Animals*, engages in causal analysis or exegesis, the paradoxographer has preferred simply to excerpt from Aristotle and others "that which is strange and extraordinary" (τὸ ξένον καὶ παράδοξον). In an essential study of the paradoxographical tradition, Guido Schepens argues that from Antigonus's perspective philosophical analysis interferes with our ability to experience wonder, and the generation of wonder is the fundamental aim of the paradoxographer: "The deliberate omission of any attempt at rational 'exegesis' must... be regarded as an intrinsic part of [Antigonus'] endeavor to inspire his readers with a sense of the marvelous."[75] Antigonus, as it were, recognizes the Aristotelian position (see Ch. 1.6, "Plato and Aristotle") that

[73] For the extensive overlap between the paradoxographical tradition and ancient meteorology, see Bakker (2016, 113–125).

[74] For the debate about the author's identity, see Dorandi (1999, xi–xxxii). Lightfoot (2021, 42 n.1) concludes that "the dating of at least the core of the collection to the third century BCE seems secure."

[75] Schepens and Delcroix (1996, 392). Lightfoot (2021, 72–78) is a more recent examination of the passage.

the knowledge of causes eradicates wonder, and he exploits it brilliantly. If ignorance of causes heightens the affective intensity of a confrontation with wondrous nature, it makes perfect sense that when Antigonus mines his sources for wondrous facts, he edits out the causal analysis. A paradoxographer like Antigonus, then, narrates, but does not analyze, nature's marvels.

With this in mind, we can turn to Sicily and Mount Aetna. The paradoxographers develop an interest in Sicily at an early date. In the third century, Nymphodorus of Syracuse compiled a wonder-book devoted to Sicily's extraordinary features.[76] Subsequent authors, including Antigonus, are similarly drawn to the island.[77] By the Roman period, Sicily had become a locus of wonder. Lucretius, for one, observes that many people consider the whole region great and wondrous and well worth seeing (1.726–727), and Seneca acknowledges the many marvels on and around the island (*NQ* 4a.1.1). Such passages take familiarity with the paradoxographical tradition for granted.

Of Sicily's many *mirabilia*, Aetna stands out.[78] Among the fragments of the paradoxographers themselves, pseudo-Aristotle, the author of *On Wondrous Things Heard* (Περὶ θαυμασίων ἀκουσμάτων), is particularly taken with the mountain's lava flow (ῥύαξ).[79] The text highlights the intermittent nature of the flow (38), its astounding breadth and the heights to which it reaches (40), and the fact that it often spreads over the entire island (105.3). Aetna's fiery stream is a recurring source of wonder, and pseudo-Aristotle scrupulously refuses to explain its causes even though those causes will have received exacting analysis at the hands of Aristotle's pupil Theophrastus, author of a lost treatise *On the Lava Flow in Sicily* (Περὶ ῥύακος τοῦ ἐν Σικελίᾳ, Diog. Laert. 5.49).[80]

The most striking example of Aetna's status as a *paradoxon* comes from *Subl.* 35.2–5, a passage the well demonstrates Longinus's own will to

[76] *On the wonders of Sicily* (Περὶ τῶν ἐν Σικελίᾳ θαυμαζομένων) may have formed part of a larger work titled *Periploi*; testimonia and fragments at Giannini (1966, 112–115).

[77] Philostephanus 8 (p. 23); Antigonus 8, 139–140 (pp. 34, 92); [Arist.] 112, 148 (pp. 278/280, 302); cf. Apollonius 2 (pp. 120/122), Paradox. Vat. 39 (p. 342), [Arist.] 130 (p. 228).

[78] The earliest witness to Aetna's marvelous status is Pind. *Pyth.* 1.26. For the imperial period, see Strab. 6.2.3 (269), with Posidonius fr. 234 Edelstein-Kidd, and 6.2.8 (274); Sil. *Pun.* 14.66; Plin. *HN* 2.236, 3.88.

[79] Third-century Peripatetic material constitutes the compendium's core, which was subsequently expanded and reworked; see Flashar (1972, 39–50).

[80] Some speculate that the *Aetna* poet may have drawn on this treatise; see Goodyear (1984, 355), Garani (2009, 120), and Verde (2020, 92–99).

sublimity.[81] Longinus begins by declaring that nature has brought mankind to a magnificent festival and produced in us a passion for greatness that drives the mind beyond the limits of the cosmos, and he asserts that the purpose of our existence is the search for the exceptional, the great, and the beautiful. At this point, Longinus adopts the role of the paradoxographer. To illustrate our vocation, he lists extraordinary natural phenomena that inspire more wonder than their everyday counterparts—the Nile, Ister, Rhine, and Ocean, the stars, and "the craters of Aetna . . . whose eruptions bring up rocks and whole masses of earth from the deep and sometimes pour forth rivers of that earthborn, self-generated fire" (*Subl.* 35.4). The description of Aetna draws on Pindar (*Pyth.* 1) and *Prometheus Bound*, where the monstrous Typhos powers the mountain.[82] In Longinus's treatise, however, Typhos does not appear and the Olympians are nowhere to be found.[83] Similarly, despite the celebration of human intellectual ambition and potency, Longinus does not hint at the physics of the mountain; there are no subterranean winds, only rivers of fire. The conclusion follows swiftly: "But in all such matters, we might invoke the familiar thought that what is useful or necessary is readily available to humans, while it is always the extraordinary that arouses wonder (ἀλλ᾽ ἐπὶ τῶν τοιούτων ἁπάντων ἐκεῖν᾽ ἂν εἴποιμεν, ὡς εὐπόριστον μὲν ἀνθρώποις τὸ χρειῶδες ἢ καὶ ἀναγκαῖον, θαυμαστὸν δ᾽ ὅμως ἀεὶ τὸ παράδοξον," *Subl.* 35.5, trans. based on Halliwell 2022). The basic proposition that *paradoxa* inspire wonder sums up the entire rationale of the paradoxographical tradition, and Longinus's most compelling argument in favor of its validity is the very idea of Aetna in a state of eruption. The mountain is the quintessentially sublime *paradoxon*.[84]

Aetna's lava flow has another claim to the extraordinary—the story of the pious brothers of Catania. The exploit was a widely known exemplum that demonstrated the value of piety; it is attested in the fourth century, appears in multiple texts of the early imperial period, and was even featured on

[81] The passage draws on a mixture of Stoic, Epicurean, and popular ideas and images; see the commentaries of Russell (1964) and Halliwell (2022). My point here does not depend on assuming that Longinus works from a lost paradoxographical source; for this possibility, however, see Porter (2007, 172–174; 2016, 450–454). For further discussion of *Subl.* 35.2–4, see Ch. 1.6 ("Longinus and the Rhetorical Tradition").

[82] In addition to *Pyth.* 1.22–23 and *PV* 367–368, Russell (1964, 167) cites Heraclit. *Quaestiones Homericae* 26.

[83] Well emphasized by Billault (2004, 203). The adjective "earthborn" (γηγενής) is the only trace of a monster like Typhos, but the epithet has been demythologized.

[84] Cf. Innes (1979, 168), who points out that Aetna fits one of Hermogenes's topics suitable for the production sublimity, "movements of the earth and sea" (*De ideis* 1.6, pp. 243–244 Rabe). For context, see Ch. 1.6 ("Longinus and the Rhetorical Tradition").

coins.[85] Among the paradoxographers, the anecdote appears in pseudo-Aristotle (154):

> When Aetna's craters had been broken open and [the lava flow] was being carried hither and thither over the earth like a torrent, the divine power (τὸ δαιμόνιον) honored the race of the pious. When they been surrounded by the stream because they were carrying their aged parents on their shoulders and were trying to save them, the stream of fire approached them, was split, diverted part of the blaze this way and part of it that way, and kept the young men safe and sound along with their parents.

The appeal to a divine power adds a new dimension to pseudo-Aristotle's well-documented interest in Aetna's lava flows, and it's a dimension that makes the mountain appear altogether less comprehensible. Who, after all, is τὸ δαιμόνιον? And what is its connection with the mountain?[86] Another treatise attributed to Aristotle, *De mundo*, contains the same narrative in almost the same language (6.400a–b). There, however, the action of τὸ δαιμόνιον makes sense, since it names the quasi-Platonic, quasi-Aristotelian, quasi-Stoic divinity who steers the cosmos and whom the author hymns at length. If the paradoxographer took his passage from *De mundo*, he has turned the original significance of the divine power upside down, transforming a symbol of cosmic order and rationality into an unexpected and inexplicable mystery.[87]

The paradoxographical tradition offers a way to write about nature's marvels that simultaneously activates the sublime and refuses to engage in causal analysis. This makes paradoxography a striking counterpoint specifically to those authors who locate the thrill of the sublime and its concomitant expansion of mind precisely in the apprehension of nature's causes. We have already observed, apropos of the Stoics' list of representative marvels (Diog. Laert. 7.123, *SVF* 3.642), the considerable overlap between the philosophical study of nature and the topics that interest the paradoxographers. This overlap figures prominently in the Latin philosophical tradition itself.

[85] Fourth century: Lycurg. *Leoc.* 95–96. Early imperial texts: Strab. 6.2.3 (629), Conon fr. 1.43 *FGrH*, Val. Max. 5.4 ext. 4, Sen. *Ben.* 3.37, 6.36, Sil. *Pun.* 14.196–197, Mart. 7.24; cf. Hyg. *Fab.* 254.4. For discussion, including of the coins, see Taub (2009, 137–141), Giampiccolo (2010), and Santelia (2012, 24–43).

[86] Contrast *Mir. ausc.* 152, where the connection between a marvelous spring, oath breakers, and Zeus *horkios* is conceptually, if not mechanistically, comprehensible.

[87] *De mundo* is likely Hellenistic in date; see Porter (2016, 474 with n.267).

As Sarah Myers observes, in the Roman period natural philosophers used paradoxographical collections "to provide examples of the ability of the laws of natural philosophy to explain even the most remarkable in nature."[88] Lucretius and Seneca, for instance, devote considerable energy to explaining phenomena that appear or might have appeared in paradoxographical treatises and lists. If paradoxographical writings privilege the narration of wonders over causal analysis, Lucretius's explanation of the poisonous exhalations around Lake Avernus or Seneca's account of the relationship between Arethusa and the Alpheus reassert the primacy of analytical thinking.[89] The same point applies to the *Aetna* poet himself. Because Aetna's explosivity is deeply embedded in the paradoxographical tradition, the poet's elucidation of its causes strikes back against the paradoxographical notices of an author like pseudo-Aristotle. In this, the *Aetna* poet's analyses perform the same function as Lucretius's own explanation of Aetna's volcanism (6.639–702). However, given the mountain's status as a recognizable *paradoxon*, when our author pivots to the story of the *pii fratres*, he conspicuously adopts the role of the paradoxographer and engages with the sublime from the perspective not of the student of nature, but of the compiler of a defiantly nonanalytical wonder-book.

Natura artifex

The prominent placement of *natura artifex* in the buildup to the *pii fratres* narrative signals this change in perspective. Besides the poets of *Aetna* and *Octauia* (see Ch. 5.3), the only first-century authors to use the expression are Valerius Maximus and the Elder Pliny, neither of whom has any vested interest in the Stoa.[90] In both Valerius and Pliny, *natura artifex* points to the incomprehensibility of nature's grandeur and the baffling obscurity of her plans. It is a concept to be deployed when the simple narration of wonders

[88] Myers (1994, 148). Myers's basic point does not apply only to those who write in Latin; see, e.g., Meeusen (2016, 237–257) on Plutarch's philosophical interest in *mirabilia*.

[89] Lake Avernus: Lucr. 6.738–739 vs. Antigonus 152b (p. 98), [Arist.] 102 (p. 270). Arethusa and the Alpheus: Sen. *NQ* 3.26.5–6 vs. Antigonus 140 (p. 92). For *mirabilia* and the *NQ*, see Garani (2020b).

[90] The *HN* is in certain ways "inspired by Stoicism" (Paparazzo 2011, 103), and Pliny might be characterized as "sympathetic to Stoicism" (Fear 2011, 26), but he is fundamentally "an eclectic and non-specialist author" (Beagon 1992, 26), not a committed or self-identified Stoic, and is perfectly capable of adopting rather un-Stoic positions (e.g., the soul is destroyed at death, 7.188).

trumps causal analysis, and it can be closely connected with the paradoxographical tradition.

Valerius writes of nature the demiurge near the end of his chapter "On wonders" (*De miraculis*, 1.8). Many of the stories involve the apparition or intervention of a divine figure in human affairs, which tends to fall outside the parameters of the paradoxographers (Schepens and Delcroix 1996, 382). At 1.8 ext. 12–17, Valerius turns to "nature's playthings" (*ludibria naturae*), by which he means primarily instances of extraordinary human physiology. These anecdotes lead into the following declaration: "Why she played these particular games with the children of very powerful kings or a most famous leader or a poet of distinguished talent or very learned men or a man of no note, not even nature herself could explain, the fertile demiurge of all matter good and bad" (*quapropter haec potissimum aut in liberis potentissimorum regum aut in duce clarissimo aut in uate ingenii florentis aut in uiris eruditissimis aut in homine sortis ignotae luserit, ne ipsa quidem, omnis bonae malaeque materiae fecunda artifex, rationem rerum natura reddiderit*, 1.8 ext. 18). A *ratio* exists that explains the logic of nature's wonders, but it is entirely impenetrable. Valerius expresses this idea through a striking paradox: not even the *artifex* herself can render an account of the rules of her creation. The expression *natura artifex* names the divine creator who does not simply scorn but is incapable of causal analysis, so vast and unwieldy is her fecundity.

At this point, Valerius lists several other phenomena that demonstrate nature's opacity: the unusual behavior of Cretan goats, the unusual behavior of Cephallanian goats, a strange temple of Juno, and waters that have an intoxicating effect. The two sets of goats and the intoxicating waters all appear in more than one collection of wonders,[91] and the temple of Juno is mentioned by Pliny in a list of unusual fires (2.240). Valerius is working from a paradoxographical source. Ultimately, he advises us to record such items, rather than wonder at them, "since we know that the highest degree of freedom is rightly claimed by the one in whose hands rests the infinite work of creating all things" (*cum sciamus recte ab ea plurimum licentiae uindicari penes quam infinitus cuncta gignendi labor constitit*, 1.8 ext. 18). An account of the unfettered freedom with which nature the demiurge

[91] Cretan goats: Antigonus 30 (p. 48), [Arist.] 4 (p. 224); cf. Plin. 8.97. Cephallanian goats: Alexander 2 (p. 165), [Arist.] 9 (p. 226). Intoxicating water: Antigonus 164 (p. 102), Paradox. Flor. 20 (p. 320); cf. Ov. *Met.* 15.329–331, Sen. *NQ* 3.20.5–6, Plin. *HN* 2.230, with Garani (2020b, 207–208).

carries out a task without end, implies Valerius, can only be a record of facts without explanation. In the following chapter, which is the last of the book, Valerius refers to these phenomena as "things that exceed ordinary reason" (*supra usitatam rationem excedentia*, 1.8 ext. 19). The contemplation of *natura artifex* takes the reader beyond the boundaries of everyday life and thought, where *ratio* itself falters.

For Pliny, knowledge tends to consist "more of a collection of extraordinary things than an accumulation of scientifically proved or studied evidence" (Naas 2011, 59). Indeed, he can write that his task is simply to point out the wonders of nature, not to investigate uncertain causes (*nobis propositum est naturas rerum manifestas indicare, non causas indagare dubias*, 11.8).[92] Given this basic outlook, it is not surprising that Pliny occasionally appeals to *natura artifex*; we will consider one prominent example, and we will return to Pliny in the Epilogue to explore further what his encyclopedia can tell us about the history of the scientific sublime in the early Flavian period.

In the opening chapters of book 2, Pliny grapples with the spatial and temporal dimensions of the *mundus*. It is infinite, eternal, and boundless. To study its dimensions or to attempt to investigate that which lies beyond it is sheer madness. Similarly mad is to posit the existence of countless universes, whether one categorizes them under countless systems of nature or just one, in which case it will still be necessary to posit countless suns and moons and stars, which are already measureless and innumerable in the one universe that we inhabit. Why should anyone speculate about an infinity of universes if our own universe is already infinite? "It's as if the same issues were not always going to arise at the end of the thought experiment due to our need for some kind of limit, or as if, should this limitlessness be able to be assigned to nature the demiurge of all things, the same thing were not more easily understood in one work, especially one so great" (*quasi non eaedem quaestiones semper in termino cogitationi sint occursurae desiderio finis alicuius aut, si haec infinitas naturae omnium artifici possit adsignari, non idem illud in uno facilius sit intellegi, tanto praesertim opere*, 2.3). As Pliny struggles to articulate the limits of the limitless, he reaches for *natura artifex*. Although her defining quality is boundlessness, Pliny must acknowledge her boundaries, and while it may be "easier" to understand her infinity within the confines of our one *mundus*, it is not "easy." The idea *natura artifex* points to an irreconcilable tension at the very heart of the cosmos;

[92] See Beagon (2011, 84–85) on Pliny's inconsistent stance on causes.

to borrow Valerius's expression, she is a byword for "things that exceed ordinary reason."

As Valerius and Pliny demonstrate, the concept *natura artifex* is bound up with the general ethos and sometimes the actual compilations of the paradoxographers. In the first century CE, it tends to signal an approach to nature that elevates the narration of wonders over causal analysis and that locates the sublime in the opacity of and a principled indifference to nature's causes. Although other names and epithets can express the teleology of nature or personify the god who has constructed and is immanent in the *mundus*, they generally do not activate the same confounding tension.[93] Manilius, for instance, writes of "nature, the beginning and guardian of things that lie hidden" (*principium rerum et custos natura latentum*, 3.47). But rather than emphasize nature's obscurity, Manilius describes the mechanical process through which she created the world, bringing harmony and order to its parts, and declares that she made human life dependent on the stars so that nothing would be excluded from the *ratio* of the whole (3.48–60). Without discussing causes, Manilius reassures the reader that an explanation exists for everything; *natura* is sublime but not inscrutable. Nor does Seneca's comparison of the study of nature to mystic initiation (see Ch. 3.4 on *NQ* 7.30.6) set up the same kind of baffling condundrum as *natura artifex*. The point in Seneca's comparison is that progress is measured in generational time; the present age has only received preliminary instruction. The contradiction that is *natura artifex*, by contrast, exists beyond the tools of ordinary reason, and its identity and characteristics remain fundamentally unknowable—what does it look like? how does it work? how is it related to the Olympian gods? Such questions are beside the point, for nature the demiurge exceeds our capacities, axiomatically. Equipped with a better understanding of the paradoxographical tradition and the enigmatic conception of *natura artifex*, we can return to the *Aetna*.

The *pii fratres*

Throughout the *Aetna*, the two halves of the dual imperative to know the mountain and to call attention to its wonders operate in tandem. From lines

[93] One exception is *natura opifex* at Plin. *HN* 31.1. On Lucretius's *natura creatrix* (1.629, 2.1117, 5.1362), Beagon (1992, 31 n.13) writes that it is "no more than a figurative way of describing the impersonal mechanics of the atomic system."

1 through 567, the emphasis falls on the investigation of causes, and this exposition stages the scientific sublime. In lines 568 to 645, however, the poet conspicuously subordinates causal analysis to the narration of an inexplicable marvel. In so doing, he adopts the role of the paradoxographer and casts the mountain as a *paradoxon*. If the poem had stopped at line 567, the tension would be slight, perhaps even unnoticeable. The poem's excess, then, only enhances and enriches its sublimity.[94]

From lines 568 to 598, the poet engages in a moral diatribe against those of his contemporaries who travel the world to visit the famous cities of mythology and to see spectacular works of art that depict mythological figures. The journey over land and sea is harrowing, bringing us close to death (*per proxima fatis*, 570), and the rewards are sublime: it is a thrill (*iuuat*, 573) to behold the walls of Thebes,[95] and we wonder (*miramur*, 578) at its sights and history; Sparta and its backstory hold us captive (*detinet*, 579); the Athens of Cecrops, Minerva, and Theseus is a spectacle (*spectantur*, 581), and Philomela's saga summons us (*euocat*, 587); we wonder (*miramur*, 589) at Troy, with all its associations; the mythological paintings and statues of the Greeks immobilize viewers (*quin etiam Graiae fixos tenuere tabellae / signaue*, 593–594); a thousand works of art, a whole crowd of them, hold us back (*mille manus operum turbaeque morantur*, 598). The entire passage encodes the sublime, again and again, but it is a sublimity grounded in falsehood, "the lies of ancient rumor" (*ueteris mendacia famae*, 571).[96]

In contrast to this debased experience of sublimity, the poet summons the reader to contemplate something truly magnificent. The text of the exhortation, unfortunately, is deeply corrupt (599–604):

> haec uisenda putas terrae dubiusque marisque?
> artificis naturae ingens opus aspice, nulla
> †cum tanta humanis phoebus spectacula cernes

[94] Note that the text's will to sublimity draws added vitality from numerous echoes of *Aen.* 2, where Troy burns and Aeneas must carry his own aged father to safety. The connection between Aeneas and the pious brothers of Catania, however, predates our poem (e.g., Sen. *Ben.* 3.37, 6.36).

[95] For *iuuat* and the sublime, see Ch. 5.3 on *Octauia* 385–387 and Ch. 4.2 ("Nero-Phaethon").

[96] Volk (2005a, 89–90) and Welsh (2014, 127) gesture toward the importance of the sublime here. Thus, Welsh is not wrong when he writes that "the objects [that these tourists] long to see are simply less exalted and ennobling than the volcano."

. . . .

praecipueque uigil feruens ubi Sirius ardet.
insequitur miranda tamen sua fabula montem,
nec minus ille pio quam forti nobilis igni est.[97]

Do you think that these things must be seen, as you move between land and sea?[98] Behold the massive work of nature the demiurge, no...human...you will behold spectacles...and especially watchful when fiery Sirius burns. Nevertheless, its own wondrous story follows the mountain, and it is not less famous for pious than for powerful fire.

The invocation of *natura artifex* sets the tone: we should marvel at the incomprehensible wonder that is Aetna. What follows is unclear; the author seems to contrast the mountain with other spectacles and to declare its superiority, but the text remains muddled, and no convincing explanation for the reference to Sirius has been proposed. Following the criticism of sightseeing tourists, we might have expected the poet to urge the reader to travel to Sicily. We know from Seneca and Strabo that people could and did ascend Aetna in a spirit of natural inquiry, and Strabo even reports what a group of explorers learned from their trip to the mountain's summit.[99] For the *Aetna* poet, no such real-world expedition is desirable—the mountain's causes are entirely beside the point. Even though the mountain is a preeminent wonder here and now, nevertheless (*tamen*) it is a wondrous story (*miranda...fabula*) that demonstrates its truly exceptional character. Given the context and the associations of the *pii fratres* narrative, the phrase *miranda fabula* is virtually a calque for *historia paradoxos* (cf. Antigonus's Ἱστοριῶν παραδόξων συναγωγή). Not to leave the matter in doubt, at the end of the narrative we learn that "the songs of the poets marvel" at the brothers (*illos mirantur carmina uatum*, 642).

[97] Goodyear (1966) prints line 604 thus: *nec minus ille pio †quamquam sors† nobilis igni est.* Goodyear (1965, 208 n.4), considering the merits of the emendation *quam forti*, observes that the "magnitude and power of the volcano is in many parts of the poem the object of admiration and interest"; he cites lines 2, 443–444, 554–564, and 565. The point is entirely apposite, and the reading fits perfectly into the poem's thematic economy. The reading, however, does not affect my argument.

[98] The expression *terrae dubiusque marisque* is odd, but its meaning is explained by reference to *traducti maria et terras* at the beginning of the passage (line 570). Hine (2012) translates "swithering between land and sea."

[99] Sen. *Ep.* 79.1–4, Strab. 6.2.8 (274). For discussion, see Hine (2002, 62–63). Later, it is said that Hadrian climbed Aetna to see the sunrise (*SHA* Had. 13.3).

The *paradoxon* appears to the carry the familiar moral lesson, which unexpectedly situates the mountain within a new moral context. Amid the eruption, those who attempt to salvage their own goods are represented as thieves and criminals who weigh themselves down with their plunder: "The fire consumes those who delay and everywhere scorches the greedy: it follows those who think that they've escaped and burns the plunder together with captured owners. The conflagration feeds on these things, intending to spare none or only the pious" (621–624). Whereas pseudo-Aristotle's divine power honors the race of the pious, *natura artifex* uses the eruption to punish the avaricious while sparing the brothers.[100] Subsequently, the fire grants them safe passage (632), feels shame at the prospect of touching them (634), makes way wherever they go (635), and restrains itself as they move through a burning landscape (639–640). The expected moral lesson soon emerges: "O *pietas*, greatest thing in the world and rightly considered the virtue that gives most protection to humankind" (*o maxima rerum / et merito pietas homini tutissima uirtus*, 632–633). The danger that the mountain poses elsewhere in the poem falls away as the brothers escape unscathed and receive a special dwelling place and rights in the underworld (643–645). The eruption is evidence of nature's justice.

At the same time, the story inverts the relative weight that the poet elsewhere gives to causal analysis and the narration of wonders. At the outset, the poet promises to explain the mountain's causes (*quae tam fortes uoluant incendia causae* [sc. *carmen erit*], 2), rejects mythological aitia that debase the gods (*non est tam sordida diuis / cura*, 32–33) or are not grounded in physical evidence (*turpe et sine pignore*, 40), and scoffs at those who claim to have seen (*uiderunt*, 77) that which does not exist below the earth and in the heavens. "Such license," he writes, "is owed to poems, but all my concern is for truth" (*debita carminibus libertas ista, sed omnis / in uero mihi cura*, 91–92).[101] Of course, it is not always possible to determine specific causes, but that does not bother the poet. Reflecting on his inability to explain the origins of the hollow spaces within the earth, the poet states that "it is not a cause for sorrow as long as the effect is consistent with the cause" (*non*

[100] Ellis (2008, xc–xcv) notes parallels, some more convincing than others, between *De mundo* and the *Aetna*; cf. Goodyear (1984, 354 with n.34).

[101] Cf. De Vivo (1985, 267–269) and Lassandro (1993, 326–327) on the centrality of *causae* here.

hic est causa dolendi, / dum stet opus causae, 116–117).[102] In the finale, the principles of causal analysis and careful observation fade into the background, but a commitment to the grandeur of the divine and the supremacy of the truth remains. If traditional mythology makes a mockery of anthropomorphized gods, the *paradoxon* reveals the magnificence of *natura artifex*. If an *impia fabula* (42) locates a monster beneath the slopes of the mountain, the *miranda fabula*, which is historically factual, reveals the ultimate reality of nature's plan, however impervious to mechanistic explanation.[103] Previously, knowledge of the earth was "the greatest reward in the world" (*haec rerum maxima merces*, 271), while now *pietas* claims a special place in nature's construction (*maxima rerum... pietas*, 632–633). The author declares the day of the eruption a blessed one (*felix illa dies*, 636), and the poem ends on a note of celebratory ecstasy. While causal analysis of the mountain is an electrifying thrill, the mountain is all the more breathtaking for being, at least partially, incomprehensible. But rather than a stale lesson about *pietas*, it is the mountain's spectacular grandeur, its status as an emblem of *natura artifex*, its undeniable sublimity, that matters most. A poet who denigrates farming (260–269) does not try to instill in his readers deep respect for traditional Roman values—a tired exemplum has become an opportunity to fill ourselves with the majesty of nature the demiurge.[104]

The establishment of a moral framework and the inversion of the relative weights of causal analysis and sheer wonder contribute fundamentally to the poem's volcanic aesthetics and simultaneously to its poetics of novelty. The "mind" of the poem (*mens carminis haec est*, 28) swerves suddenly, like a Lucretian atom, and ends up somewhere quite unexpected. The result is a mountain and a poem that are discursively dissonant, variegated, and altogether new.

[102] Goodyear (1965) prints *docenda* for *dolendi*, which is the reading of most manuscripts. At 373, the poet is unperturbed that the *causa* of periods of inactivity remains hidden.

[103] On the importance of historicity to the authors of the paradoxographical tradition, see Schepens and Delcroix (1996, 382–389) and Geus and King (2018, 434, 441–443). Many scholars characterize the *pii fratres* narrative as myth (e.g., Effe 1977, 214; Pingoud 2008, 209), but note Kruschwitz (2015, 81, 86, 92).

[104] Wolff (2004, 81) compares the poem to an epic in which the hero is Aetna or perhaps "la nature, célébrée avec enthousiasme pour sa beauté et sa puissance grandioses qu'il faut contempler guidé par la raison." For different approaches to the moral lesson, see De Vivo (1989, 83), Taub (2009, 135–137), and Williams (2020, 118–119).

The closest parallel for the poem's construction of the mountain is *Subl.* 35.2–4, where, as we have seen, a "lightly personified" nature (Halliwell 2022, 372) invites mankind to a magnificent festival and summons us to the contemplation of the cosmos, which leads into to a list of *paradoxa* that culminates in an eruption of Aetna. In this context, it is worth observing that Longinus later writes of "nature the demiurge of humankind" (τὴν δημιουργήσασαν φύσιν τὸν ἄνθρωπον, *Subl.* 43.5).[105] Are the two texts related? Fernand Delarue cautiously hypothesizes a literary coterie that formed around Seneca and Lucan and included the authors of *On the Sublime* and *Aetna*.[106] Given the uncertainty of the dates and our general ignorance about mid-first-century literary culture, we have no way of knowing. What is beyond doubt, however, is that by the middle of the first century a pantheistic worldview has found a powerful means of self-expression through the image of Aetna in state of eruption, which is framed as a spectacular show.[107]

Throughout the *Aetna*, both halves of the dual imperative are operative; the poet insists that we come to know the earth and draw attention to its wonders. Generally, this process unfolds according to the logic of the scientific sublime: the study of the mountain draws attention to one of the wonders of the earth, the experience is of unparalleled intensity, and the student comes into contact with the divine. In such passages, the author works in the tradition of Lucretius, Manilius, Seneca, and Lucan, even if he conspicuously differentiates himself on multiple fronts. At times, however, the pursuit of knowledge and the celebration of wonder diverge: the divine power that supervises the *mundus* is without witness (193–196), and the poet has no desire to probe the causes of *natura artifex* (600–645). By the end of the poem, causal analysis has yielded entirely to the celebration of wonder in a way that recalls the value system and modus operandi of the paradoxographers. In these moments, the *Aetna* poet resembles Longinus or the Elder

[105] Halliwell (2022, 372) connects the two passages. The context in Longinus, however, is quite different: in our endeavor to avoid using sordid and contemptable words, we should imitate nature, the demiurge of mankind, who placed our "dishonorable parts" out of sight as much as possible.

[106] Delarue (2000, 22–33), with the group's shared interest in Aetna at 28–29; the argument expands on Cizek (1972, 60–64), and Delarue (2006) returns to it. Effe (1977, 210–211) connects the *Natural Questions*, *On the Sublime*, and *Aetna* but insists on their shared Stoicism. Billault (2004, 203) summarizes Longinus's views of the divine and the marvelous in language that could easily be applied to the *Aetna*.

[107] Cf. Volk (2005a, 81), who identifies in the poem "eine Art kosmischen Pantheismus"; she applies the idea to the *Astronomica*, *Natural Questions*, and *Aetna*.

Pliny, who similarly invokes nature the demiurge, eschews the investigation of causes, and tirelessly catalogs the marvels of the earth (*terrae miracula*, 2.206). At the beginning of the Flavian era, a shift has occurred, at least in some intellectual quarters. Under certain circumstances, it is better for the student of nature not to know too much, not to ask questions that are too precise, but simply to revel in the inconceivable pleasures of the infinite abundance of nature. To get a better sense of this broader cultural shift and what it can tell us about the history of the scientific sublime at Rome, we turn again to Pliny.

The Scientific Sublime in Imperial Rome: Manilius, Seneca, Lucan, and the Aetna. Patrick Glauthier, Oxford University Press. © Oxford University Press 2025. DOI: 10.1093/9780197787588.003.0006

Epilogue

The Scientific Sublime and the Case of the Elder Pliny

The scientific sublime implies a set of values—above all, a commitment to the study of the causes and first principles of the natural world, the belief that such study has the power to expand the mind and ennoble the spirit, and a bold willingness to adopt different and often discordant modes of thought as one rushes exuberantly ahead on a tumultuous and uncertain journey toward a reality that appears truer and more potent than that which meets the eyes. This set of values figures prominently in a wide range of Latin authors, including Manilius, Seneca, Lucan, and the *Aetna* poet. Each of these authors inflects the scientific sublime differently, moving the reader from the boundless expanses of the heavens to the elemental level of atmospheric phenomena, from the very ends of the earth to the subterranean spaces that rumble and gape below our feet. With each recalibration, the scientific sublime takes on new dimensions—it figures the experience of infinite and unending empire, it functions as an antidote to the corrupting influences of the seamy present, it collapses under the weight of its own pretensions, it makes way for the appreciation of wonders that we cannot comprehend, the spectacle of nature. Each author who engages with this tradition forges his own scientific sublime, putting it in dialogue with or using it to critique diverse areas of inquiry, competing literary traditions, and specific cultural practices and institutions, many of which make, or can be imagined to make, their own claims to sublimity. Put differently, the continuities and disagreements that have emerged over the preceding chapters are evidence of a culturally significant and widely diffused discourse on sublimity and attest to the multitude of ways in which our authors come close to theorizing the sublime. While these texts do not explicitly address the nature of sublimity, they repeatedly juxtapose competing modes of sublimity and push readers to think about their relative merits and social functions. On this reading, the scientific sublime is a set of values and a medium of philosophical and literary communication and debate, with both the present and the past, and it fuels a vital current of Latin literary production.

The four texts on which I have focused allow us to trace the history of attitudes toward the study of nature in Latin literature of the early imperial period. The positions that each chapter explores reflect, in broad strokes, contemporary cultural and political developments. Manilius's positivistic triumphalism echoes the elite messaging of the late Augustan Principate, Seneca's cautious reappraisal and Lucan's damning indictment of the scientific sublime mirror the growing sense of instability and righteous indignation that permeate the final years of Nero's reign. The *Aetna* poet's commitment to both the study of the earth and the spectacle of nature reflects a desire to turn the page on the chaos of the 60s CE and to find elevation in displays of imperial power that are at once comprehensible and tantalizingly mysterious. This narrative, however, runs the risk of flattening the history of Roman attitudes toward the study of nature and the scientific sublime, with shifts at the top of the political pyramid constraining the possibilities of thought and expression within narrow limits; this a conclusion that we should resist. The differing tones of Seneca and Lucan readily illustrate the potential for contemporaneous authors to adopt contrasting stances toward natural inquiry and the sublime, and other texts within the Senecan corpus might reveal additional perspectives. Similarly, a figure such as Germanicus is less invested in the sublimity of celestial inquiry than his contemporary Manilius (see Ch. 2.2), and there is no shortage of early imperial texts that are downright hostile to astrology. At any given moment, it is possible to conceptualize the study of nature in more ways than one, and we should not assume that the positions that Manilius, Seneca, Lucan, and the *Aetna* poet formulate are the only ones available to their readers or that parallels do not exist elsewhere, in earlier or later literature. These four authors are part of a conversation, but there are other voices.

This brings us to the Elder Pliny's *Natural History*. Pliny constitutes a necessary coda for the present study because he complicates the history of the relationship between the study of nature, the sublime, and empire, and he does so at roughly the same time as the *Aetna* poet—the *Natural History* is dedicated to Titus, and Pliny himself famously died during the eruption of Vesuvius in 79 CE. As we saw in Ch. 2.6, the *Natural History* is animated by the same imperialistic drive as the *Astronomica*, whereby the seemingly infinite accumulation of knowledge figures the seemingly infinite accumulation of empire. The narratives of Chapters 3–4 problematized that harmonious coupling of natural inquiry and political power, but similar perspectives will have remained available and attractive to many. In point of fact, Chapter 5 showed that the *Aetna* poet's deployment of the mechanical sublime ties the rush of imperial spectacle to the experience of studying the earth. We can make a similar point about the complex historical relationship between sublimity and Roman *imperium* itself. Seneca and Lucan might view the sublimity of empire as dangerous and destructive, but others will have thrilled to its rhetoric, iconography, rituals, building projects, and more. Martial's *Liber spectaculorum*, for one, vividly illustrates how displays of imperial power can take one's breath away.[1] In a similar vein, Pliny consistently construes the evidence of empire as a sublime spectacle even as he repeatedly deplores the moral degeneracy of the age. The histories of Roman attitudes toward the sublime and the study of nature, on the one hand, and toward the sublime and imperial power, on the other, are recursive rather than linear.

At quite a different level, however, Pliny repeatedly distances himself specifically from the scientific sublime. Pliny frames the *Natural History* as a catalog of 20,000 facts that are worthy of attention ($\overline{XX}$ *rerum dignarum cura*, praef. 17). Within this catalog, he does not place a high premium on understanding nature's causes, refuses to allow his mind to soar beyond the ramparts of the world, to borrow Lucretian terminology, and suggests that both he and his contemporaries are incapable of fully accessing the vitality and thrill that the systematic study of nature used to inspire. To be sure, Pliny does discuss nature's causes, and Stoic principles buttress or inform his discussions of various phenomena in ways that are not always immediately obvious.[2] What qualifies as a causal explanation, however, is not always likely to satisfy the demands of the philosophers. When Pliny explains the perpetual discord that brings Indian elephants and serpents into conflict, he reaches for the language of imperial spectacle: "What other cause for such great discord might one adduce than that nature is putting on a show for herself and matching combatants?" (*quam quis aliam tantae discordiae causam attulerit nisi naturam spectaculum sibi ac paria componentem*, 8.34).[3] Nature produces a gladiatorial show in her great amphitheater, and when

[1] For the *Liber spectaculorum* and the sublime, see Chomse (2018).

[2] As I point out in Ch. 5.5 ("*Natura artifex*"), with references, Pliny is not a committed Stoic. Beagon (2011, 84–86) discusses Pliny's attitude toward causation.

[3] Mayhoff (1892–1909) does not print *ac*.

we catch a glimpse of the action, we gain access, in some vague way, to her inner workings.

The implication of such passages is that the natural world is spectacular and, frequently, sublime. More than anything, Pliny is interested in that which arouses wonder in all aspects of life and all areas of the world. As Mary Beagon puts it, Pliny uses wonder as a stimulus to kick-start the reader into a state of alertness, and precisely because his narratives refuse to offer satisfying causal explanations, wonder leads to more wonder. "For Pliny," writes Beagon (2011, 86),

> perpetual change and variation in the very fabric of terrestrial nature can make explanation itself inexact and open-ended. Wonders in such an environment will, literally, never cease. By this [Pliny] is not suggesting that humanity should allow itself to be dazzled and stunned by the world around it to the extent of abandoning all attempt at rational understanding. Instead he suggests that the key to understanding the sub-lunar world is to recognize its volatility and variety.... To understand nature for Pliny is to understand that wonder and explanation can knit together in a never-ending circle of intellectual curiosity, rather than presenting the inquirer with a simple and finite one-way journey from wonder to explanation. This is the essence of reality, of consciousness, of being truly alive.

The outlook that Beagon describes resembles the ethos of the scientific sublime in its configuration of the study of nature as an ongoing and often inconclusive adventure through an infinitely complex and ever-changing landscape. Significantly, however, Pliny's relentless focus on *mirabilia*, with minimal attention to causal analysis, and his frequent use of identifiable paradoxographical sources align him with the paradoxographical tradition.[4] Pliny's alienation from the scientific sublime, on the one hand, and his infatuation with the wonders of the natural world, on the other, make it possible to read him alongside Lucan and the *Aetna* poet, respectively. That the *Aetna* exhibits a similar combination of contradictory impulses toward causal analysis and paradoxography in an explicitly earth-focused context reveals a commonality of vision that suggests a shared cultural dissatisfaction

[4] For Pliny's connections with the paradoxographical tradition, see Ziegler (1949, 1165–1166), Schepens and Delcroix (1996, esp. 433–439), and Naas (2002, 243–292), suggestively subtitled "L'*Histoire naturelle*, encyclopédie paradoxographique?"

with the scientific sublime, especially in its aspirations for celestial elevation. Grouping these texts together potentially shifts the balance of power in the history of the scientific sublime at Rome and alters our sense of the center of gravity of the Latin tradition that writes about the natural world.

The following discussion briefly explores the distance that Pliny establishes between himself and the scientific sublime. Although the study of nature continues to thrill, the details and methods of that study look new, and when Pliny does encode its sublimity, the experience suggests something more akin to Kant's dynamically sublime than to the scientific sublime as we have come to know it. Because Pliny thinks about the scientific sublime and the study of the natural world in ways that resemble the perspectives of Lucan and the *Aetna* poet and that ultimately sideline the scientific sublime altogether, the *Natural History* takes us away from the main topic of the present study. For precisely this reason, however, Pliny's text constitutes an important element of the conversation that the present study documents. Moreover, as we will see, Pliny's stance on the scientific sublime itself reveals points of contact with other first-century authors, one of whom is none other than Longinus.[5]

Pliny's alienation from the scientific sublime is apparent from the very outset of the treatise's first expository book. Following book 1, in which Pliny outlines his sources and topics, book 2 begins thus (2.1–4):

> The *mundus*—this thing, whatever it is, that people have been pleased to call by another name, the heavens, under whose vault all life progresses—it is reasonable that it is believed to be a divinity (*numen*), eternal, immeasurable, neither born nor ever going to perish. To investigate what lies beyond it is not relevant to humankind, nor can the reach of the human mind take it in (*huius extera indagare nec interest hominum nec capit humanae coniectura mentis*). It is sacred, eternal, immeasurable, wholly within the whole, or rather it is itself the whole, infinite and like the finite, fixed in all its details and resembling something random, embracing within itself all that is without and within, and at the same time the work of nature and nature herself. It is madness (*furor*) that certain individuals have studied and dared to publish its measurement, and that others ... have handed down that there are countless worlds.... It is madness, sheer

[5] My discussion here is intended to be suggestive. I plan to treat Pliny's combination of natural inquiry with the sublime in more detail in the future.

> madness (*furor est profecto, furor*), to go out of it and, as if its whole contents were already clearly known, to probe what lies beyond it, as if anyone could really take the measurement of any object when he does not know his own, or the mind of man could see that for which the world does not have room.

The *Natural History* will be a study of something that might reasonably be called god, and the contradictions through which Pliny initially strives to articulate its dimensions bear witness to the inspirational ineffability of the divine; this is the language of sublimity.

But just as he prepares the reader's mind to fly beyond the limits of cosmos, like Lucretius's Epicurus, Pliny stops us in our tracks. To place limits on "the reach of the human mind" in the context of cosmic exploration is to deny the possibility of the scientific sublime. Longinus (*Subl.* 35.3), for one, excitedly celebrates the fact that "not even the whole universe is enough for the reach of human contemplation and understanding" (τῇ θεωρίας καὶ διανοίας τῆς ἀνθρωπίνης ἐπιβολῇ οὐδ᾽ ὁ σύμπας κόσμος ἀρκεῖ) and that "our intuitions often go beyond even the boundaries of the world."[6] Similarly, Lucretius (2.1044–1047) declares that "since the totality of space is infinite outside beyond walls of this world, the mind seeks an explanation of what there is far out there, where the spirit wishes continually to gaze and where the unrestrained cast of the mind itself flies on" (*animi iactus liber quo peruolet ipse*). Pliny's *coniectura*, Longinus's ἐπιβολή, and Lucretius's *iactus* all figure the flight of the mind and the thrilling adventure of the scientific sublime.[7] For Pliny, however, the journey is emphatically madness (*furor*), which makes such investigators exemplars of a dark and dangerous sublime subjectivity. This is not primarily a rejection of atomism or Lucretius, nor does Pliny simply think that we ought to strive to know ourselves before we turn to the natural world. Pliny's outrage is triggered first by those who have dared to measure the *mundus* internally. The world is axiomatically unmeasurable, and to appreciate and to keep intact its grandeur, we must not know it with too much precision. This is the mindset of the

[6] The text is problematic; I follow Halliwell (2022) and slightly modify his translation.

[7] For the connections here between Lucretius and Longinus, see Porter (2007, 174; 2016, 453–454). A similar yearning for mental expansion animates Petronius's Eumolpus when he declares that the author of a poem on civil war must propel his free spirit headlong (*praecipitandus est liber spiritus, Sat.* 118.6) through the twists and turns of the mythological and the fabulous. For different perspectives on this passage and the sublime, see Conte (1996, 69–72), Hardie (2009, 226–227), Hunter (2009, 133–134), and Ripoll (2019, 131–133).

paradoxographer, and, as we have already seen (Ch. 5.5, "*Natura artifex*"), Pliny invokes *natura artifex* in this very context. At the opening of his first expository book, Pliny establishes distance between himself and the scientific sublime; it is morally indefensible hybris.

The opening of *HN* 2 points to Pliny's alienation from one major strain of the scientific sublime. Later in the same book, Pliny discusses two figures, Endymion and Hipparchus, whose stories suggest that an impossible distance separates Pliny's Rome from the electrifying thrill of celestial inquiry as it used to be practiced in the past.

Through Endymion, Pliny comes close to encoding the scientific sublime. The context is the order of the planets, moving from that which is furthest from the earth to that which is closest (2.41, 43):

> sed omnium admirationem uincit nouissimum sidus, terris familiarissimum et in tenebrarum remedium ab natura repertum, lunae. multiformis haec ambigua torsit ingenia contemplantium et proximum ignorare sidus maxime indignantium....quae singula in ea deprehendit hominum primus Endymion; ob id amor eius fama traditur. non sumus profecto grati erga eos qui labore curaque lucem nobis aperuere in hac luce, miraque humani ingeni peste sanguinem et caedes condere annalibus iuuat, ut scelera hominum noscantur mundi ipsius ignaris.
>
> But the amazement of all is overcome by the last star, the one most familiar to those on earth and devised by nature to be a remedy for darkness—namely, the moon. With her multiplicity of forms, she tortured the wavering intellects of those who gazed at her and were especially resentful that they were ignorant about the star closest to them. [At this point, Pliny lists various lunar phenomena that are difficult to explain.] The first to discover these individual facts about her was Endymion; on account of this, his love for her is handed down by tradition. To be sure, we feel no gratitude toward those who with hard work and care have shed light for us on this light, and thanks to an astonishing disease of the human mind it is thrilling to record bloodshed and slaughter in history books, so that the crimes of men may become known to those who are ignorant of the world itself.

The moon fills those on earth with sublime stupefaction. What follows is a dramatization, with a twist, of the Aristotelian position on wonder and philosophy (see Ch. 1.6, "Plato and Aristotle"). Once upon a time, an experience of sublimity, inspired by the moon, led to philosophical inquiry, but the complexity of lunar phenomena ended in angry aporia rather than enlightenment. Enter Endymion. Pliny rationalizes Endymion's romance with the moon, transforming his mythological *amor* into the eros that drives philosophy and that ought to dispel mental *tenebrae*.[8] Sadly, Endymion's discoveries do not free mankind from intellectual darkness or emotional disturbance. Even now, laments Pliny, we prefer the sadistic rush (*iuuat*) that comes from writing and reading military and political history to the sublime *admiratio* that summoned our ancestors to study the stars. As we have seen (Ch. 4.2, "Nero-Phaethon"), the impersonal verb *iuuat* regularly names the thrill of the scientific sublime. In the *Natural History*, this energy has been redirected and morally compromised. Endymion's passion for nature does not offer a paradigm that we can hope to reenact.

Pliny's treatment of the Hellenistic astronomer Hipparchus (fl. c. 125 BCE) works similarly. Pliny is especially impressed with Hipparchus's star catalog, which contained the names and ecliptic coordinates of over 800 stars.[9] Once upon a time, Hipparchus witnessed something in the heavens (most likely a nova) that led him to question one of the most basic facts of imperial-era cosmology—namely, that the fixed stars are fixed. As a result, he set out to catalog the stars and discovered precession, thereby demonstrating that, from the perspective of an earthly observer, the stars are not quite as fixed as everyone tends to think.[10] We have seen Seneca suggest that the fixed stars might not actually be fixed (Ch. 3.4 on *NQ* 2.32.8). Seneca construes such daring thought experiments as sublime, and Pliny frames the matter similarly (2.95):

> idem Hipparchus numquam satis laudatus, ut quo nemo magis adprobauerit cognationem cum homine siderum animasque nostras partem esse caeli, nouam stellam et aliam in aeuo suo genitam deprehendit eiusque motu, qua fulsit, ad dubitationem est adductus, anne hoc saepius fieret mouerenturque et eae, quas putamus adfixas, ideoque ausus rem etiam deo improbam, adnumerare posteris stellas ac sidera ad nomen

[8] At Nonnus, *Dion.* 41.379–381, Endymion is said to be the first to have worked out the phases of the moon. Roller (2022, 45) suggests that we are dealing with "a faint tradition" according to which Endymion was the first astronomer.

[9] The catalog is lost, but some scholars have argued for a connection between it and Ptolemy's star catalog in *Almagest* 7–8; see Evans (1998, 247–250, 264–274).

[10] For precession in antiquity, see Evans (1998, 245–250, 259–262).

> expungere...caelo in hereditate cunctis relicto, si quisquam, qui cretionem eam caperet, inuentus esset.
>
> The same Hipparchus—who cannot be praised enough, seeing that no one more than he has proved that mankind is related to the stars and that our souls are a part of heaven—discovered that a new and different kind of star had been born in his lifetime. Because of its motion, where it shone, he was led into uncertainty about whether this happened more often and whether the stars that we consider fixed also move, and so he dared to do something that would be improper even for a god: to tally the stars for posterity and to mark off the heavenly bodies by name on a list...thereby bequeathing the heavens as a legacy for all, if anyone had been found to take up the inheritance!

Hipparchus's willingness to cross boundaries leads to an act of astronomical daring that elevates him to the level of a god or, perhaps, beyond the level of a god (*ausus rem etiam deo improbam*). At the same time, his extraordinary achievement demonstrates the divinity of human beings, and since to catalog the stars is to know and hence to control them, Hipparchus can bequeath the heavens to posterity, thereby giving us the opportunity to elevate ourselves. On multiple levels, then, Hipparchus is a sublime subject who, like Lucretius's Epicurus, makes us equal to heaven. Tragically, however, in Pliny's day, no one had continued Hipparchus's project. Once again, Pliny's contemporaries cannot relive the adventures of their intellectual forebears. Such paradigms are of limited use, which means that the scientific sublime has become a relic of the past.[11]

The Endymion and Hipparchus narratives suggest that Pliny values the scientific sublime and longs for it once again to become accessible. These figures throw into sharp relief the intellectual and ethical degradation that Pliny identifies in contemporary Rome; in contrast to Epicurus in *De rerum natura* or Phaethon in the *Astronomica* and *Natural Questions*, Endymion and Hipparchus do not summon readers of the *Natural History* to greatness. Like Lucan, then, Pliny senses that a historical rupture has caused old models for achieving sublimity to become inoperable; elevation must be sought elsewhere, in political engagement or contemplation of nature's terrestrial

[11] Some critics (e.g., Beagon 1992, 44) detect a certain ambivalence in Pliny's portrait of Hipparchus and parallel figures; my reading here offers one way to make sense of this tension. For Pliny's praise of other astronomers, see Herbert-Brown (2007, 118–119).

wonders. The criteria by which he identifies terrestrial wonders are more generous than those of the *Aetna* poet, but his turn toward the earth represents a similar suspicion of that which is up above.

There is another relevant context within which we should situate Pliny's unfulfilled desire for the scientific sublime. In his anxious longing for an inaccessible mode of sublimity, Pliny expresses a frustration that Gian Biagio Conte (1996, 37–72) has identified in Longinus and Petronius and that we have observed while thinking about moral depravity in Seneca's *Natural Questions* (Ch. 3.6), where, for the most part, knowledge of the heavens is also out of reach. In *Subl.* 44, Longinus purports to record a conversation that he had with an unnamed philosopher. Both parties agree that truly sublime literature and oratory are only rarely encountered any longer. The philosopher ascribes this decline to the loss of political freedom under the imperial system, whereas Longinus identifies the cause in moral decay. Conte (1996, 44) argues that this lament over the disappearance of true sublimity constitutes "a fundamental theme" of Petronius's *Satyrica*, "to the point that Petronius... gives his protagonist [i.e., Encolpius] powerful words that express all the yearning for that sublime literature which the giants of the past could create." Pliny's yearning is more specific; it's a deep desire for the sublimity of celestial inquiry, which represents the scientific sublime in its most exalted form. In assessing the causes of historical change, however, Pliny implicitly sides with Longinus himself; contemporary Rome, Neronian Rome, Flavian Rome, cannot answer the call to this particular mode of sublimity. From multiple perspectives, then, thinking about Pliny's relationship with the scientific sublime reveals that he is well attuned to the concerns of his times.

By way of conclusion, we can turn to the mode of sublimity that Pliny's study of the sublunar world does offer. Pliny everywhere exhorts the reader to marvel at nature's infinite variety and unbounded power, what he calls "the force and grandeur of nature" (*naturae... rerum uis atque maiestas*, 7.7). The phrase resembles Kant's formulation of the dynamically sublime, through which he treats "nature as a power,"[12] but Pliny recognizes nature's force and grandeur at every level of the natural world, not just in those objects and phenomena that pose a threat to human equanimity and/or life. Consider the gemstone, which is the subject of book 37, the last of the

[12] Kant (2000, 143, from *Critique* §28). For further discussion, see the Introduction.

work.[13] In the first chapter, Pliny discusses how in the gemstone "nature's grandeur is gathered together within the narrowest limits" (*in artum coacta rerum naturae maiestas*): the variety, colors, texture, and elegance of such stones make them "more wondrous" (*mirabilior*) in the eyes of many than any other part of nature's creation; some consider it sacrilegious to violate them with engravings; they seem to be beyond price and to defy evaluation in terms of human wealth; and consequentially "very many people find that a single gemstone alone is enough to provide them with a supreme and perfect vision of nature" (*plerisque ad summam absolutamque naturae rerum contemplationem satis sit una aliqua gemma*). The compressed dimensions of an individual gemstone paradoxically bring into focus the vastness of the natural world, and it is that dynamic tension between the unfathomably large and the incredibly small that quivers with the energy of the sublime. Such objects lie beyond all ordinary categories of human reckoning, literally crystallizing that which is truly magnificent and thereby bringing us into contact with something greater than ourselves.

Implicit in Pliny's hymn to the gemstone is the idea that such contemplation leads to *megalopsuchia* or *megalophrosunē*. The grandeur that the gemstone allows us to grasp and to cultivate stands in stark contrast to the small-mindedness with which Pliny judges his contemporaries and so many historical figures to have fetishized precious stones, especially those that have been engraved and worn as jewelry; consequently, Pliny's narrative is full of moral outrage. Immediately following his initial celebration of gemstones, Pliny describes a widespread fascination with gems that has blazed to great heights (*a quibus initiis in tantum admiratio haec exarserit*, 37.2), and locates in precious stones a cultural authority that has been elevated into an almost erotic passion (*his initiis coepit auctoritas in tantum amorem elata*, 37.3); such language implies profoundly flawed priorities and sets the stage for Pliny's moral judgment. Pliny's hymn to the gemstone, then, offers readers an opportunity to catch sight of nature's true magnificence and to turn away from the petty obsessions of contemporary culture, which ultimately degrades and misconstrues the natural world.[14]

In demonstrating proper and improper ways to marvel at gemstones, Pliny negotiates competing modes of sublimity, but the narrative itself offers

[13] I discuss *HN* 37.1 in a slightly different context at Glauthier (2023, 20–21).

[14] Naas (2011, 57) writes that for Pliny *mirabilia* "exemplify the power of Nature and force Man to regain the respect he has lost for her." I take this to be a complementary way of conceptualizing the point I am making here.

history lessons and catalogs rather than causal analysis and an experience of the scientific sublime. The account of *adamas*, the diamond (37.55–61), shows the extent to which this process unfolds in proud and defiant opposition to any systematic analysis of causes. Near the outset of the passage, Pliny frames the *Natural History* itself as an exploration of two principles, *discordia* and *concordia*, which might suggest the charged energy of Empedoclean Strife and Love as much as Stoic thinking; either way, the combination points to the sublimity of the study of nature.[15] In the *Astronomica*, this combination of forces (*discordia concors*, 1.142) highlights the interconnected vitality of a cosmos that is ruled by mathematical *ratio* (see Ch. 2.1). In the *Natural Questions*, the same combination (*tota haec mundi concordia ex discordibus constat*, 7.27.4) leads to reflections on the infinite variety with which nature enhances her own grandeur, but it is precisely this variety that the human mind struggles to comprehend as it finds itself forced to rely on conjecture rather than sure reason (see Ch. 3.1 on 7.27.4, 3.2 on 7.29.3, and 3.3. on 7.27.5–6.). From the perspective of the *Natural History*, the interrelationship and overlap between discord and harmony is simultaneously more destabilizing and more inspiring (37.59–60):

> nunc quod totis uoluminibus his docere conati sumus de discordia rerum concordiaque, quam antipathian Graeci uocauere ac sympathian, non aliter clarius intellegi potest, siquidem illa inuicta uis, duarum uiolentissimarum naturae rerum ferri igniumque contemptrix, hircino rumpitur sanguine, neque aliter quam recenti calidoque macerata et sic quoque multis ictibus, tunc etiam praeterquam eximias incudes malleosque ferreos frangens. cuius hoc inuento quoue casu repertum? aut quae fuit coniectura experiendi rem immensi pretii in foedissimo animalium? numinum profecto talis inuentio est et hoc munus omne, nec quaerenda ratio in ulla parte naturae, sed uoluntas!
>
> Now, in all these books of mine I have tried to teach about the discord and harmony of things, what the Greeks have called antipathy and sympathy. Nowhere else can this be more clearly understood, seeing that the invincible force [of the diamond], which despises nature's two most violent

[15] Note the juxtaposition of *concordia discors* with *Empedocles* at Hor. *Epist.* 1.12.19–20 (*quid uelit et possit rerum concordia discors, / Empedocles an Stertinium deliret acumen*). Paparazzo (2011, 100–101), by contrast, reads this passage from a Stoic perspective. For other parallels, see Vottero (1989, 716) on Sen. *NQ* 7.27.4.

> materials, iron and fire, is broken apart by goat's blood, so long as it has been steeped in blood that is fresh and warm, and even under these circumstances, it may need many blows and still may break all but the best anvils and iron hammers. By whose research or by what accident was this discovered? What guesswork led to testing a material of boundless value on the foulest of animals? To be sure, so great a discovery and every such gift belong to divine powers. We should not look for *ratio* in any part of nature, but will!

The most shocking example of nature's discordant harmony or harmonious discord is the incongruous relationship that brings together two opposites, the diamond ("a material of boundless value") and the goat ("the foulest of animals"). The ability of the impossibly low to shatter the limitless power of the impossibly high is so unexpected that no systematic process of investigation could have led to it; it is evidence of the divine will that pervades the world, and it is accessible only by random guesswork. This is the mindset of the paradoxographer, the adherent of *natura artifex*, and Pliny drives the point home by drawing a sweeping lesson that is meant to astound the reader every bit as much as the simple fact under consideration: in nature, do not look for reason or regularity (*ratio*), but rather an expression of potency, power, will (*uoluntas*). At the heart of nature, lies an indefatigable will to sublimity.

The worldview of the Hellenistic paradoxographers never disappeared—the *nil admirari* tradition continually protests against it—but the early years of the Flavian period elevate it within Latin literary culture. Like the wonder-books themselves, but on a massively grander scale, Pliny's text often appears disjointed, structured more by associative thinking than a careful masterplan. Attempting to follow the unexpected twists and turns of his narrative induces "referential dizziness."[16] His deep dives into rich and sometimes shocking sensory worlds, his vertiginous pivots from the heights of Roman culture to the fleshy realities of animal bodies, create a "surréalisme naturaliste" that is both disorienting and revelatory.[17] It is a textual universe that refuses closure and defies the critic to search for symmetry, proportion, and balance. Pliny explicitly compares it to sculptures and paintings that

[16] Murphy (2004, 30); cf. 49 on Pliny's "aesthetic flirtation with chaos."

[17] Michel (1987, 67). Manolaraki (2018) explores connections between the senses and "numinous experience" in the *HN*; many of her examples could be analyzed through sublimity.

bear provisional titles "as if art were always being brought into being and incomplete" (*tamquam inchoata semper arte et imperfecta*, praef. 26); these are the kinds of works at which we cannot wonder enough (*quae mirando non satiamur*, praef. 26). Put differently, this encyclopedic maze embodies the basic values of flawed sublimity that Longinus and Pliny's own nephew so enthusiastically embrace (see Ch. 1.5, "Acrobats, Heroes, and Flawed Sublimity"). In the present context, though, we might construe each of the 20,000 noteworthy facts that comprise the text of the *Natural History* as individual gems, as so many potential sites of revelation and rapture.[18] Pliny's wonder-book is a virtual collection of precious stones, a textual *dactyliotheca* (37.11), and each item is meant to sparkle and to captivate.

The dazzling brilliance of the *Natural History* shines through powerfully in a final passage that beautifully encodes the sublimity of Pliny's conception of the study of nature and his own turn toward the earth. In book 18, in the very middle of work, Pliny declares that nature, in her unbelievable generosity, has given mankind a sign for sowing a reaping: fields of nighttime fireflies (*cicindelae*, 18.250). The seasonal appearance of fireflies serves the same function as the annual rising of the star-group Vergiliae (18.251–253):

> iam uergilias in caelo notabiles caterua fecerat [*sc.* natura]; non tamen his contenta terrestres fecit alias ueluti uociferans: "cur caelum intuearis, agricola? cur sidera quaeras, rustice? iam te breuiore somno fessum premunt noctes. ecce tibi inter herbas tuas spargo peculiares stellas easque uespera et ab opere disiungenti ostendo ac, ne possis praeterire, miraculo sollicito. uidesne ut fulgor igni similis alarum compressu obtegatur, secumque lucem habeant et nocte?... cur etiamnum altius spectes ipsumque caelum scrutere? habes ante pedes tuos ecce uergilias." incertis hae diebus proueniunt durantque, sed esse sideris huiusce partum eas certum est. proinde quisquis aestiuos fructus ante illas seuerit, ipse frustrabitur sese.
>
> [Nature] had already formed the Vergiliae, which are easy to observe in the sky because they make a cluster, but not content with these, she has made others on earth, crying aloud, as it were: "Why should you contemplate the heavens, farmer? Why search for the stars, rustic? Already with

[18] Since at least the time of Posidippus (fl. *ca.* 250 BCE), engraved gemstones could figure exquisite literary craftsmanship. For wonder in Posidippus's *lithika*, see frr. 5, 13, 15 Austin-Bastianini with Smith (2004). For gemstone as poem, see Elsner (2014).

> too short sleep the nights weigh on you in your exhaustion. Behold! I scatter special stars for you among your plants, and I display them to you in the evening as you unyoke and leave off work, and I rouse you with a marvel so that you won't be able to pass them by—do you see how their fire-like brilliance is covered by the folds of their wings, and how they carry daylight with them even at night?... Why then should you look higher up and probe the heavens themselves? Behold! you have before your feet your Vergiliae." The days on which these appear and their duration are not fixed, but it is certain that they are the offspring of this heavenly body. Accordingly, whoever sows his summer seeds before them will cheat himself.

As she reveals the presence of the divine on earth, Nature issues a dramatic exhortation that stages the sublime (*ecce! uidesne? ecce!*), while her real-time narrative encodes the affective intensity with which she stops her addressee in his tracks (*ne possis praeterire, miraculo sollicito*). This is *natura artifex* come to life, and her primary concern is to admonish the ideal Roman, the simple farmer, not to study the stars.[19] It is striking that the language of this admonition echoes a well-known bit of proverbial wisdom to which Ennius, in his *Iphigenia*, gave lasting form: "What is before one's feet no one watches, they probe the expanses of heaven" (*quod est ante pedes nemo spectat, caeli scrutantur plagas*, fr. 187 Jocelyn, 82 Goldberg-Manuwald). This is the same line that the *Aetna* poet invokes when he prioritizes the study of the earth over that of the heavens (Ch. 5.2 on *Aetna* 251–256); it is hard to believe that there is not a direct connection. But Pliny goes further. He feels no need to understand the heavens in detail, nor is it important to explain the physical or causal mechanisms that relate the morning rising of the Vergiliae to the seasonal appearance of *cicindelae*. Causal analysis diminishes the immediate impact of the *uis atque maiestas* of nature.

Significantly, though, Nature's conventional moralizing coincides with a vibrant injection of wonder into the quotidian world, which suddenly becomes a sublime spectacle. We, as readers, are not allowed to pass by the dazzling array of tiny lights without acknowledging that they embody nature's potency and plan and symbolize the centrality of humankind to the *mundus*. At the same time, the likes of Endymion and Hipparchus give way

[19] Beagon (1992, 169–173) demonstrates that technical knowledge does play an important role in Pliny's approach to agriculture. Pliny is making a philosophical, rather than agricultural, point.

to a mysterious goddess and the practically minded Italian farmer who is free to answer Nature's call. At this point, we recall that Pliny has assembled the *Natural History* in his spare time at night (*subsiciuisque temporibus ista curamus, id est nocturnis*, praef. 18). The farmer, then, figures the author and, by extension, the reader, and the dazzling array of fireflies become the seemingly infinite series of facts that sparkle on the scroll of papyrus. A farmer at night gazing in awe at a field of celestial fireflies—it's a striking image for the experience of learning about nature through the vast expanses of the *Natural History*. The brilliance of the image makes it one that we might expect from Lucretius, whose Nature also rebukes the reader in the middle of his exposition (3.931–977). Similarly, the tableaux would fit nicely in the works of Manilius, Seneca, or the *Aetna* poet, with his ecstatic enthusiasm for nature the demiurge, but it's all the more arresting for appearing in a text that is so thoroughly imbued with the spirit of the paradoxographical tradition. For Pliny, the process of studying nature is sublime, but it's a form of study that records facts and remains stationary, with both eyes fixed on the ground, rather than asking questions and soaring through the heavens. This paradigm constitutes a kind of intellectual retrenchment after the turbulence and uncertainty of the 60s and early 70s CE. In a dangerous world where political power changes hands quickly and Pliny himself has official responsibilities, it seems preferable to stay close to home and not to cross certain boundaries. The study of nature looks noticeably different than it does in the *Astronomica*, *Natural Questions*, *Civil War*, and *Aetna*, but despite these changes, the experience remains thrillingly sublime.

The Scientific Sublime in Imperial Rome: Manilius, Seneca, Lucan, and the Aetna. Patrick Glauthier, Oxford University Press. © Oxford University Press 2025. DOI: 10.1093/9780197787588.003.0007

Bibliography

Abry, J.-H. 1993. "Manilius et Germanicus: Une énigme historique et littéraire." *RÉL* 71: 179–202.

———. 1998. "Les *anaphorai* des signes du zodiaque dans les écrits astrologiques." In *Sciences exactes et sciences appliquées à Alexandrie*, ed. G. Argoud and J.-Y. Guillaumin, 305–324. Saint-Étienne: Publications de l'Université de Saint-Étienne.

———. 2002. "*Inueniunt et in astra uias*... (Manilius, *Astr.*, 4, 152–161)." *Pallas* 59: 65–73.

———. 2006. "L'*excursus* sur les latitudes (Manilius, *Astronomiques*, 3, 301–384)." *Pallas* 72: 149–170.

———. 2008. "Les vers 5, 710–745 sont-ils l'épilogue des *Astronomiques*?" In *Commencer et finir: débuts et fins dans les littératures grecque, latine et néolatine*, ed. B. Bureau and C. Nicolas, 2.653–667. Paris: De Boccard.

———. 2011. "*Cosmos* and *imperium*: Politicized Digressions in Manilius' *Astronomica*." In Green and Volk, eds., 222–234.

Ahl, F.M. 1976. *Lucan: An Introduction*. Ithaca and London: Cornell University Press.

Alexander, W.H. 1948. "Seneca's *Naturales Quaestiones*: The Text Emended & Explained." *University of California Publications in Classical Philology* 13: 241–332.

Alston, R., and E. Spentzou. 2011. *Reflections of Romanity: Discourses of Subjectivity in Imperial Rome*. Columbus: Ohio State University Press.

Althoff, J. 2005. "Senecas *Naturales Quaestiones*, Buch 2, und Lukrez." In *Seneca: Philosophus et Magister*, ed. T. Baier, G. Manuwald, and B. Zimmermann, 9–34. Freiburg: Rombach.

Ambühl, A. 2015. *Krieg und Bürgerkrieg bei Lucan und in der griechischen Literatur: Studien zur Rezeption der attischen Tragödie und der hellenistischen Dichtung im* Bellum Civile. Berlin and Boston: De Gruyter.

Armisen-Marchetti, M. 1986. "Imagination et méditation chez Sénèque: l'exemple de la praemeditatio." *RÉL* 64: 185–195.

———. 1989. *Sapientiae facies: étude sur les images de Sénèque*. Paris: Les Belles Lettres.

———. 1990. "Pline le Jeune et le sublime." *RÉL* 68: 88–98.

Asper, M. 1997. *Onomata allotria: zur Genese, Struktur und Funktion poetologischer Metaphern bei Kallimachos*. Stuttgart: F. Steiner.

Aujac, G., ed. 1975. *Géminos: Introduction aux phénomènes*. Paris: Les Belles Lettres.

Aumont, J. 1968. "Sur l'épisode des reptiles dans la *Pharsale* de Lucain (IX, 587–937)." *Bulletin de l'Association Guillaume Budé* 1: 103–119.

Aygon, J.-P. 2019. "Fulgurances «sublimes» chez Sénèque: le choix de Phaéthon et le silence d'Œdipe." *Cahiers des Études Anciennes* 56: 189–205.

Baccani, D. 1992. *Oroscopi greci: documentazione papirologica*. Messina: Sicania.

Bailey, C., ed. 1926. *Epicurus: The Extant Remains*. Oxford: Clarendon Press.

———, ed. 1947. *Titi Lucreti Cari De Rerum Natura Libri Sex*. 3 vols. Oxford: Clarendon Press.

Bajoni, M.G. 2004. "Gli *Astronomica* di Manilio come rappresentatzione politica dello spazio celeste." *Latomus* 63: 98–107.

Bakhouche, B. 2002. "Lucain (X, 194–331): les crues du Nil ou la négation de la science." *Vita Latina* 165: 34–44.

Bakker, F.A. 2016. *Epicurean Meteorology: Sources, Method, Scope and Organization*. Leiden and Boston: Brill.

Baldini Moscadi, L. 1981. "Il poeta fra storia e ideologia: Manilio e le guerre civili." In *Cultura e ideologia da Cicerone a Seneca*, ed. A. Ronconi, 37–69. Florence: Le Monnier.

Barbara, S. 2008. "Science, mythe et poésie dans le «Catalogue des serpents» de Lucain (*Phars.* IX, 700–733)." *Pallas* 78: 257–277.
Barchiesi, A. 2009. "Phaethon and the Monsters." In *Paradox and the Marvellous in Augustan Literature and Culture*, ed. P.R. Hardie, 163–188. Oxford: Oxford University Press.
Barrenechea, F. 2004. "The Star Signs at Brundisium: Astral Symbolism in Lucan 2.691–692." *CQ* 54: 312–317.
——. 2010. "Didactic Aggressions in the Nile Excursus of Lucan's *Bellum Civile*." *AJP* 131: 259–284.
Barrière, F. 2019. "L'ombre de Sénèque: Lucain entre philosophie et poésie de la nature." In *La Philosophie des non-philosophes dans l'empire romain du Ier au IIIe siècle*, ed. S. Aubert-Baillot, C. Guérin, and S. Morlet, 67–82. Paris: Éditions de Boccard.
——. 2020. "Les astres de Lucain: représentations érudites et poétiques." In *Poésie et astronomie: de l'antiquité au romantisme*, ed. F. Barrière and C. Bertonèche, 41–59. Grenoble: UGA Éditions.
Barton, T. 1994. *Ancient Astrology*. London and New York: Routledge.
Bartsch, S. 1997. *Ideology in Cold Blood: A Reading of Lucan's* Civil War. Cambridge, MA, and London: Harvard University Press.
——. 2006. *The Mirror of the Self: Sexuality, Self-Knowledge, and the Gaze in the Early Roman Empire*. Chicago: The University of Chicago Press.
Batinski, E.E. 1991. "Cato and the Battle with the Serpents." *Syllecta Classica* 3: 71–76.
Beagon, M. 1992. *Roman Nature: The Thought of Pliny the Elder*. Oxford: Clarendon Press.
——. 2011. "The Curious Eye of the Elder Pliny." In Gibson and Morello, eds., 71–88.
Beaujeu, J. 1979. "L'astronomie de Lucain." In *L'astronomie dans l'antiquité classique*, ed. G. Aujac and J. Soubiran, 209–223. Paris: Les Belles Lettres.
Beck, R. 2007. *A Brief History of Ancient Astrology*. Malden, MA: Blackwell.
Benson, G.C. 2014. "Archimedes the Poet: Generic Innovation and Mathematical Fantasy in the *Cattle Problem*." *Arethusa* 47: 169–196.
Berno, F.R. 2002. "Ostio Quadra allo specchio: riflessioni speculari e speculative su *Nat. Quaest.* 1, 16–17." *Athenaeum* 90: 214–228.
——. 2003. *Lo specchio, il vizio e la virtù: studio sulle* Naturales Quaestiones *di Seneca*. Bologna: Pàtron.
——. 2012. "Non solo acqua: elementi per un diluvio universale nel terzo libro delle *Naturales Quaestiones*." In *Seneca e le scienze naturali*, ed. M. Beretta, F. Citti, and L. Pasetti, 49–68. Firenze: Olschki.
——. 2019. "Apocalypses and the Sage: Different Endings of the World in Seneca." *Gerión* 37: 75–95.
Berti, E., ed. 2000. *M. Annaei Lucani Bellum Civile, Liber X.* Firenze: Le Monnier.
Billault, A. 2004. "Volcanisme et esthétique: à propos du traité «Du sublime» 35, 4." In Foulon, ed., 193–203.
Bömer, F., ed. 1969. *P. Ovidius Naso: Metamorphosen, Buch I-III.* Heidelberg: Carl Winter.
Bouché-Leclercq, A. 1899. *L'astrologie grecque*. Paris: Leroux.
Boyle, A.J., ed. 2008. *Octavia: Attributed to Seneca*. Oxford: Oxford University Press.
——, ed. 2017. *Thyestes: Seneca*. Oxford: Oxford University Press.
Bramble, J.C. 1982. "Lucan." In *The Cambridge History of Classical Literature. II. Latin Literature*, ed. E.J. Kenney and W.V. Clausen, 533–557. Cambridge: Cambridge University Press.
Braund, S.H., trans. 1992. *Lucan: Civil War*. Oxford: Clarendon Press.
Brind'Amour, P. 1983. "Manilius and the Computation of the Ascendant." *CPh* 78: 144–148.
Brink, C.O., ed. 1971. *Horace on Poetry: The Ars Poetica*. Cambridge: Cambridge University Press.
——, ed. 1982. *Horace on Poetry: Epistles Book II, The Letters to Augustus and Florus*. Cambridge: Cambridge University Press.
Brown, R.D. 1990. "The Structural Function of the Song of Iopas." *HSCP* 93: 315–334.

Buckley, E., and M.T. Dinter, eds. 2013. *A Companion to the Neronian Age*. Chichester: Wiley-Blackwell.

Buffière, F. 1956. *Les mythes d'Homère et la pensée grecque*. Paris: Les Belles Lettres.

Burke, E. 1998. *A Philosophical Enquiry into the Origin of Our Ideas of the Sublime and Beautiful, and Other Pre-Revolutionary Writings*. Ed. D. Womersley. London: Penguin.

Burkert, W. 1987. *Ancient Mystery Cults*. Cambridge, MA, and London: Harvard University Press.

Buxton, R. 2016. "Mount Etna in the Greco-Roman *imaginaire*: Culture and Liquid Fire." In *Valuing Landscape in Classical Antiquity: Natural Environment and Cultural Imagination*, ed. J. McInerney and I. Sluiter, 25–45. Leiden and Boston: Brill.

Caldini Montanari, R. 1981. "Virgilio, Manilio e Germanico: memoria poetica e ideologia imperiale." In *Cultura e ideologia da Cicerone a Seneca*, no editor, 71–114. Florence: Le Monnier.

Cazzaniga, I. 1957. "L'episodio dei serpi libici in Lucano e la tradizione dei *Theriaka* Nicandrei." *Acme* 10: 27–41.

Celotto, G. 2018. "The Metapoetic Function of Magic: Ovid's Orpheus and Lucan's Erichto." *Latomus* 77: 628–645.

Champlin, E. 2003. *Nero*. Cambridge, MA: Harvard University Press.

Chaudhuri, P. 2014. *The War with God: Theomachy in Roman Imperial Poetry*. New York: Oxford University Press.

Chomse, S. 2018. "Instability and the Sublime in Martial's *Liber Spectaculorum*." In *After 69 CE: Writing Civil War in Flavian Rome*, ed. L.D. Ginsberg and D.A. Krasne, 387–409. Berlin: De Gruyter.

———. "Building the Sublime Emperor in 'A More Vertical Rome.'" In *Luoghi dell'Abitare, Immaginazione Letteraria e Identità Romana*, ed. M. Citroni, M. Labate, and G. Rosati, 161–184. Pisa: Edizioni ETS.

Cizek, E. 1972. *L'époque de Néron et ses controverses idéologiques*. Leiden: Brill.

Clausen, W., ed. 1994. *A Commentary on Virgil,* Eclogues. Oxford: Clarendon Press.

Codoñer Merino, C., ed. 1979. *L. Annaei Senecae Naturales Quaestiones*. 2 vols. Madrid: Consejo Superior de Investigaciones Científicas.

Coleman, K.M. 1990. "Fatal Charades: Roman Executions Staged as Mythological Enactments." *JRS* 80: 44–73.

———. 1993. "Launching into History: Aquatic Displays in the Early Empire." *JRS* 83: 48–74.

Coleman, R., ed. 1977. *Vergil: Eclogues*. Cambridge: Cambridge University Press.

Connors, C. 2015. "In the Land of the Giants: Greek and Roman Discourses on Vesuvius and the Phlegraean Fields." *ICS* 40: 121–137.

Conte, G.B. 1994. "Instructions for a Sublime Reader: Form of the Text and Form of the Addressee in Lucretius's *De Rerum Natura*." In *Genres and Readers: Lucretius, Love Elegy, Pliny's Encyclopedia*, trans. G.W. Most, 1–34. Baltimore and London: The Johns Hopkins University Press.

———. 1996. *The Hidden Author: An Interpretation of Petronius' Satyricon*. Trans. E. Fantham. Berkeley, Los Angeles, London: University of California Press.

Cullyer, H.C. 1999. "Greatness of Soul from Aristotle to Cicero: The Genealogy of a Virtue." Diss. Yale.

D'Alessandro Behr, F. 2007. *Feeling History: Lucan, Stoicism, and the Poetics of Passion*. Columbus: Ohio State University Press.

Day, H.J.M. 2013. *Lucan and the Sublime: Power, Representation and Aesthetic Experience*. Cambridge: Cambridge University Press.

de Angelis, F. 2014. "Sublime Histories, Exceptional Viewers: Trajan's Column and Its Visibility." In *Art and Rhetoric in Roman Culture*, ed. J. Elsner and M. Meyer, 89–114. Cambridge: Cambridge University Press.

de Bolla, P. 1989. *The Discourse of the Sublime: Readings in History, Aesthetics, and the Subject*. Oxford: Basil Blackwell.

de Jonge, C.C. 2012. "Dionysius and Longinus on the Sublime: Rhetoric and Religious Language." *AJP* 133: 271–300.

——— . 2020. "Ps.-Longinus on Ecstasy: Author, Audience, and Text." In *Experience, Narrative, and Criticism in Ancient Greece*, ed. J. Grethlein, L. Huitink, and A. Tagliabue, 148–171. Oxford: Oxford University Press.

De Lacy, P. 1943. "The Philosophy of the Aetna." *TAPA* 74: 169–178.

de Nadaï, J.-C. 2000. *Rhétorique et poétique dans la* Pharsale *de Lucain: la crise de la représentation dans la poésie antique*. Louvain and Paris: Peeters.

De Vivo, A. 1985. "Motivi proemiali nell'*Aetna*." *Vichiana* 14: 259–278.

——— . 1989. "Considerazioni sull'*Aetna*: rapporti con Seneca, epoca della composizione." *Vichiana* 18: 63–85.

——— . 1992a. "Il proemio dell'*Aetna* Pseudo-Virgiliano." In *Prefazioni, prologhi, proemi di opere tecnico-scientifiche latine, Volume II*, ed. C. Santini and N. Scivoletto, 661–681. Rome: Herder.

——— . 1992b. *Le parole della scienza: sul trattato "De Terrae Motu" di Seneca*. Salerno: P. Laveglia.

——— . 1998. *Costruire la memoria: ricerche sugli storici latini*. Naples: Loffredo.

——— . 2009. "Il volo di Fetonte da Ovidio a Seneca." *GIF* 61: 123–137.

——— . 2012. "Seneca e i terremoti: (*Questioni Naturali*, Libro VI)." In *Seneca e le scienze naturali*, ed. M. Beretta, F. Citti, and L. Pasetti, 93–106. Firenze: Olschki.

Degl'Innocenti Pierini, R. 1990. "Seneca, Ovidio e il diluvio." In *Tra Ovidio e Seneca*, 177–210. Bologna: Pàtron.

Delarue, F. 2000. *Stace, poète épique: originalité et cohérence*. Louvain and Paris: Peeters.

——— . 2006. "Sénèque lecteur d'Ovide et le *Traité du Sublime*." *Interférences: Ars Scribendi* 4. http://ars-scribendi.ens-lyon.fr/spip.php?article44.

Delle Donne, C.G. 2021. "Ulysses *proficiens*: A Reassessment of Seneca's Reading." *Rónai* 9: 4–21.

Di Giovine, C. 1981. "La polemica con Virgilio in *Aetna* 260 sgg." *RFIC* 109: 298–303.

Diels, H. 1886. *Seneca und Lucan*. Berlin: Akademie der Wissenschaften.

Dinter, M.T. 2012. *Anatomizing Civil War: Studies in Lucan's Epic Technique*. Ann Arbor: University of Michigan Press.

Domenicucci, P. 2013. *Il cielo di Lucano*. Pisa: ETS.

Donini, P. 1979. "L'eclettismo impossibile: Seneca e il platonismo medio." In *Modelli filosofici e letterari: Lucrezio, Orazio, Seneca*, ed. P. Donini and G.F. Gianotti, 149–300. Bologna: Pitagora.

Doody, A. 2013. "Literature of the World: Seneca's *Natural Questions* and Pliny's *Natural History*." In Buckley and Dinter, eds., 288–301.

Doran, R. 2015. *The Theory of the Sublime from Longinus to Kant*. Cambridge: Cambridge University Press.

Dorandi, T., ed. 1999. *Antigone de Caryste: Fragments*. Paris: Les Belles Lettres.

Dover, K.J., ed. 1968. *Aristophanes: Clouds*. Oxford: Clarendon Press.

Dross, J. 2019. "Le sublime, langage de la philosophie? Quelques réflexions sur *La tranquillité de l'âme* et les *Lettres à Lucilius* de Sénèque." *Cahiers des Études Anciennes* 56: 159–175.

Duret, L. 1988. "Néron-Phaéthon ou la témérité sublime." *RÉL* 66: 139–155.

D'Urso, V. 2019. *Vivit post proelia Magnus: commento a Lucano,* Bellum Civile VIII. Naples: Loffredo.

Economou, G.D. 2002. *The Goddess Natura in Medieval Literature*. Notre Dame: University of Notre Dame Press.

Edmonds, R.G. 2019. *Drawing down the Moon: Magic in the Ancient Greco-Roman World*. Princeton: Princeton University Press.

Edwards, C. 1997. "Self-Scrutiny and Self-Transformation in Seneca's Letters." *Greece and Rome* 44: 23–28.

Effe, B. 1977. *Dichtung und Lehre: Untersuchungen zur Typologie der antiken Lehrgedichts*. Munich: C.H. Beck.

Eldred, K.O. 2000. "Poetry in Motion: The Snakes of Lucan." *Helios* 27: 63–74.
Ellis, R. 1899. "The Literary Relations of 'Longinus' and Manilius." *CR* 13: 294.
——— , ed. 2008. *Aetna*. Ed. K. Volk. Reprint of the 1901 edition with new introduction and bibliography. Exeter: Bristol Phoenix Press.
Elsner, J. 2014. "Lithic Poetics: Posidippus and his Stones." *Ramus* 43: 152–172.
Endt, J., ed. 1969. *Adnotationes super Lucanum*. Leipzig: B.G. Teubner.
Erler, M. 2002. "Epicurus as *Deus Mortalis*: *Homoiosis Theoi* and Epicurean Self-Cultivation." In Frede and Laks, eds., 159–181.
Esposito, P. 2007. "I segnali della tempesta nella riscrittura lucanea (*Phars.* 5, 540–550)." In Landolfi and Monella, eds., 83–110.
Evans, J. 1998. *The History and Practice of Ancient Astronomy*. New York: Oxford University Press.
Evans, J., and J.L. Berggren, eds. 2006. *Geminos's* Introduction to the Phenomena*: A Translation and Study of a Hellenistic Survey of Astronomy*. Princeton: Princeton University Press.
Faggin, G. 1967. "Il misticismo astrale di Seneca." *Atene e Roma* 12: 23–37.
Fantham, E., ed. 1992a. *Lucan: De Bello Civili, Book II*. Cambridge: Cambridge University Press.
———. 1992b. "Lucan's Medusa-Excursus: Its Design and Purpose." *MD* 29: 95–119.
Fanti, G. 2019. "Didactic Dismissals in Manilius' *Astronomica*." In *Il vero condito: caratteri e ambiti della poesia didascalica nel mondo antico*, ed. R. Colombo, F. Gasti, M. Gay, and F. Sorbello, 59–73. Pavia: Pavia University Press.
Fear, F.T. 2011. "The Roman's Burden." In Gibson and Morello, eds., 21–34.
Feeney, D.C. 1991. *The Gods in Epic: Poets and Critics of the Classical Tradition*. Oxford: Clarendon Press.
Feldherr, A. 2021. "Caesar or Cleopatra? Lucan's Tragic Queen." In *Identities, Ethnicities and Gender in Antiquity*, ed. J. Fabre-Serris, A. Keith, and F. Klein, 135–152. Berlin and Boston: De Gruyter.
Feraboli, S., E. Flores, and R. Scarcia, eds. 1996–2001. *Manilio: il poema degli astri (Astronomica)*. 2 vols. Milan: A. Mondadori.
Festugière, A.J. 1944–1954. *La révélation d'Hermès Trismégiste*. 4 vols. Paris: J. Gabalda.
Flashar, H., ed. 1972. *Aristoteles: Mirabilia*. Berlin: Akademie-Verlag.
Flemming, R. 2003. "Empires of Knowledge: Medicine and Health in the Hellenistic World." In *A Companion to the Hellenistic World*, ed. A. Erskine, 449–463. Malden, MA: Blackwell.
Flinterman, J.-J. 2014. "Pythagoreans in Rome and Asia Minor around the Turn of the Common Era." In *A History of Pythagoreanism*, ed. C.A. Huffman, 341–359. Cambridge: Cambridge University Press.
Flores, E. 1991. "L'urbs e la cosmopoli in Man. 5,734ss. e un confronto con un'antica concezione astrologica cinese." In *Sýnesis: studi su forme del pensiero storico e politico greco e romano*, 113–124. Naples: Liguori.
Ford, A. 1992. *Homer: The Poetry of the Past*. Ithaca, NY: Cornell University Press.
Foulon, É, ed. 2004. *Connaissance et représentations des volcans dans l'antiquité*. Clermont-Ferrand: Presses Universitaires Blaise Pascal.
Fowler, D.P. 1989. "Lucretius and Politics." In *Philosophia Togata: Essays on Philosophy and Roman Society*, ed. M.T. Griffin and J. Barnes, 120–150. Oxford: Clarendon Press.
Fratantuono, L.M. 2012. "Andromeda, Perseus, and the End of the *Astronomica*." *Maia* 64: 305–315.
Frede, D., and A. Laks, eds. 2002. *Traditions of Theology: Studies in Hellenistic Theology, Its Background and Aftermath*. Leiden and Boston: Brill.
French, R. 1994. *Ancient Natural History: Histories of Nature*. London and New York: Routledge.
Gale, M.R. 1994. *Myth and Poetry in Lucretius*. Cambridge: Cambridge University Press.

——— . 2000. *Virgil on the Nature of Things: The* Georgics*, Lucretius, and the Didactic Tradition.* Cambridge: Cambridge University Press.

——— . 2011. "Digressions, Intertextuality, and Ideology in Didactic Poetry: The Case of Manilius." In Green and Volk, eds., 205–221.

——— . 2022. "Plagues and the Limits of Didactic Authority: Lucretius and Others." In *Teaching through Images: Imagery in Greco-Roman Didactic Poetry*, ed. J. Strauss Clay and A. Vergados, 205–230. Leiden and Boston: Brill.

Galli Milić, L. 2016. "Manilius et l'éloge de Néron (Lucan. 1,33–66): quelques considérations intertextuelles sur le *proemium* du *Bellum Civile*." In *Lucan and Claudian: Context and Intertext*, ed. V. Berlincourt, L. Galli Milić, and D.P. Nelis, 107–125. Heidelberg: Universitätsverlag Winter.

Garani, M. 2009. "Going with the Wind: Visualizing Volcanic Eruptions in the Pseudo-Vergilian *Aetna*." *BICS* 52: 103–121.

——— . 2020a. "Seneca as Lucretius' Sublime Reader (*Naturales Quaestiones* 3 Praef.)." In *Lucretius Poet and Philosopher: Backgrounds and Fortunes of* De Rerum Natura, ed. P.R. Hardie, V. Prosperi, and D. Zucca, 105–126. Berlin and Boston: De Gruyter.

——— . 2020b. "Seneca on Pythagoras' *Mirabilia Aquarum* (*NQ* 3.20–1, 25–6; Ovid *Met.* 15.270–336)." In *Intertextuality in Seneca's Philosophical Writings*, ed. M. Garani, A.N. Michalopoulos, and S. Papaioannou, 198–224. London and New York: Routledge.

——— . 2021. "Seneca's Medical Imagery in the Eschatological Flood (*Naturales Quaestiones* 3.27–30)." *Lucius Annaeus Seneca* 1: 151–182.

——— . 2022. "Keep up the Good Work: (Don't) Do It like Ovid (Sen. *QNat.* 3.27–30)." In *Philosophy in Ovid, Ovid as Philosopher*, ed. K. Volk and G.D. Williams, 145–163. New York: Oxford University Press.

——— . 2023. "Seneca's Platonism Revisited: Myth and the Sublime in Plato, Ovid, and Seneca (*Q. Nat.* 3)." *Lucius Annaeus Seneca* 2: 191–224.

Gardner, H.H. 2019. *Pestilence and the Body Politic in Latin Literature.* Oxford: Oxford University Press.

Garelli-François, M.-H. 2007. "Présence du sublime dans les tragédies du Sénèque: hauteurs et précipices." In *La littérature et le sublime*, ed. P. Marot, 59–77. Toulouse: Presses Universitaires du Mirail.

Gauly, B.M. 2004. *Senecas* Naturales Quaestiones*: Naturphilosophie für die römische Kaiserzeit.* Munich: C.H. Beck.

——— . 2018. "Grenzen des Reiches und Grenzen des Wissens in Lucans Bürgerkriegsepos." In *Konzeptionen des Unendlichen: eine europäische Kulturkonstante?* Ed. C. Böttigheimer and R. Dausner, 81–101. Würzburg: Königshausen und Neumann.

Gee, E. 2013. *Aratus and the Astronomical Tradition.* New York: Oxford University Press.

Getty, R.J., ed. 1940. *M. Annaei Lucani De Bello Civili, Liber I.* Cambridge: Cambridge University Press.

Geus, K., and C.G. King. 2018. "Paradoxography." In *The Oxford Handbook of Science and Medicine in the Classical World*, ed. P.T. Keyser and J. Scarborough, 431–444. New York: Oxford University Press.

Giampiccolo, E. 2010. "La *miranda fabula* dei *Pii Fratres*: la leggenda, le monete, la statua." *RCCM* 52: 71–79.

Giannini, A., ed. 1966. *Paradoxographorum Graecorum Reliquiae.* Milan: Istituto Editoriale Italiano.

Gibson, R.K., and R. Morello, eds. 2011. *Pliny the Elder: Themes and Contexts.* Leiden and Boston: Brill.

Gigon, O. 1991. "Senecas *Naturales Quaestiones*." In *Sénèque et la prose latine: neuf exposés suivis de discussions*, ed. P. Grimal, 313–346. Geneva: Fondation Hardt.

Gildenhard, I. 2004. "Confronting the Beast: From Virgil's Cacus to the Dragons of Corenlias van Haarlem." *Proceedings of the Virgil Society* 25: 27–48.

Giuliese, C. 2001. "Divina animi voluptas... cognoscere terram: *osservazioni su* Aetna *219–281*." *Invigilata Lucernis* 23: 111–120.

Gladhill, B. 2016. *Rethinking Roman Alliance: A Study in Poetics and Society*. Cambridge: Cambridge University Press.

Glauthier, P. 2011. "*Census* and *commercium*: Two Economic Metaphors in Manilius." In Green and Volk, eds., 188–201.

———. 2017. "Repurposing the Stars: Manilius, *Astronomica* 1, and the Aratean Tradition." *AJP* 138: 267–303.

———. 2018. "Playing the Volcano: *Prometheus Bound* and Fifth-Century Volcanic Theory." *CPh* 113: 255–278.

———. 2020. "*Bugonia* and the Aetiology of Didactic Poetry in Virgil, *Georgics* 4." *CQ* 69: 745–763.

———. 2022. "An Image Sublime: The Milky Way in Aratus and Manilius." In *Teaching through Images: Imagery in Greco-Roman Didactic Poetry*, ed. J. Strauss Clay and A. Vergados, 82–104. Leiden and Boston: Brill.

———. 2023. "The Classical Sublime." In *The Cambridge Companion to the Romantic Sublime*, ed. C. Duffy, 17–28. Cambridge: Cambridge University Press.

———. 2024. "Time Stood Still, and It Was Sublime: *Proto-Gospel of James* 18." In *Making Time for Greek and Roman Literature*, ed. K. Gilhuly and J.P. Ulrich, 205–225. London and New York: Routledge.

Goodyear, F.R.D., ed. 1965. *Incerti Auctoris Aetna*. Cambridge: Cambridge University Press.

———, ed. 1966. "Aetna." In *Appendix Vergiliana*, ed. W.V. Clausen, F.R.D. Goodyear, E.J. Kenney, and J.A. Richmond. Oxford: Clarendon Press.

———. 1984. "The *Aetna*: Thought, Antecedents, and Style." *ANRW* 2.32.1: 344–363.

Goold, G.P. 1961. "A Greek Professorial Circle at Rome." *TAPA* 92: 168–192.

———, ed. 1997. *Manilius: Astronomica*. 2nd ed. Cambridge, MA: Harvard University Press.

———, ed. 1998. *M. Manilii Astronomica*. 2nd ed. Stuttgart and Leipzig: Teubner.

Gowers, E. 2011. "The Road to Sicily: Lucilius to Seneca." *Ramus* 40: 168–197.

Graver, M. 2000. "Commentary on Inwood." In *Proceedings of the Boston Area Colloquium in Ancient Philosophy 15*, ed. J.J. Cleary and G.M. Gurtler, 44–56. Leiden and Boston, Köln: Brill.

Graziano, M.R. 2021. "Cose innumerabili e onniscienza divina : Lucan. 5, 181–182." *Bollettino di Studi Latini* 51: 530–536.

Green, C.M.C. 1991. "*Stimulos dedit aemula virtus*: Lucan and Homer Reconsidered." *Phoenix* 45: 230–254.

Green, S.J. 2014. *Disclosure and Discretion in Roman Astrology: Manilius and His Augustan Contemporaries*. Oxford: Oxford University Press.

Green, S.J., and K. Volk, eds. 2011. *Forgotten Stars: Rediscovering Manilius'* Astronomica. Oxford: Oxford University Press.

Griffin, M.T. 1984. *Nero: The End of a Dynasty*. London: Batsford.

Gross, N. 1989. *Senecas Naturales Quaestiones: Komposition, naturphilosophische Aussagen und ihre Quellen*. Stuttgart: F. Steiner.

Guillemin, A.-M. 1954. "Sénèque directeur d'âmes, III, les théories littéraires." *RÉL* 32: 250–274.

Gunderson, E. 2015. *The Sublime Seneca: Ethics, Literature, Metaphysics*. Cambridge: Cambridge University Press.

Habinek, T. 2007. "Probing the Entrails of the Universe: Astrology as Bodily Knowledge in Manilius' *Astronomica*." In *Ordering Knowledge in the Roman Empire*, ed. J. König and T. Whitmarsh, 229–240. Cambridge: Cambridge University Press.

———. 2011. "Manilius' Conflicted Stoicism." In Green and Volk, eds., 32–44.

Hadot, P. 1995. *Philosophy as a Way of Life: Spiritual Exercises from Socrates to Foucault*. Ed. A.I. Davidson. Trans. M. Chase. Malden, MA: Blackwell.

Halliwell, S. 2011. *Between Ecstasy and Truth: Interpretations of Greek Poetics from Homer to Longinus*. Cambridge: Cambridge University Press.

——— , ed. 2021. *Sul Sublime*. Trans. L. Lulli. Milan: Mondadori.
——— , ed. 2022. *Pseudo-Longinus:* On the Sublime. Oxford: Oxford University Press.
Hannah, R. 1996. "Lucan *Bellum Ciuile* 1.649–65: The Astrology of P. Nigidius Figulus Revisited." *Papers of the Leeds International Latin Seminar* 9: 175–190.
Hannay, M. 2016. "*Magnanimus Phaethon*: The Sublime in Ovid's *Metamorphoses*." MA Thesis. Leiden.
Hardie, P.R. 1986. *Virgil's* Aeneid*: Cosmos and Imperium*. Oxford: Clarendon Press.
——— . 2009. *Lucretian Receptions: History, the Sublime, Knowledge*. Cambridge: Cambridge University Press.
——— . 2013a. "Flavian Epic and the Sublime." In *Flavian Epic Interactions*, ed. G. Manuwald and A. Voigt, 125–138. Berlin and Boston: De Gruyter.
——— . 2013b. "Lucan's *Bellum Civile*." In Buckley and Dinter, eds., 225–240.
——— . 2022a. *Celestial Aspirations: Classical Impulses in British Poetry and Art*. Princeton: Princeton University Press.
——— . 2022b. "The Ovidian Sublime. Antiquity and After." *Dictynna* 19. https://journals.openedition.org/dictynna/2793?lang=en.
Harrison, S.J., ed. 1991. *Vergil: Aeneid 10*. Oxford: Clarendon Press.
Heath, M. 2012. "Longinus and the Ancient Sublime." In *The Sublime: From Antiquity to the Present*, ed. T.M. Costelloe, 11–23. Cambridge: Cambridge University Press.
——— . 2013. *Ancient Philosophical Poetics*. Cambridge: Cambridge University Press.
Henderson, J. 1987. "Lucan/The Word at War." *Ramus* 16: 122–164.
Herbert-Brown, G. 2007. "Scepticism, Superstition, and the Stars: Astronomical Angst in Pliny the Elder." In *Vita Vigilia Est: Essays in Honour of Barbara Levick*, ed. E. Bispham, G. Rowe, and E. Matthews, 113–126. London: Institute of Classical Studies.
Heslin, P. 2007. "Augustus, Domitian and the So-Called Horologium Augusti." *JRS* 97: 1–20.
Hinds, S. 1987. "Generalizing about Ovid." *Ramus* 16: 4–31.
Hine, H.M., ed. 1981. *An Edition with Commentary of Seneca Natural Questions, Book Two*. New York: Arno Press.
——— , ed. 1996a. *L. Annaei Senecae Naturalium Quaestionum Libros*. Stuttgart: Teubner.
——— . 1996b. *Studies in the Text of Seneca's Naturales Quaestiones*. Stuttgart: Teubner.
——— . 2002. "Seismology and Vulcanology in Antiquity?" In *Science and Mathematics in Ancient Greek Culture*, ed. C.J. Tuplin and T.E. Rihll, 56–75. Oxford: Oxford University Press.
——— . 2006. "Rome, the Cosmos, and the Emperor in Seneca's *Natural Questions*." *JRS* 96: 42–72.
——— , trans. 2010. *Seneca: Natural Questions*. Chicago and London: The University of Chicago Press.
——— , trans. 2012. "*Aetna*: A New Translation Based on the Text of F.R.D. Goodyear." *Studies in History and Philosophy of Science* 43: 316–325.
Holzberg, N. 2020. "Einführung." In *Appendix Vergiliana: Lateinisch-Deutsch*, ed. F. Zogg, 11–42. Berlin and Boston: De Gruyter.
Hosius, C. 1892. "Lucanus und Seneca." *Neue Jahrbücher für Philologie und Pädagogik. Abteilung 1. Jahrbücher für classische Philologie* 145: 337–356.
Housman, A.E., ed. 1903–1930. *M. Manilii Astronomicon Libri*. 5 vols. London: Richards.
——— , ed. 1926. *M. Annaei Lucani Belli Civilis Libri Decem*. Oxford: Blackwell.
Hübner, W. 1984. "Manilius als Astrologe und Dichter." *ANRW* 2.32.1: 126–320.
——— , ed. 2010. *Manilius,* Astronomica, *Buch V*. 2 vols. Berlin and New York: De Gruyter.
Hunink, V.J.C. 1989. "Ondergronde stromen: Lucretius en de Aetna-dichter." *Lampas* 22: 22–35.
——— , ed. 1992. *M. Annaeus Lucanus, Bellum Civile, Book III*. Amsterdam: Gieben.
Hunter, R.L. 2008. "Written in the Stars: Poetry and Philosophy in the *Phainomena* of Aratus." In *On Coming After: Studies in Post-Classical Greek Literature and Its Reception*, 153–188. Berlin and New York: De Gruyter.

——. 2009. *Critical Moments in Classical Literature: Studies in the Ancient View of Literature and Its Uses.* Cambridge: Cambridge University Press.
Innes, D.C. 1979. "Gigantomachy and Natural Philosophy." *CQ* 29: 165–171.
Inwood, B. 1995. "Seneca in His Philosophical Milieu." *HSCP* 97: 63–76.
——. 2000. "God and Human Knowledge in Seneca's *Natural Questions*." In *Proceedings of the Boston Area Colloquium in Ancient Philosophy 15*, ed. J.J. Cleary and G.M. Gurtler, 23–43, 55–56. Leiden and Boston, Köln: Brill.
——. 2002. "God and Human Knowledge in Seneca's *Natural Questions*." In Frede and Laks, eds., 119–157.
——. 2009. "Why Physics?" In *God and Cosmos in Stoicism*, ed. R. Salles, 201–223. Oxford: Oxford University Press.
Iodice, M.G., ed. 2002. *Appendix Vergiliana*. Milan: Mondadori.
Johnson, W.R. 1987. *Momentary Monsters: Lucan and His Heroes.* Ithaca, NY: Cornell University Press.
Johnston, P.A. 1996. "Under the Volcano: Volcanic Myth and Metaphor in Vergil's *Aeneid*." *Vergilius* 42: 55–65.
Kant, I. 2000. *Critique of the Power of Judgment.* Ed. P. Guyer. Trans. P. Guyer and E. Matthews. Cambridge: Cambridge University Press.
Kaufman, D.H. 2020. "Lucan's Cato and Popular (Mis)Conceptions of Stoicism." In Zientek and Thorne, eds., 133–149. London and New York: Bloomsbury.
Kennedy, D.F. 2011. "Sums in Verse or a Mathematical Aesthetic?" In Green and Volk, eds., 165–187.
——. 2013a. *Antiquity and the Meanings of Time: A Philosophy of Ancient and Modern Literature.* Cambridge: Cambridge University Press.
——. 2013b. "The Political Epistemology of Infinity." In *Lucretius: Poetry, Philosophy, Science*, ed. D. Lehoux, A.D. Morrison, and A.R. Sharrock, 51-67. Oxford: Oxford University Press.
Kenney, E.J., ed. 2014. *Lucretius: De Rerum Natura, Book III.* 2nd ed. Cambridge: Cambridge University Press.
Kersten, M. 2018. *Blut auf pharsalischen Feldern: Lucans Bellum Ciuile und Vergils Georgica.* Göttingen: Vandenhoeck und Ruprecht.
Keyser, P.T. 1994. "On Cometary Theory and Typology from Nechepso-Petosiris through Apuleius to Servius." *Mnemosyne* 47: 624–651.
——. 2013. "The Name and Nature of Science: Authorship in Social and Evolutionary Context." In *Writing Science: Medical and Mathematical Authorship in Ancient Greece*, ed. M. Asper and A.-M. Kanthak, 17–61. Berlin and Boston: De Gruyter.
Kidd, I.G, ed. 1988. *Posidonius, Voume II: The Commentary.* 2 vols. Cambridge: Cambridge University Press.
Kim, L.Y. 2010. *Homer Between History and Fiction in Imperial Greek Literature.* Cambridge: Cambridge University Press.
Kirwan, J. 2005. *Sublimity: The Non-Rational and the Irrational in the History of Aesthetics.* New York and London: Routledge.
Klotz, A., ed. 1927. *C. Iuli Caesaris Commentarii. Vol. III. Commentarii Belli Alexandrini, Belli Africi, Belli Hispaniensis. Accedunt C. Iuli Caesaris et A. Hirtii fragmenta.* Stuttgart and Leipzig: Teubner.
Knoche, U. 1935. *Magnitudo Animi: Untersuchungen zur Entstehung und Entwicklung eines römischen Wertgedankens.* Leipzig: Dieterich.
König, J., and T. Whitmarsh, eds. 2007. *Ordering Knowledge in the Roman Empire.* Cambridge: Cambridge University Press.
Kreuzwieser, C. 2016. *Der Begriff* natura *und seine ethische Relevanz in Senecas Prosaschriften.* Göttingen: V&R Unipress.
Kruschwitz, P. 2015. "Getting on Top of Things: Form and Meaning in the Pseudo-Vergilian Aetna." *Habis* 46: 75–97.

Kuhn, T.S. 1957. *The Copernican Revolution: Planetary Astronomy in the Development of Western Thought*. Cambridge, MA: Harvard University Press.

Kuhn-Treichel, T. 2020. "Was leistet Lehrdichtung? Neue Perspektiven auf das Motiv der Himmelsreise in Manilius' *Astronomica*." *Hermes* 148: 492–504.

Küppers, J. 1996. "'Kosmosschau' und *virtus* in den Philosophica Senecas." *A&A* 42: 57–75.

Kyriakidis, S. 2015. "Rome and the *fata Asiae* (Manilius, *Astr.* 1.512)." In *Lemmata: Beiträge zum Gedenken an Christos Theodoridis*, ed. M. Tziatzi, M. Billerbeck, F. Montanari, and K. Tsantsanoglou, 265–285. Berlin and Boston: De Gruyter.

——— . 2018. "*Deflexus solito cursu*: Phaethon between Ovid and Manilius." *Myrtia* 33: 109–153.

La Penna, A. 2001. "Tra Fetonte e Icaro: ardimento o amore della scienza?" *Maia* 53: 535–563.

Lagière, A. 2017. *La Thébaïde de Stace et le Sublime*. Brussels: Latomus.

Lamberton, R. 1986. *Homer the Theologian: Neoplatonist Allegorical Reading and the Growth of the Epic Tradition*. Berkeley: University of California Press.

Landolfi, L. 1990. "Manilio e gli eroi della Via Lattea: tra doctrina e ideologia." *GIF* 42: 87–98.

——— . 1991. "Manilio e la gerarchia delle stelle (*Astr.* 5. 734–745)." *Prometheus* 17: 247–258.

——— . 2003. Integra Prata*: Manilio, i proemi*. Bologna: Pàtron.

——— . 2007. "Stratigrafie multiple e suggestioni dotte: l'esempio di Luc. *Phars.* 9, 700–733." In Landolfi and Monella, eds., 111–149.

Landolfi, L., and P. Monella, eds. 2007. *Doctus Lucanus: aspetti dell'erudizione nella Pharsalia di Lucano*. Bologna: Pàtron.

Lapidge, M. 1979. "Lucan's Imagery of Cosmic Dissolution." *Hermes* 107: 344–370.

——— . 1989. "Stoic Cosmology and Roman Literature, First to Third Centuries A.D." *ANRW* 2.36.3: 1379–1429.

Lassandro, D. 1993. "*Nosse fidem rerum*: l'elogio della scienza nell'*Aetna*." *Bollettino di Studi Latini* 23: 320–328.

Lausberg, M. 1990. "Epos und Lehrgedicht: ein Gattungsvergleich am Beispiel von Lucans Schlangenkatalog." *WJA* 16: 173–203.

Le Blay, F. 2006. "Le poème latin sur l'Etna: témoin d'un savoir oublié." In *Musa Docta: recherches sur la poésie scientifique dans l'antiquité*, ed. C. Cusset, 335–361. Saint-Étienne: Publications de l'Université de Saint-Étienne.

Lehoux, D. 2011. "Myth and Explanation in Manilius." In Green and Volk, eds., 45–56.

——— . 2012. *What Did the Romans Know? An Inquiry into Science and Worldmaking*. Chicago and London: The University of Chicago Press.

Leigh, M. 1997. *Lucan: Spectacle and Engagement*. Oxford: Clarendon Press.

——— . 1999. "Lucan's Caesar and the Sacred Grove: Deforestation and Enlightenment in Antiquity." In *Interpretare Lucano: Miscellanea di Studi*, ed. P. Esposito and L. Nicastri, 167–205. Salerno and Naples: Arte Tipografica.

——— . 2000. "Lucan and the Libyan Tale." *JRS* 90: 95–109.

——— . 2006. "Statius and the Sublimity of Capaneus." In *Epic Interactions: Perspectives on Homer, Virgil, and the Epic Tradition Presented to Jasper Griffin by Former Pupils*, ed. M.J. Clarke, B.G.F. Currie, and R.O.A.M. Lyne, 217–241. Oxford: Oxford University Press.

Leitão, D.D. 1998. "Senecan Catoptrics and the Passion of Hostius Quadra (*Sen. Nat. 1*)." *MD* 41: 127–160.

Leroux, V. 2004. "La représentation de l'Etna dans l'épopée latine." In Foulon, ed., 57–78.

Leventhal, M. 2015. "Counting on Epic: Mathematical Poetry and Homeric Epic in Archimedes' Cattle Problem." *Ramus* 44: 200–221.

——— . 2017. "Eratosthenes' Letter to Ptolemy: The Literary Mechanics of Empire." *AJP* 138: 43–84.

Lightfoot, J. 2021. *Wonder and the Marvellous from Homer to the Hellenistic World*. Cambridge: Cambridge University Press.

Limburg, F.J.G. 2007. "*Aliquid ad mores*: The Prefaces and Epilogues of Seneca's Naturales Quaestiones." Diss. Leiden.
Littlewood, C.A.J. 2004. *Self-Representation and Illusion in Senecan Tragedy*. Oxford: Clarendon Press.
———. 2017. "*Hercules Furens* and the Senecan Sublime." *Ramus* 46: 153–174.
Liuzzi, D. 1988. "Echi degli *Aratea* di Cicerone negli *Astronomica* di Manilio." *Rudiae* 1: 115–159.
Lombardo, G. 2018. "Le sublime et la poétique cosmologique." In *Le sublime: poétique, esthétique, philosophie*, ed. P.-H. Frangne, C. Flécheux, and D. Laroque, 51–64. Rennes: Presses universitaires de Rennes.
Long, A.A. 1982. "Astrology: Arguments pro et Contra." In *Science and Speculation: Studies in Hellenistic Theory and Practice*, ed. J. Brunschwig, M.F. Burnyeat, and M. Schofield, 165–192. Cambridge: Cambridge University Press.
Loupiac, A. 1998. *La poétique des éléments dans la Pharsale de Lucain*. Brussels: Latomus.
Lowe, D. 2014. "Heavenly and Earthly Elements in Manilius' *Astronomica*." *Ramus* 43: 45–66.
———. 2015. *Monsters and Monstrosity in Augustan Poetry*. Ann Arbor: University of Michigan Press.
Lühr, F.-F. 1969. "*Ratio* und *Fatum*: Dichtung und Lehre bei Manilius." Diss. Frankfurt.
———. 1971. "Die Kritik des Aetna-Dichters an Manilius." *Hermes* 99: 141–149.
MacGregor, A.P. 2004. "Which Art in Heaven: The Sphere of Manilius." *ICS* 29: 143–157.
———. 2005. "Was Manilius Really a Stoic?" *ICS* 30: 41–65.
Mader, G.J. 1983. "Some Observations on the Senecan Götterdämmerung." *Acta Classica* 26: 61–71.
Magnavacca, A. 2020. "«Caesareae domus series»: Cesare, Augusto e… Germanico nel *Bellum ciuile* di Lucano." *Maia* 72: 336–361.
Malamud, M. 2003. "Pompey's Head and Cato's Snakes." *CPh* 98: 31–44.
Mann, W.-R. 2011. "On Two Stoic 'Paradoxes' in Manilius." In Green and Volk, eds., 85–103.
Mannering, J. 2013. "Seneca's Philosophical Writings: *Naturales Quaestiones, Dialogi, Epistulae Morales*." In Buckley and Dinter, eds., 188–203.
Manolaraki, E. 2013. *Noscendi Nilum Cupido: Imagining Egypt from Lucan to Philostratus*. Berlin and Boston: De Gruyter.
———. 2018. "Senses and the Sacred in Pliny's *Natural History*." *ICS* 43: 207–233.
Mansfeld, J., and D.T. Runia, eds. 2023. *Aetius: Placita*. Cambridge, MA, and London: Harvard University Press.
Marti, B.M. 1975. "Lucan's Narrative Techniques." *La Parola del Passato* 30: 74–90.
Martin, P.-M. 2019. "Archéologie du mot *sublimis*." *Cahiers des Études Anciennes* 56: 37–60.
Maruotti, A. 2019. "Sortir de soi: le parcours de Lucilius dans les *Lettres* de Sénèque." *Cahiers des Études Anciennes* 56: 177–187.
Master, J. 2015. "The Shade of Sallust: History-Writing in the *Natural Questions* of Seneca." *CPh* 110: 333–352.
Masters, J. 1992. *Poetry and Civil War in Lucan's* Bellum Civile. Cambridge: Cambridge University Press.
Matthews, M., ed. 2008. *Caesar and the Storm: A Commentary on Lucan, De bello civili, Book 5, Lines 476–721*. Bern and Frankfurt am Main: Lang.
Mayhoff, C., ed. 1892–1909. *C. Plini Secundi Naturalis Historiae Libri XXXVII*. Stuttgart and Leipzig: Teubner.
Mazzoli, G. 1968. "Due note anneane, II. L'*inclitum carmen* di Vagellio." *Athenaeum* 46: 363–368.
———. 1970. *Seneca e la poesia*. Milan: Ceschina.
———. 1990. "Seneca e il sublime." In *Dicibilità del sublime*, ed. T. Kemeny and E.C. Cotta Ramusino, 89–97. Udine: Campanotto.
———. 1991. "Seneca e la poesia." In *Sénèque et la prose latine: neuf exposés suivis de discussions*, ed. P. Grimal, 177–209. Geneva: Fondation Hardt.

Mazzucchi, C.M., ed. 2010. *Dionisio Longino: Del Sublime.* 2nd ed. Milan: V&P.
McCarter, S. 2015. *Horace between Freedom and Slavery: The First Book of* Epistles. Madison, WI: The University of Wisconsin Press.
McCartney, E.S. 1960. "Vivid Ways of Indicating Uncountable Numbers." *CPh* 55: 79–89.
McCloskey, P., and E. Phinney, Jr. 1968. "Ptolemaeus Tyrannus: The Typification of Nero in the Pharsalia." *Hermes* 96: 80–87.
McConnell, S. 2017a. "*Magnitudo Animi* and Cosmic Politics in Cicero's *De Re Publica.*" *CJ* 113: 45–70.
——— . 2017b. "The Epicurean Virtue of Μεγαλοψυχία." *CPh* 112: 175–199.
McRoberts, S. 2015. "Pompey as Phaethon." *Syllecta Classica* 26: 51–75.
Meeusen, M. 2016. *Plutarch's Science of Natural Problems: A Study with Commentary on* Quaestiones Naturales. Leuven: Leuven University Press.
Meyer, A.-S. 2023. *Naturphänomene in Lucans* Bellum Civile. Basel: Schwabe.
Michel, A. 1969. "Rhétorique, tragédie, philosophie: Sénèque et le sublime." *GIF* 21: 245–257.
——— . 1987. "L'esthétique de Pline l'Ancien." *Helmantica* 38: 55–67.
Migliorini, P. 1997. *Scienza e terminologia medica nella letteratura latina di età neroniana: Seneca, Lucano, Persio, Petronio.* Bern and Frankfurt am Main: Lang.
Miller, P. 1965. *The Life of the Mind in America, from the Revolution to the Civil War.* New York: Harcourt, Brace and World, Inc.
Montanari, F. 2015. *The Brill Dictionary of Ancient Greek.* Co-authored by I. Garofalo and D. Manetti. English edition by M. Goh and C. Schroeder. Leiden: Brill.
Morel, W. 1928. "Iologica." *Philologus* 83: 345–389.
Morford, M.P.O. 1967. *The Poet Lucan: Studies in Rhetorical Epic.* New York: Barnes and Noble.
Morgan, L. 1998. "Assimilation and Civil War: Hercules and Cacus, *Aeneid* 8." In *Vergil's* Aeneid*: Augustan Epic and Political Context*, ed. H.-P. Stahl. London: Duckworth.
——— . 2003. "Child's Play: Ovid and His Critics." *JRS* 93: 66–91.
Most, G.W. 2012. "The Sublime, Today?" In *Dynamic Readings: Studies in the Reception of Epicureanism*, ed. B. Holmes and W.H. Shearin, 239–266. New York: Oxford University Press.
Muecke, F. 2006. "A Literary Voyage Round Mt Etna." *Classicum* 32: 13–23.
Murphy, T.R. 2004. *Pliny the Elder's* Natural History*: The Empire in the Encyclopedia.* Oxford: Oxford University Press.
Musso, S. 2012. *La Via Lattea dei Greci e dei Romani: Manilio,* Astronomica, *I 666–804.* Vercelli: Mercurio.
Myers, K.S. 1994. *Ovid's Causes: Cosmogony and Aetiology in the* Metamorphoses. Ann Arbor: University of Michigan Press.
Naas, V. 2002. *Le projet encyclopédique de Pline l'Ancien.* Rome: École française de Rome.
——— . 2011. "Imperialism, *mirabilia* and Knowledge: Some Paradoxes in the *Naturalis Historia.*" In Gibson and Morello, eds., 57–70.
Narducci, E. 1979. *La provvidenza crudele: Lucano e la distruzione dei miti augustei.* Pisa: Giardini.
——— . 1983. "Pauper Amyclas (modelli etici e poetici in un episodio della *Pharsalia*)." *Maia* 35: 183–194.
——— . 2001. "Pompeo in Cielo (*Pharsalia* IX 1–24; 186–217), un verso di Dante (*Parad.* XXII 135) e il senso delle allusioni a Lucano in due epigrammi di Marziale (IX 34; XI 5)." *MH* 58: 70–92.
——— . 2002. *Lucano: un'epica contro l'impero: interpretazione della* Pharsalia. Rome: Laterza.
Nelis, D.P. 2011. "Praising Nero (Lucan, *De Bello Civili* 1, 33–66)." In *Dicere laudes: elogio, comunicazione, creazione del consenso*, ed. G. Urso, 253–264. Pisa: ETS.
Netz, R. 2009. *Ludic Proof: Greek Mathematics and the Alexandrian Aesthetic.* Cambridge: Cambridge University Press.

Neuburg, M. 1993. "Hitch Your Wagon to a Star: Manilius and His Two Addressees." *MD* 31: 243–283.

Newlands, C.E. 2002. *Statius'* Silvae *and the Poetics of Empire*. Cambridge: Cambridge University Press.

Nicolet, C. 1991. *Space, Geography, and Politics in the Early Roman Empire*. Ann Arbor: University of Michigan Press.

Nicolson, M.H. 1959. *Mountain Gloom and Mountain Glory: The Development of the Aesthetics of the Infinite*. Ithaca, NY: Cornell University Press.

Nye, D.E. 1994. *American Technological Sublime*. Cambridge, MA, and London: The MIT Press.

O'Gorman, F. 2021. "Endless Breath? The Pipe Organ and Immortality." In *The Life of Breath in Literature, Culture and Medicine: Classical to Contemporary*, ed. D. Fuller, C. Saunders, and J. Macnaughton, 305–324. Cham: Palgrave Macmillan.

O'Higgins, D. 1988. "Lucan as Vates." *CA* 7: 208–226.

O'Rourke, D. 2019. "Knowledge Is Power: Dynamics of (Dis)Empowerment in Didactic Poetry." In *Didactic Poetry of Greece, Rome and Beyond: Knowledge, Power, Tradition*, ed. L.G. Canevaro and D. O'Rourke, 21–52. Swansea: Classical Press of Wales.

Osorio, P. 2020. "Vergil's Physics of *Bugonia* in *Georgics* 4." *CPh* 115: 27–46.

Ozanam, A.-M. 1990. "Le mystère et le sacré dans le stoïcisme romain à l'époque néronienne." *Bulletin de l'Association Guillaume Budé* 3: 275–288.

Page, D.L., ed. 1981. *Further Greek Epigrams: Epigrams before A.D. 50 from the Greek Anthology and Other Sources, Not Included in "Hellenistic Epigrams" or "The Garland of Philip."* Cambridge: Cambridge University Press.

Paisley, P.B., and D.R. Oldroyd. 1979. "Science in the Silver Age: *Aetna*, a Classical Theory of Volcanic Activity." *Centaurus* 23: 1–20.

Palmer, J.A. 2009. *Parmenides and Presocratic Philosophy*. Oxford and New York: Oxford University Press.

Papaioannou, S. 2005. "Epic Transformation in the Second Degree: The Decapitation of Medusa in Lucan, *BC* 9, 619–889." In Walde, ed., 216–236.

Paparazzo, E. 2011. "Philosophy and Science in the Elder Pliny's *Naturalis Historia*." In Gibson and Morello, eds., 89–111.

Parroni, P. 2000. "Le *Naturales Quaestiones* fra scienza e morale." In *Seneca e il suo tempo*, ed. P. Parroni, 433–444. Rome: Salerno.

———, ed. 2002. *Seneca: Ricerche Sulla Natura*. Milan: A. Mondadori.

Payne, M. 2016. "*Aetna* and Aetnaism: Schiller, Vibrant Matter, and the Phenomenal Regimes of Ancient Poetry." *Helios* 43: 89–108.

Pease, A.S., ed. 1955–1958. *M. Tulli Ciceronis De Natura Deorum Libri III*. 2 vols. Cambridge, MA: Harvard University Press.

———, ed. 1979. *M. Tulli Ciceronis De Divinatione*. Reprint of 1920 edition. 2 vols. New York: Arno Press.

Pépin, J. 1986. "Cosmic Piety." In *Classical Mediterranean Spirituality: Egyptian, Greek, Roman*, ed. A.H. Armstrong, 408–435. London: Routledge and Kegan Paul.

Phillips, O.C. 1995. "Singing Away Snakebite: Lucan's Magical Cures." In *Ancient Magic and Ritual Power*, ed. M.W. Meyer and P.A. Mirecki, 391–400. Leiden and New York: Brill.

Pichon, R. 1912. *Les sources de Lucain*. Paris: Leroux.

Pingoud, J. 2008. "La science rattrapée par le mythe: le poète de l'*Aetna* et l'hermétisme d'Héraclite." *Pallas* 78: 207–219.

Porter, J.I. 2007. "Lucretius and the Sublime." In *The Cambridge Companion to Lucretius*, ed. S. Gillespie and P.R. Hardie, 167–184. Cambridge: Cambridge University Press.

———. 2016. *The Sublime in Antiquity*. Cambridge: Cambridge University Press.

———. 2020. "Living on the Edge: Self and World *in extremis* in Roman Philosophy." *CA* 39: 225–283.

Postl, B. 1970. "Die Bedeutung des Nil in der römischen Literatur mit besonderer Berücksichtigung der wichtigsten griechischen Autoren." Diss. Wien.

Preisendanz, K., and A. Henrichs, eds. 1973. *Papyri Graecae Magicae*. 2nd ed. Stuttgart: Teubner.

Quadlbauer, F. 1958. "Die genera dicendi bis Plinius d. J." *WS* 71: 55–111.

Quint, D. 1993. *Epic and Empire: Politics and Generic Form from Virgil to Milton*. Princeton: Princeton University Press.

———. 2004. "Fear of Falling: Icarus, Phaethon, and Lucretius in Paradise Lost." *Renaissance Quarterly* 57: 847–881.

———. 2020. "Erictho and Demogorgon: Poetry against Metaphysics." *Arion* 28: 1–40.

Ramelli, I. 2014. "Manilius and Stoicism." In *The Philosophizing Muse: The Influence of Greek Philosophy on Roman Poetry*, ed. M. Garani and D. Konstan, 161–186. Newcastle upon Tyne: Cambridge Scholars Press.

Raschle, C.R., ed. 2001. *Pestes Harenae: die Schlangenepisode in Lucans Pharsalia (IX, 587–949)*. Bern and Frankfurt am Main: Lang.

Reeh, A. 1973. "Interpretationen zu den Astronomica des Manilius mit besonderer Berücksichtigung der philosophischen Partien." Diss. Marburg.

Reitz, C. 2020. "Caesar and the Supernatural: An Interpretation of Lucan. V 504–549." *Maia* 72: 404–411.

Reydams-Schils, G. 2016. "Teaching Pericles: Cicero on the Study of Nature." In *Roman Reflections: Studies in Latin Philosophy*, ed. G.D. Williams and K. Volk, 91–108. New York: Oxford University Press.

———. 2017. "'Becoming like God' in Platonism and Stoicism." In *From Stoicism to Platonism*, ed. T. Engberg-Pedersen, 142–158. Cambridge: Cambridge University Press.

Richards, A. 1999. "Automatic Genius: Mozart and the Mechanical Sublime." *Music & Letters* 80: 366–389.

Richardson, J.S. 2008. *The Language of Empire: Rome and the Idea of Empire from the Third Century BC to the Second Century AD*. Cambridge: Cambridge University Press.

Rihll, T.E. 1999. *Greek Science*. Oxford: Oxford University Press.

Ripoll, F. 2019. "Le *Bellum ciuile* de Pétrone: un Exemple de «sublime raté»?" *Cahiers des Études Anciennes* 56: 129–157.

Rix, R.W. 2023. "The Arctic Sublime." In *The Cambridge Companion to the Romantic Sublime*, ed. C. Duffy, 154–165. Cambridge: Cambridge University Press.

Roby, C.A. 2014. "Seneca's Scientific Fictions: Models as Fictions in the *Natural Questions*." *JRS* 104: 155–180.

Roche, P.A. 2005. "Righting the Reader: Conflagration and Civil War in Lucan's *De Bello Civili*." *Scholia* 14: 52–71.

———, ed. 2009. *Lucan:* De Bello Civili, *Book 1*. Oxford: Oxford University Press.

Roller, D.W. 2022. *A Guide to the Geography of Pliny the Elder*. Cambridge: Cambridge University Press.

Roloff, D. 1970. *Gottähnlichkeit, Vergöttlichung und Erhöhung zu seligem Leben: Untersuchungen zur Herkunft der platonischen Angleichung an Gott*. Berlin: De Gruyter.

Romm, J.S. 1992. *The Edges of the Earth in Ancient Thought: Geography, Exploration, and Fiction*. Princeton: Princeton University Press.

Rosenmeyer, T.G. 2000. "Seneca and Nature." *Arethusa* 33: 99–119.

Rossetti, M. 2017. "Elementi protrettici nel finale del IV libro degli *Astronomica* di Manilio." *Vichiana* 54: 75–91.

Rossi, A. 2005. "*sine fine*: Caesar's Journey to Egypt and the End of Lucan's *Bellum Civile*." In Walde, ed., 237–260.

Rostagni, A. 1933. *Virgilio minore: saggio sullo svolgimento della poesia virgiliana*. Turin: Chiantore.

———, ed. 1979. *Suetonio* De poetis *e biografi minori*. Reprint of the 1944 edition. New York: Arno Press.

Rowe, C.J., ed. 1993. *Plato: Phaedo*. Cambridge: Cambridge University Press.
Rubincam, C. 2003. "Numbers in Greek Poetry and Historiography: Quantifying Fehling." *CQ* 53: 448–463.
Runia, D.T. 2002. "The Beginnings of the End: Philo of Alexandria and Hellenistic Theology." In Frede and Laks, eds., 281–316.
Russell, D.A., ed. 1964. *"Longinus": On the Sublime*. Oxford: Clarendon Press.
Rutherford, R.B. 1989. *The Meditations of Marcus Aurelius: A Study*. Oxford: Oxford University Press.
Salemme, C. 2000. *Introduzione agli* Astronomica *di Manilio*. 2nd ed. Naples: Loffredo.
Santelia, S. 2012. *La miranda fabula dei pii fratres in* Aetna *603–645*. Bari: Cacucci.
Schepens, G., and K. Delcroix. 1996. "Ancient Paradoxography: Origin, Evolution, Production and Reception." In *La letteratura di consumo nel mondo greco-latino*, ed. O. Pecere and A. Stramaglia, 373–460. Cassino: Università degli Studi di Cassino.
Schiesaro, A. 2003. *The Passions in Play: Thyestes and the Dynamics of Senecan Drama*. Cambridge: Cambridge University Press.
———. 2014. "*Materiam superabat opus*: Lucretius Metamorphosed." *JRS* 104: 73–104.
Schindler, C. 2000. *Untersuchungen zu den Gleichnissen im römischen Lehrgedicht: Lucrez, Vergil, Manilius*. Göttingen: Vandenhoeck und Ruprecht.
Schmidt, M.G. 1986. *Caesar und Cleopatra. Philologischer und historischer Kommentar zu Lucan 10,7–171*. Frankfurt: Lang.
Schmidt, W., ed. 1899. *Heronis Alexandrini Opera Quae Supersunt Omnia. Volumen I: Pneumatica et Automata*. Stuttgart: B.G. Teubner.
Schotes, H.-A. 1969. *Stoische Physik, Psychologie und Theologie bei Lucan*. Bonn: Habelt.
Schrijvers, P.H. 1983. "Le chant du monde: remarques sur *Astronomica* I 1–24 de Manilius." *Mnemosyne* 36: 143–150.
———. 2005. "The 'Two Cultures' in Lucan: Some Remarks on Lucan's *Pharsalia* and Ancient Sciences of Nature." In Walde, ed., 26–39.
———. 2006. "Silius Italicus and the Roman Sublime." In *Flavian Poetry*, ed. R.R. Nauta, H.-J. van Dam, and J.J.L. Smolenaars, 97–111. Leiden: Brill.
Schwemmler, F. 1916. "De Lucano Manilii imitatore." Diss. Giessen.
Scott, J. 1999. "The Ethics of the Physics in Seneca's *Natural Questions*." *Classical Bulletin* 75: 55–68.
Seidman, J. 2017. "A Poetic Caesar in Lucan's *Pharsalia*." *CJ* 113: 72–95.
Seitz, K. 1965. "Der pathetische Erzählstil Lucans." *Hermes* 93: 204–232.
Setaioli, A. 1985. "Seneca e lo stile." *ANRW* 2.32.2: 776–858.
———. 1991. "Due messe a punto senecane, 1: la tradizione dei versi di Cleante a Zeus e al fato; 2: Seneca e il sublime." *Prometheus* 17: 137–154.
———. 2000. *Facundus Seneca: aspetti della lingua e dell'ideologia senecana*. Bologna: Pàtron.
———. 2007. "Seneca and the Divine: Stoic Tradition and Personal Developments." *International Journal of the Classical Tradition* 13: 333–368.
Shackleton Bailey, D.R., ed. 1997. *M. Annaei Lucani De Bello Civili Libri X*. Stuttgart and Leipzig: B.G. Teubner.
Shaw, P. 2017. *The Sublime*. 2nd ed. London and New York: Routledge.
Sherwin-White, A.N. 1966. *The Letters of Pliny: A Historical and Social Commentary*. Oxford: Clarendon Press.
Shuger, D.K. 1984. "The Grand Style and the 'Genera Dicendi' in Ancient Rhetoric." *Traditio* 40: 1–42.
Sklenář, R. 2003. *The Taste for Nothingness: A Study of* Virtus *and Related Themes in Lucan's* Bellum Civile. Ann Arbor: The University of Michigan Press.
Smith, M. 2004. "Elusive Stones: Reading Posidippus' *Lithika* through Technical Writing on Stones." In *Labored in Papyrus Leaves: Perspectives on an Epigram Collection Attributed to Posidippus (P.Mil.Vogl. VIII 309)*, ed. B. Acosta-Hughes, E. Kosmetatou, and M. Baumbach, 105–117. Cambridge, MA: Harvard University Press.

Stahl, W.H. 1962. *Roman Science: Origins, Development, and Influence to the Later Middle Ages.* Madison, WI: The University of Wisconsin Press.

Staley, G.A. 2010. *Seneca and the Idea of Tragedy.* Oxford: Oxford University Press.

Star, C. 2021. *Apocalypse and Golden Age: The End of the World in Greek and Roman Thought.* Baltimore: The Johns Hopkins University Press.

Stewart, Z. 1959. "The Song of Silenus." *HSCP* 64: 179–205.

Stoneman, R. 2020. "Giants or Science: Cosmic Strife, Mount Etna, and *Aetna*." In *Conflict and Competition: Agon in Western Greece*, ed. H.L. Reid, J. Serrati, and T. Sorg, 187–201. Sioux City, IA: Parnassos Press-Fonte Aretusa.

Sudhaus, S., ed. 1898. *Aetna*. Leipzig: B.G. Teubner.

Taub, L.C. 2003. *Ancient Meteorology.* London and New York: Routledge.

———. 2008. *Aetna and the Moon: Explaining Nature in Ancient Greece and Rome.* Corvallis, OR: Oregon State University Press.

———. 2009. "Explaining a Volcano Naturally: *Aetna* and the Choice of Poetry." In *Authorial Voices in Greco-Roman Technical Writing*, ed. L.C. Taub and A. Doody, 125–141. Trier: Wissenschaftlicher Verlag.

Taylor, J.C. 2020. "Even *natura* Nods: Lucan's Alternative Explanations of the Syrtes (9.303–318)." In Zientek and Thorne, eds., 91–109.

Tesoriero, C. 2005. "Trampling over Troy: Caesar, Virgil, Lucan." In Walde, ed., 202–215.

Theiler, W. 1957. "Demiurgos." *RAC* 3: 694–711.

Tomcik, M. 2021. "Looking for LVCAN: Weather Signs and Wordplay in the Pharsalia (5, 541–550)." *MH* 78: 234–241.

Toohey, P. 1996. *Epic Lessons: An Introduction to Ancient Didactic Poetry.* London and New York: Routledge.

———. 2004. *Melancholy, Love, and Time: Boundaries of the Self in Ancient Literature.* Ann Arbor: University of Michigan Press.

Torre, C. 2003. "Sublime del potere, potere del sublime in Seneca." In *Potere sovrano: simboli, limiti, abusi*, ed. S. Simonetta, 235–251. Bologna: Il Mulino.

———. 2007a. "«Alia temptanda est via»: alcune riflessioni sui recenti sviluppi della questione dei 'due' Seneca (morale e tragico)." *Acme* 60: 37–84.

———. 2007b. "Tra Ovidio e Seneca: la traccia dell'epos di Pitagora nel programma filosofico delle *Naturales Quaestiones*." In *La poesia filosofica*, ed. A. Costazza, 45–61. Milan: Cisalpino.

Toulze-Morisset, F. 2004. "Sénèque s'étonne-t-il dans les «Questions naturelles»?" In *Conceptions et représentations de l'extraordinaire dans le monde antique*, ed. O. Bianchi, O. Thévenaz, and P. Mudry, 199–220. Bern and Frankfurt am Main: Lang.

Tracy, J. 2010. "*Fallentia Sidera*: The Failure of Astronomical Escapism in Lucan." *AJP* 131: 635–661.

———. 2014. *Lucan's Egyptian Civil War.* Cambridge: Cambridge University Press.

Truitt, E.R. 2015. "Between Art and Nature: *Natura artifex*, Neoplatonism, and Literary Automata." In *Medieval Robots: Mechanism, Magic, Nature, and Art*, 40–68. Philadelphia: University of Pennsylvania Press.

Tutrone, F. 2017. "Seneca On the Nature of Things: Moral Concerns and Theories of Matter in *Natural Questions* 6." *Latomus* 76: 765–789.

Uden, J. 2011. "A Song from the Universal Chorus: The Perseus and Andromeda Epyllion." In Green and Volk, eds., 235–252.

Verde, F. 2020. "Epicurean Meteorology, Lucretius, and the *Aetna*." In *Lucretius Poet and Philosopher: Background and Fortunes of* De Rerum Natura, ed. P.R. Hardie, V. Prosperi, and D. Zucca, 83–101. Berlin and Boston: De Gruyter.

Volk, K. 2001. "Pious and Impious Approaches to Cosmology in Manilius." *MD* 47: 85–117.

———. 2002. *The Poetics of Latin Didactic: Lucretius, Vergil, Ovid, Manilius.* Oxford: Oxford University Press.

——— . 2003. "Manilius' Solitary Chariot-Ride (*Astronomica* 2.138–140)." *CQ* 53: 628–33.

——— . 2004. "'Heavenly Steps': Manilius 4.119–121 and Its Background." In *Heavenly Realms and Earthly Realities in Late Antique Religions*, ed. R. Boustan and A.Y. Reed, 34–46. Cambridge: Cambridge University Press.

——— . 2005a. "*Aetna* oder Wie man ein Lehrgedicht schreibt." In *Die Appendix Vergiliana: Pseudepigraphen im literarischen Kontext*, ed. N. Holzberg, 68–90. Tübingen: Narr.

——— . 2005b. "Lehrgedicht oder «Naturgedicht»? Naturwissenschaft und Naturphilosophie in der Lehrdichtung von Hesiod bis zur *Aetna*." In *Wissensvermittlung in dichterischer Gestalt*, ed. M. Horster and C. Reitz, 155–173. Stuttgart: Steiner.

——— . 2009. *Manilius and His Intellectual Background.* Oxford: Oxford University Press.

——— . 2010. "Literary Theft and Roman Water Rights in Manilius' Second Proem." *MD* 65: 187–197.

——— . 2011. "Manilian Self-Contradiction." In Green and Volk, eds., 104–119.

Vottero, D., ed. 1989. *Lucio Anneo Seneca: Questioni Naturali.* Turin: Unione Tipografico-Editrice Torinese.

——— . 1998. "Seneca e la natura." In *L'uomo antico e la natura*, ed. R. Uglione, 291–303. Turin: Celid.

Waiblinger, F.-P. 1977. *Senecas Naturales Quaestiones. Griechische Wissenschaft und römische Form.* Munich: Beck.

Walde, C., ed. 2005. *Lucan im 21. Jahrhundert. Lucan in the 21st Century. Lucano nei primi del XXI Secolo.* Munich: Saur.

——— . 2007. "Per un'idrologia poetica: fiumi e acque nella *Pharsalia* di Lucano." In Landolfi and Monella, eds., 13–47.

Walsh, P.G., trans. 2006. *Pliny the Younger: Complete Letters.* Oxford: Oxford University Press.

Walters, J. 1998. "Making a Spectacle: Deviant Men, Invective, and Pleasure." *Arethusa* 31: 355–367.

Warren, J. 2000. "Epicurean Immortality." *OSAP* 18: 231–261.

Wehrli, F. 1946. "Der erhabene und der schlichte Stil in der poetisch-rhetorischen Theorie der Antike." In *Phyllobolia für P. von der Mühll zum 60. Geburtstag*, ed. O. Gigon, K. Meuli, W. Theiler, F. Wehrli, and B. Wyss, 9–34. Basel: Schwabe.

Weiner, J., and T.A. Benz. 2018. "Detroit and the Classical Sublime, or in Defense of 'ruin porn.'" In *Landscapes of Dread in Classical Antiquity: Negative Emotion in Natural and Constructed Spaces*, ed. D. Felton, 279–302. London: Routledge.

Welsh, J.T. 2014. "How to Read a Volcano." *TAPA* 144: 97–132.

Wessels, A. 2016. "Das Paradox der Katastrophe: zu Senecas Schrift über das Erdbeben (*Naturales quaestiones*, Buch 6)." In In *Erdbeben in der Antike: Deutungen, Folgen, Repräsentationen*, ed. J. Borsch and L. Carrara, 209–224. Tübingen: Mohr Siebeck.

West, D.A. 1969. *The Imagery and Poetry of Lucretius.* Edinburgh: Edinburgh University Press.

Wheeler, S. 2002. "Lucan's Reception of Ovid's *Metamorphoses*." *Arethusa* 35: 361–380.

——— . 2009. "Into New Bodies: The Incipit of Ovid's Metamorphoses as Intertext in Imperial Latin Literature." *MD* 61: 147–160.

Wick, C., ed. 2004. *M. Annaeus Lucanus, Bellum Civile, Liber IX.* 2 vols. Munich: K.G. Saur.

Wiener, C. 2016. "*ratio terrorem prudentibus excutit* - Zur Methode und Zielsetzung der Evaluierung von Erdbebentheorien in Senecas *Naturales Quaestiones*." In *Erdbeben in der Antike: Deutungen, Folgen, Repräsentationen*, ed. J. Borsch and L. Carrara, 189–208. Tübingen: Mohr Siebeck.

Wildberger, J. 2005. "*Quanta sub nocte iaceret nostra dies* (Lucan. 9, 13f.) - Stoizismus als Mittel der Verfremdung bei Lucan." In Walde, ed., 56–88.

——— . 2006. *Seneca und die Stoa: der Platz des Menschen in der Welt.* 2 vols. Berlin and Boston: De Gruyter.

Williams, G.D. 2012. *The Cosmic Viewpoint: A Study of Seneca's Natural Questions.* New York: Oxford University Press.

——— . 2016. "Minding the Gap: Seneca, the Self, and the Sublime." In *Roman Reflections: Studies in Latin Philosophy*, ed. G.D. Williams and K. Volk, 172–191. New York: Oxford University Press.

——— . 2017a. "A Passion for Nature: Seneca's *Natural Questions* and *Hippolytus* in His *Phaedra*." *Maia* 69: 312–325.

——— . 2017b. *Pietro Bembo on Etna: The Ascent of a Venetian Humanist*. New York: Oxford University Press.

——— . 2020. "Volcanic Wonder: A Starry-Eyed View of the *Aetna*." In *Constructing Authors and Readers in the Appendices Vergiliana, Tibulliana, and Ouidiana*, ed. T.E. Franklinos and L. Fulkerson, 112–130. New York: Oxford University Press.

Wilson, A.M. 1985. "The Prologue to Manilius 1." *Papers of the Liverpool Latin Seminar* 5: 283–298.

Wilson, E.R. 2014. *The Greatest Empire: A Life of Seneca*. New York: Oxford University Press.

Wilson, K. 2020. "The Astronomer-Poet at Night: The Evolution of a Motif." In *The Values of Nighttime in Classical Antiquity: Between Dusk and Dawn*, ed. J. Ker and A. Wessels, 131–149. Leiden and Boston: Brill.

Wolff, É. 2004. "L'image de l'Etna dans l'anonyme *Aetna*." In Foulon, ed., 79–84.

Wolkenhauer, A. 2011. *Sonne und Mond, Kalender und Uhr: Studien zur Darstellung und poetischen Reflexion der Zeitordnung in der römischen Literatur*. Berlin and New York: De Gruyter.

Wooten, C.W., trans. 1987. *Hermogenes'* On Types of Style. Chapel Hill and London: The University of North Carolina Press.

Yunis, H., ed. 2011. *Plato: Phaedrus*. Cambridge: Cambridge University Press.

Zetzel, J.E.G., ed. 1995. *Cicero: De Re Publica. Selections*. Cambridge: Cambridge University Press.

Ziegler, K. 1949. "Paradoxographoi." *RE* 18.3: 1137–1166.

Zientek, L. 2017. "The Auditory Sublime from Vergil to Lucan." *The Classical Outlook* 92: 35–43.

——— . 2020. "Mining and Morality in Lucan and Seneca." In Zientek and Thorne, eds., 73–90.

Zientek, L., and M. Thorne, eds. 2020. *Lucan's Imperial World: The* Bellum Civile *in Its Contemporary Contexts*. London and New York: Bloomsbury.

Zwierlein, O. 1974. "Cäsar und Kleopatra bei Lucan und in späterer Dichtung." *A&A* 20: 54–73.

Index Locorum

Since the index has been created to work across multiple formats, indexed terms for which a page range is given (e.g., 52–53, 66–70, etc.) may occasionally appear only on some, but not all of the pages within the range.

Aeschylus
TrGF
70 49 n.82
pseudo-Aeschylus
PV
367–368 221 n.56, 230 n.82
368–369 209
Aesop
Fab. (Perry)
40 39 n.58
Aetius
3.2.9 120 n.43
3.3 167 n.54
3.5.7–9 42 n.66
4.1 189 n.104
Aetna
1–93 202
1–4 224–225
2 237 n.97, 238–239
7–8 202 n.9
8 226
9–23 226
24 202
25 221
28 226, 239
30 210 n.25
32–33 238–239
40 238–239
42 238–239
54–73 213–214
72–73 210
77 238–239
91–92 238–239
94–176 202
116–117 238–239
135–139 210–211
139 225
156–157 211
156 216–217
168 210 n.25
171 210–211
173–174 210–211
177–384 202
177–180 211–212
183 211–212, 225
187b–187 211–212
187b 214
188 212, 214
191 212
192–198 212–214, 226–227
193–196 240–241
193–194 221
203 213–214
219–281 203–209
224–250 203–204
225 206 n.17
250 208
251–256 205–207, 256
252 214
256 221
257–269 206–207
260–269 238–239
271–281 207–208
271 238–239
272 214
273–279 213–214
273 214
282–305 217
293–300 217–220
293 224
294 221
296–297 224
319 210 n.25
326–328 220–221
328 210
330 210 n.25
334–336 221–222
351 214
366–384 211 n.28
373 239 n.102
374 210 n.25

Aetna (*Continued*)
378 210 n.25
384 132 n.75, 216–217
385–567 202
394 210 n.26
416–424 222–223
431 199 n.1
443–444 237 n.97
464 214
469 210
510–511 205 n.15
511 222
515–519 222–223
554–565 237 n.97
559–560 214
561 210 n.25
565–567 226–227
568–645 203, 226–227, 235–241
568–598 236
570 237 n.98
599–604 236–237
600–645 240–241
600–603 226–227
600 214, 214 n.39
621–624 238
632–633 238–239
636 238–239
643–645 238
642 237

Anaxagoras (25 *EGP*/59 D-K)
D4.10/A42.10 62 n.16
D53/A84 167 n.54
D63/A89 167 n.54
P35/A30.18–21 49 n.83
P40/A1 49 n.83

Anaximander (6 *EGP*/12 D-K)
D33/A23 167 n.54

Anaximenes (7 *EGP*/13 D-K)
D21/A17 167 n.54
D22 167 n.54
D23/A17 167 n.54
R11 39 n.59

Antipater of Thessalonica (Gow-Page)
91 217 n.45

Apollodorus
Bibl.
1.6.2 210 n.25
1.9.7 70 n.38

Apollonius of Rhodes
Argon.
1.494–515 15 n.9

Aratus
Phaen.
469–479 64
473–475 64
474–475 64

Aristides
Orat.
48 189 n.104

Aristophanes
Ach.
530–531 4 n.6, 52 n.89, 167 n.53
Nub.
218–238 26–27
223 26–27
227–234 26
228 27 n.31

Aristotle
Cael.
2.9.290b 71 n.41
2.14.297b–298a 26 n.29
Eth. Eud.
1.5.1216a 49 n.83
Fragments
On the Nile's Flood
686–695 Gigon 189 n.104
Protrepticus
18–19 Düring 49 n.83
Unassigned
838 Gigon 125 n.56
IA
11.711a 215 n.42
12.711a 215 n.42
Metaph.
1.2 43–44
1.2.982b 43–44
1.2.983a 43–44
Mete.
1.1.338a–339a 111
1.6.342b–343a 120 n.43
1.6.343a 121 n.48
1.14.352a 26 n.29
2.1–3.353a–359b 151 n.22
2.9 167 n.55
Part. an.
1.5.644b–645a 44–45, 205 n.16
1.5.645a 215 n.42

pseudo-Aristotle
Mund.
4.396a 151 n.22
6.400a–b 231
Mir. ausc.
4 233 n.91

9 233 n.91
38 229
40 229
102 232 n.89
105.3 229
112 229 n.77
130 229 n.77
148 229 n.77
152 231 n.86
154 230–231

Athenaeus
4.174b–f 219 n.52

M. Aurelius
Med.
1.15 46 n.72

Caesar
BGall.
6.14.6 152 n.25

Callimachus
Epigr. (Pfeiffer/Gow-Page)
23/53 39
Fragments (Pfeiffer)
1.23–24 15 n.11
1.36 210 n.25
384.31–32 189–190
407–411 227 n.72

Calvus, C. Licinius (*FLP*)
9 15 n.10

Cassius Dio
62.14.1 156 n.34
63.8.1–2 117 n.40

Celsus
5.27.3 179–180

Cicero
Amic.
24 212 n.30
Aratea
fr. 33.249 64
fr. 33.250 64
fr. 33.252 64
Diu.
1.63 48 n.78
2.91 84
Fin.
4.11–12 49–50, 204 n.11
Marcell.
28 190 n.107
Nat. D.
2.37 49 n.83
2.58 214 n.38
2.95 125 n.56
3.51 42 n.66
Off.
1.64–65 27 n.33
Orat.
15–16 52–53
Part. or.
56 52–53
Rep.
6.10 25–26
6.14 25–26
6.15 26 n.28
6.16 25–26
6.16–19 25–26
6.18 71 n.42
6.19 71 n.39, 71 n.41
Ti.
6 214 n.37
Tusc.
1.40 26 n.29
3.43 219 n.50

Ciris
1–53 54 n.93

Clement of Alexandria
Strom.
5.138.2 48 n.79

Conon
FGrH
1.43 231 n.85

Demetrius
Eloc.
53 32 n.42
75 53

Democritus (27 *EGP*/68 D-K)
D98/A91 62 n.16
D99 122 n.50
D100 122 n.50
D117/A130 167 n.54
D217/B18 16–17
D219/B17 16 n.13
D221/B21 16 n.13
D230/A169 45 n.71
D295/B216 45 n.71

Dio Chrysostom
Or.
11.37–70 196 n.120

Diodorus Siculus
1.32.1 189–190
1.37–41 189 n.104
1.37.1 189–190
1.37.6 189–190

Diodorus Siculus (*Continued*)
1.37.7 196
1.40 196
1.41.10 189–190
5.23.2 68

Diogenes of Apollonia (28 *EGP*/64 D-K)
D23/A16 167 n.54

Diogenes Laertius
1.34 39 n.58
2.4 38–39
2.10 49 n.83
5.49 229
7.123 46–47, 228, 231–232
10.79 46–47
10.85 46–47
10.135 48 n.79

Diogenes of Oenoanda (William/Chilton)
fr. 63/52, coll. iii–iv 48 n.79

Dionysius of Halicarnassus
Dem.
22 162 n.47
Lys.
13 31 n.41
Pomp.
2.15 32 n.44

Empedocles (22 *EGP*/31 D-K)
D4.4–5/B112.4–5 32–33
P29/A1 40 n.61

Ennius
Tragic fragments (Jocelyn/Goldberg-Manuwald)
169/57 34 n.49
187/82 206–207, 256

Epicurus
Ep. Hdt.
79 46–47
Ep. Men.
135 48 n.79
Ep. Pyth.
85 46–47
Gnom. Vat.
10 47–48

Erasistratus of Ceos
frr. 79–80 Garofalo 215 n.42

Eudoxus of Cnidus
frr. 287–289 Lasserre 194 n.114

Euripides
HF
908 210 n.25
TrGF
306–311 72 n.44

Eustathius
Od.
1453.20–22 Stallbaum 43 n.68

Galen (Kühn)
Ars med.
26 (1.378) 215 n.42
MM
3.3 (10.174) 215 n.42
Nat. Fac.
1.12 (2.26–30) 215 n.42
2.3 (2.87) 215 n.42
2.9 (2.131–132) 215 n.42
UP
3.10 (3.234–242) 215 n.42
3.10 (3.237.11–17) 49 n.82

Gellius
NA
14.1.11–12 125 n.55

Germanicus
Aratea
11–12 64–65
290–305 64–65
455–458 64–65
457 64–65

Heraclitus
Quaest. Hom.
26 230 n.82

Hermogenes
De ideis (Rabe)
1.5–6 (pp. 241–244) 53–54
1.6 (pp. 243–244) 150, 230 n.84

Herodotus
1.47.3 163–164
2.19–31 189–190
2.19 196
2.26 190 n.107
2.28 196
2.34 190 n.107
8.100.1 32 n.44

Heron of Alexandria
Aut. (Schmidt)
pp. 338–342 219 n.52
Pneum. (Schmidt)
p. 2 219–220
p. 28 219–220
pp. 130–136 220–221
pp. 192–202 219–220

Hesiod
Fragments (Merkelbach-West)
30 70 n.38
Theog.
38 48
116 210–211
338–339 190 n.107
740 12–13
744 13 n.4
748–757 13 n.4
780 42 n.64

Hippocrates
Flat.
3 132 n.75
Progn.
1.1 48 n.78

Hippolytus
Ref.
1.1 39 n.58

Homer
Il.
1.70 48
2.484–493 82 n.73
2.494–759 82
2.509 82
2.516 82
2.524 82
6.200–202 72 n.44
8.555–559 56 n.1
16.776 33–34
18.26–27 33–34
20.61–65 166
21.388 166
Od.
6.42–46 165–166
24.40 33–34

Horace
Ars
325–332 83 n.76
453–476 39–40
457 39–40
462–463 39–40
463–466 39–40
468–469 39–40
472 39–40
472–476 39–40
Carm.
1.11.1–3 81 n.70
4.2.1–4 34 n.49
4.14.45–46 189–190
4.14.46–48 190 n.107
Epist.
1.6.1–2 45
1.6.3–5 46
1.12.19–20 57–58, 253 n.15
2.1.208–213 73

Hyginus
Astron.
2.18 71–72
Fab.
254.4 231 n.85

John the Lydian
De mens.
4.107 189 n.105

Leucippus (27 *EGP*/67 D-K)
D117/A25 167 n.54

Livy
40.22.3 165 n.51

Longinus
Subl.
1.3 4–5
1.4 4–5, 51–52, 77–78, 167 n.53
3.2 162 n.47
3.3 32, 36–37
7.1 32 n.42
9.2 4 n.6, 4–5, 23 n.23
9.5–6 28 n.35
9.6 166
10.6 81–82
12.3 53 n.91
12.4 4 n.6, 167 n.53
13.1 61 n.13
13.2 160–161
13.3–4 57 n.4
15 147 n.7
15.4 34 n.49
15.9 51–52, 77–78
16 147 n.7
33–36 32
33.1–2 31–32, 69–70
34.3 4 n.6
34.4 167 n.53
35.2–5 229–230
35.2–4 240
35.2–3 50–51
35.2 4–5, 31–32
35.3 197, 247–248
35.4–5 51 n.86
35.4 190 n.107, 225 n.68
36.1 4–5, 23 n.23
43.5 240

Longinus (*Continued*)
44 51–52, 137 n.91, 251
44.8 24
44.9–10 51–52

Lucan
1.1 147–148
1.5 165 n.50
1.38–43 151–152
1.45–59 155–159
1.45–52 155–158
1.49 158–159
1.50 158–159
1.57–58 158–159
1.60–62 157–158
1.63–66 159
1.72 163 n.48
1.74 158–159
1.150 173 n.67
1.151–157 3–4, 147–148, 167–168
1.151 167–168
1.153–154 167–168
1.233–235 159
1.392–465 149
1.412–419 149–153
1.417 197
1.418–419 170–171
1.447–462 152–153
1.642–645 193 n.113
2.7–15 193 n.113
2.14–15 160 n.43
2.289–297 152 n.23
2.410–415 158 n.38
3.399–452 170
3.458–461 170–171
4.724–729 179–180
5.97–101 161–162, 224–225
5.119 163 n.48
5.161–169 162–163
5.165 160–161
5.169 162 n.47
5.177–182 163–164
5.188–189 163–164
5.197 163–164
5.214–218 162–163
5.224 162–163
5.515–576 171–174
5.515–537 172
5.515 171
5.536 173
5.540 172–174
5.557–559 172–174
5.568–576 173
5.580–583 177
5.638–653 173–174
5.656–659 173
5.672 174
5.681 174
6.438–491 170
6.463–465 164, 214 n.36
6.481–484 164–166
5.633 163 n.48
6.479 150 n.18
6.651 160–161
6.822–823 166–167
6.826 166–167
8.2 175
8.4–5 175
8.16–17 175
8.21–23 175
8.29–32 175
8.159–201 154–155, 174–178
8.161 175
8.166–170 175
8.167 171
8.171 175–176
8.172–176 175–177
8.173 176 n.75
8.182–183 176
8.186–192 177
8.259 177–178
8.477–478 196
9.1–11 153
9.11–18 153–155
9.14 186
9.303–318 159
9.528–543 168–169
9.604–937 178–187
9.611–618 179–180, 180 n.85
9.614 181–182
9.619–623 159, 180–182
9.619 181–182
9.630 181–182
9.724 181–182
9.734 181–182
9.734–746 182–184
9.741 184
9.744 181–182
9.758–759 183–184
9.761–762 183–184
9.723 184
9.769 184
9.779 181–182
9.787 181–182
9.790–804 185–186
9.805 181–182
9.844 181–182

9.850–851 184–185
9.853–854 184–185
9.881–889 186
9.895 186–187
9.908 186–187
9.914 186–187
9.923 186–187
9.927 186–187
9.931 186–187
9.922 181–182
9.926 181–182
9.930 181–182
10.172–331 187–198
10.185–187 190–191
10.187 194 n.114
10.188–192 191–192
10.190 197
10.196 197
10.199–218 192–193
10.202–203 194–195
10.204 150 n.18
10. 219–261 194–195
10.237–239 192–193
10.262–267 192–193
10.268–331 194
10.287 194
10.295–298 195
10.300–301 194
10.331 194
10.332–335 197
10.456–459 197

Lucian
Astr.
14–15 34 n.49
19 34 n.49
Icar.
17–18 154 n.28

Lucretius
1.38–43 65 n.26
1.62–79 23–24
1.68–69 35
1.69 23–24, 47
1.69–71 23–24
1.69–70 23–24
1.72–74 197
1.73 117
1.74 35
1.76 117
1.79 23–24
1.136–145 28 n.37
1.136–139 76
1.487 93
1.629 235 n.93
1.635–920 32–33
1.716–733 32–33
1.726 32–33
1.726–727 229
1.733 32–33
1.740–741 33–34, 39–40
1.921–930 152–153
1.927–928 156–157
1.936–950 76
1.947 76
1.988 62 n.19
1.1018 14 n.8
1.1102–1113 62 n.19
1.1103 14 n.8
2.59–61 28 n.37
2.65 14 n.8
2.105 14 n.8
2.109 14 n.8
2.662 62 n.19
2.1044–1047 247–248
2.1117 235 n.93
3.1–3 28 n.37
3.6 21–22
3.14–15 20–22
3.14 165–166
3.16–18 20
3.16–17 165–166
3.18–24 165–166
3.25–30 114
3.25–27 20
3.25 166
3.26–27 37–38, 165–166
3.28–30 21
3.28–29 204
3.211 134 n.82
3.273–281 151 n.20
3.931–977 256–257
3.939 134 n.82
3.977 134 n.82
4.1–5 152–153
4.2–3 156–157
4.11–25 76
4.22 76
5.1–54 29 n.39
5.82 134 n.82
5.95–96 62 n.19
5.392–406 34–35
5.397–398 35, 68
5.399–401 34–35
5.395 34
5.398 35
5.400 36 n.53

Lucretius (*Continued*)
5.405–406 34
5.445 204 n.11
5.1016 62 n.19
5.1204–1210 204 n.12
5.1205–1217 46
5.1218–1219 47
5.1362 235 n.93
6.47 35 n.50, 156–157
6.85 134 n.82
6.96–422 167–168
6.535–607 170 n.62
6.591–595 21 n.21
6.603–607 37–38
6.608–737 190 n.107
6.647–679 26 n.29
6.712–737 189 n.104
6.738–739 232 n.89
6.1011 21 n.20
6.1098–1102 181–182

Lycurgus
Leoc.
95–96 231 n.85

Manilius
1.1–4 70
1.1–24 49 n.82
1.11 67–68
1.13–17 156–157
1.13–15 67
1.13–14 128–129, 157 n.36, 206–207
1.17 67 n.31
1.19 77–78
1.22–23 70–71
1.66–112 95–96
1.68–69 73–74
1.142 57–58, 102, 252–253
1.255–262 89
1.259 75–76
1.294–295 176 n.75
1.474–482 176
1.529 73 n.49, 90 n.82
1.539–804 59–60
1.670 75–76
1.684–804 59–66
1.701–702 60, 211 n.29
1.713 42 n.66, 61
1.714–717 60–61, 64–65
1.714–715 98–99
1.714 61 n.14
1.715 74 n.53
1.716 68
1.718–804 61–64
1.718–722 62–63
1.735–743 67–68
1.737 69
1.738 90 n.82
1.740–741 71–72
1.744–745 69
1.748–749 69
1.749 69 n.36
1.750 69–70
1.755–757 98–99
1.758–804 153 n.27
1.760 63–64
1.802 63–64
1.803–804 63–64
1.806 73 n.49, 90 n.82
1.809–926 126–127
1.867–873 120 n.43
2.25–27 70
2.49–52 68 n.35
2.58–59 67–68
2.60–135 58 n.7
2.60–61 71, 79
2.99–104 61 n.13
2.107–108 79 n.65
2.109 95–96, 208–209
2.136–140 73–74
2.141–142 73–74
2.453–465 59 n.9
2.693 78 n.62
2.701–706 80–81
2.701 204
2.722–723 88
2.765 97
2.766–767 77–78
3.1–3 202 n.9
3.29–30 75, 88–89
3.31–34 75–76
3.38 76, 88–89
3.43–159 93–94
3.47–60 235
3.47 151 n.21
3.142–144 59 n.9
3.149–155 59 n.9
3.203–509 83–91
3.206–207 83–84
3.207 88
3.211–215 86
3.218–246 85
3.218 88–89
3.219–220 86–87
3.255 88
3.247–482 85
3.295–300 87

3.301–384 89
3.356–361 89–90
3.365–374 90
3.393–394 87–88
3.396 87–88
3.483–509 86
3.483 88–89
3.486–487 87–88
4.1–121 58 n.7
4.120–121 73–74
4.152–157 76–77
4.152 77–78
4.158–160 77–78
4.189–196 78 n.60
4.294–407 78
4.387–389 78 n.63
4.387 78
4.389 78
4.390–392 78–79
4.390 71
4.407 79
4.408–501 79–80
4.411–414 79–80
4.414–415 79–80
4.416 79–80
4.431–432 79–80
4.439 79–80
4.585–817 93–94
4.701–710 59 n.9
4.773–774 92
4.821 79–80
4.866–935 58 n.7
4.883–884 92
4.833 151 n.21
4.934–935 93–94
5.1 96–97
5.4 96–97
5.8–11 97
5.10 156–157
5.26 97
5.67–101 71–72
5.91–101 70–72
5.91 70
5.93 70–71
5.94–96 71
5.98 71–72
5.99 71–72
5.99–100 71–72
5.569–570 90 n.82
5.599 90 n.82
5.650–652 72–73
5.652–655 73–74
5.726–733 92–93
5.733 93
5.734 93
5.738–739 93, 99
5.742–745 98–99
5.742 125 n.54

Martial
Epigr.
7.24 231 n.85
Spect.
28 218 n.48

Mela
1.50–54 189 n.104

Nicander
Ther.
4–7 185–186

Nonnus
Dion.
41.379–381 249 n.8

Octauia
385–387 157 n.37, 214–215, 236 n.95

Ovid
Am.
1.15 37 n.55
Fast.
1.295–310 65 n.27
1.301 65 n.27
1.309–310 65 n.27
Met.
1.1–2 202 n.9
1.8 204 n.11
1.32–33 204 n.11
1.747–2.332 36–37
1.751 36 n.53
2.56–58 36
2.111 36 n.53
2.161–168 68
2.137 158–159
2.180–181 108
2.191 36
2.193 36
2.200 36, 108
2.201–209 68
2.254–255 189–190
2.264 135 n.86
2.299 158–159
2.321–322 36
2.327–328 36–37, 108
7.533–535 182 n.90
15.72 151 n.21
15.146–149 157 n.36
15.147 151 n.21

Ovid (*Continued*)
15.147–148 156–157
15.329–331 233 n.91
15.848 120 n.43
Tr.
1.1.69–82 37 n.54
3.4.29–30 37 n.54
4.3.65–66 37 n.54

Papyri Graecae Magicae
4.484–485 81–82

Paradoxographorum Graecorum Reliquiae (Giannini)
Alexander
2 (p. 165) 233 n.91
Antigonus of Carystus
8 (p. 34) 229 n.77
30 (p. 48) 233 n.91
60 (p. 64) 228–229
139–140 (p. 92) 229 n.77
140 (p. 92) 232 n.89
152b (p. 98) 232 n.89
164 (p. 102) 233 n.91
Apollonius
2 (pp. 120/122) 229 n.77
pseudo-Aristotle
4 (p. 224) 233 n.91
9 (p. 226) 233 n.91
38 (p.238) 229
40 (p. 238) 229
102 (p. 270) 232 n.89
105.3 (p. 274) 229
112 (pp. 278/280) 229 n.77
130 (p. 228) 229 n.77
148 (p. 302) 229 n.77
152 (p. 304) 231 n.86
154 (p. 306) 230–231
Nymphodorus of Syracuse
pp. 112–115 229 n.76
Paradoxographus Florentinus
20 (p. 320) 233 n.91
Paradoxographus Vaticanus
39 (p. 342) 229 n.77
Philostephanus
8 (p. 23) 229 n.77

Parmenides (19 *EGP*/28D-K)
D4/B1 12–13
D4.6/B1.6 12 n.3
D4.9/B1.9 13 n.4
D4.11/B1.11 13 n.4
D4.17–18/B1.17–18 12–13
D4.18/B1.18 37–38
D4.19/B1.19 12 n.1
D11–31/A37–43a and B9–15, 21 12 n.1
D36/A44 12 n.1
D44–50 12 n.1

Petronius
Sat.
36.6 218–219
118.6 247 n.7

Pindar
Isthm.
7.44–47 72 n.44
Ol.
13.91–92 72 n.44
Pyth.
1.21–23 221 n.56
1.21–22 209
1.22–23 230 n.82
1.26 229 n.78

Plato
Ap.
19b–c 27 n.31
26d–e 27 n.31
Ion
533d–536d 17 n.14
Phd.
96a–b 42–43
99d 43
Phdr.
245a–c 17 n.14
246a–248b 39
248a–b 39
249c–250c 39
269e–270a 52 n.89
Resp.
6.494c–d 27 n.33
7.516c–518b 97 n.99
9.586a 61 n.13
Symp.
211b–d 25
212a 25
215a–216b 162 n.47
218b 162 n.47
Tht.
155c–d 41–44, 61
174a 38–39
176b 42, 205
Ti.
22b–d 108 n.26
29a 214 n.37
47b–c 49 n.83
90c–d 49 n.83

Pliny the Elder
HN
praef. 17 95, 244–245
praef. 18 256–257

praef. 26 254–255
2.1–4 246–248
2.3 234–235
2.41–43 248–249
2.94 120 n.43
2.95 249–250
2.174 26 n.29
2.206 240–241
2.230 233 n.91
2.236 229 n.78
2.240 233–234
3.88 229 n.78
5.51–59 189 n.104
6.181 117 n.40, 187–188
6.184–186 117 n.40
7.7 251–252
7.125 219 n.52
7.188 232 n.90
8.34 244–245
8.97 233 n.91
11.8 234
12.19 117 n.40
18.250–253 255–257
18.252–253 206 n.18
31.1 235 n.93
37.1–2 251–252
37.11 254–255
37.55–61 252–253
37.59–60 253–254

Pliny the Younger
Ep.
5.6.36–37 220 n.54
9.26.1 31–32
9.26.2–4 30–32
9.26.3 73
9.26.5 30–31

Plutarch
Crass.
37.2 32 n.44
Mor.
44b–c 45

Posidippus (Austin-Bastianini)
fr. 5 255 n.18
fr. 13 255 n.18
fr. 15 255 n.18

Posidonius (Edelstein-Kidd)
fr. 90 223 n.62
fr. 234 202 n.8, 229 n.78

Propertius
2.34.27–28 54 n.93
2.34.51–54 54 n.93
3.5.23–48 54 n.93

Ptolemy
Epigrams (*AP*)
9.577 13–14
Tetr.
1.2.1–3 60 n.12

Quintilian
12.10.23–24 167 n.53
12.10.64–65 167 n.53

Scholia
Adnotat. super Luc.
1.49 156 n.33
6.482 165 n.50
6.484 165 n.51
Comm. Bern. in Luc. *BC*
1.413 149 n.15
Schol. vet. in Arat. *Phaen.*
161 72 n.46

Scriptores Historiae Augustae
Hadr.
13.3 237 n.99

Seneca
Ben.
3.37 231 n.85, 236 n.94
6.22.1 108 n.25
6.36 231 n.85, 236 n.94
7.17.1 133 n.80
Dial.
1.3.1 136 n.89
1.4.4 108–109
1.5.4 108 n.25
1.5.10–11 108–109
4.12.4 31 n.40
5.6.1 120 n.44
6.21.2 26 n.29
7.20.2 109
7.20.5 109
8.5.3–8 49 n.81
9.17.10–11 225 n.67
10.15.5 48 n.78
12.8.3 214–215
Ep.
16.5 204 n.11
48.11 110–111
51.1 224 n.66
65.9 120 n.44
65.19 120 n.44
72.6 109 n.28
75.13 109 n.28
76.34–35 133 n.80
79.1–4 237 n.99
79.5 224–225
79.7 224–225

Seneca (*Continued*)
88.22 223 n.62
90.42–43 102 n.8
108.13 112 n.33
121.12 127 n.62
Fragments (Vottero)
19 116 n.39, 189 n.105
NQ
1 praef. 5–6 105–106
1 praef. 6 110–111, 128–129
1 praef. 6–13 128–129, 154 n.28
1 praef. 7–10 26 n.29
1 praef. 7 128–129, 154 n.28, 156–157, 206–207, 215 n.41
1 praef. 8 143
1 praef. 11–17 49 n.81
1 praef. 11–13 140
1 praef. 12 129
1 praef. 13 105–106
1 praef. 17 105–106, 110–111
1.2.2 118–119
1.3–8 42 n.66
1.16 139–142
1.16.1 139–140, 142
1.16.2–5 141–142
1.16.2 139, 141–142
1.16.7 141–142
1.16.8–9 139–140
1.16.8 141–142
1.16.9 141–142
1.17.2–4 141–142
2.1.1 112
2.6–11 130–133
2.6.4–5 130–132
2.6.6 131–132, 132 n.78
2.7.6 120 n.44
2.8 130–131
2.9.4 132 n.78
2.12.5 167–168
2.16 220 n.54
2.26.2 202 n.7
2.26.6 113 n.35
2.32.8 127–128, 249
2.45 132–133, 214–215
3 praef. 1–4 103–112
3 praef. 1–2 104, 106–107
3 praef. 3–4 104–110
3 praef. 10 105–106, 110–111, 123–124, 136–137
3 praef. 18 105–106
3.18 137–139
3.18.1 137–139
3.18.2 137–138
3.18.3 137–139
3.18.4 137–138
3.18.5 137–138
3.18.7 137–139
3.20.5–6 233 n.91
3.22 189–190, 193–194
3.26.5–6 232 n.89
3.27–30 134–137
3.27.12 135, 158–159
3.27.13 135 n.86
3.28 151 n.22
3.29.5–6 136–137
3.30.7–8 136–137
4a praef. 1 143
4a praef. 10 204 n.10
4a praef. 20–22 142–144
4a praef. 21–22 142–143
4a praef. 21 143
4a.1–2 189 n.104
4a.1.1 143–144, 229
4a.2.3 189–190
4a.2.6 115–116
4b.3 120 n.44
4b.11 143
4b.11.4 26 n.29
5.15.1–2 113–115
5.15.4 114
6.2.1 133–137, 177–178
6.2.2 134
6.2.3 133
6.2.9 106–108
6.3.2 123–124, 138–139
6.8.3–5 116–118, 189 n.105
6.8.3 187–188, 191–192
6.9.1 167 n.54
6.12–21.1 170 n.62
6.16 132
6.16.4 132 n.75
6.17.3 113 n.35
6.31.3 133–134
6.32.4 108
7.1.1 125–126
7.1.5–2.1 119–120
7.1.4–5 125–126
7.3 124
7.3.2 121–122
7.4.1 121 n.47
7.6.1 123 n.51
7.12.2 121–122
7.12.4 121–122
7.12.8 121–122, 124–125
7.13.1–2 120–121
7.13.1 121–122
7.14.4 120–121
7.17–18 121

7.17.2 121 n.47
7.18.1 124–125
7.18.2 121, 123
7.19.2 120 n.43
7.22–25 119–120
7.22.1 123
7.23.1 123
7.23.3 124
7.24.1 124–125
7.24.2–3 124–125
7.24.2 119–120
7.25.1 127–128
7.25.2 127–129
7.25.7 121–122
7.27.4 102, 252–253
7.27.5–6 124–125, 252–253
7.29.3 111, 125–126, 252–253
7.30.2 123 n.52
7.30.4–6 127–128
7.30.6 235

Servius
In Aen.
6.154 116 n.39, 189 n.105

Sextus Empiricus
Math.
5.68 83–84
9.332 63 n.21

Silius Italicus
Pun.
8.495–510 185–186
14.66 229 n.78
14.196–197 231 n.85

Sophocles
TrGF
10c 70 n.38
537–539 70 n.38

Stoicorum Veterum Fragmenta
1.102 48 n.80
1.160 204 n.11, 214 n.38
1.172 214 n.38
1.216 49 n.81
1.239 46 n.73
2.35 48 n.80
2.42 106 n.20
2.524 63 n.21
2.599 214 n.38
2.1022 48 n.80
3.264–265 49 n.81
3.269–270 49 n.81
3.274–275 49 n.81
3.642 46–47, 228–229, 231–232

Strabo
1.3.16 (57) 46
6.2.3 (269) 202 n.8, 229 n.78, 231 n.85
6.2.6 (273) 223 n.62
6.2.8 (274) 213 n.35, 229 n.78, 237 n.99
10.3.9 (467) 213 n.35
17.1.1–5 (785–791) 189 n.104

Suetonius
Aug.
94.12 84 n.77
Claud.
21.6 218
Ner.
24.2 156 n.34
41.2 218–219
54 218–219

Thales (5 *EGP*/11 D-K)
D9/A16 189–190
P12/A9 39 n.58

Thucydides
1.10.3–5 82
2.48.3 181–182

Tibullus
1.7.23–24 189–190

Vagellius (*FLP*)
2 106–107

Valerius Maximus
1.8 ext. 12–19 233–234
5.4 ext. 4 231 n.85

Virgil
Aen.
1.122–123 62–63
1.261–277 96
1.278–279 96
1.494 222 n.59
1.495 221–222
1.740–747 188 n.103
2.355–360 134 n.84
2.355 134
3.574–576 209
3.578 210 n.25
3.619–620 209
3.622 209
3.632 209
5.13–25 173 n.68
6.585–594 70 n.38
6.586 71 n.39
6.589–590 70
7.750–760 185–186
8.184–279 27–29
8.219 27–28

Virgil (*Continued*)
8.228 27–28
8.230 27–28
8.242 27–28
8.256–258 37–38
8.247 27–28
8.262 27–28
8.264–265 27–28
8.265–267 29
10.160–161 177
Ecl.
6.1–12 15
6.27–86 14–15
6.27–30 14–15
6.29–30 74 n.51
6.31–33 14–15
6.31–40 14–15
6.31–86 14–15
6.47 15 n.10
6.52 15 n.10
6.64–73 15 n.10
G.
1.1 181–182
1.351–463 171–172
1.511–514 156
2.475–494 54 n.93
3.206–207 107
3.291–293 157 n.37
3.414–422 185–186
3.419 182 n.90
4.457–459 184 n.93
Vita Donatiana Vergilii
19 199 n.1
Vitruvius
De arch.
1.6.2 22–23, 132 n.75
10.7.4 220–221
10.8 219–220

General Index

Since the index has been created to work across multiple formats, indexed terms for which a page range is given (e.g., 52–53, 66–70, etc.) may occasionally appear only on some, but not all of the pages within the range.

Aetna, Mount
as paradigm of sublimity 200, 209, 224–225
and the paradoxographical tradition 227–232
as symbol for poet, poetry, poetic tradition 32–33, 161–162, 224–226
See also *Aetna* poet
Aetna poet, the
Aetna as poem in 224–226
Aetna as imperial spectacle in 10–11, 210–214, 216–226
assimilation to god in 204, 205
astronomical study rejected by 203–209, 214–215
divinity of Aetna in 208–216, 237–238
dual imperative in 203–208, 235–236, 240–241
Jupiter in 213–214, 216, 225
lapis molaris in 201–202, 222–223
and Longinus 240–241
and Lucan 204, 208–209, 214–215, 224–225, 240–241
and Lucretius 200, 204, 208, 224–225, 231–232, 240–241
and Manilius 200, 204, 206–209, 211 n.29, 216, 240–241
natura artifex in 200–201, 210, 212–216, 225–227, 236–241
and *Octauia* 214–215
Olympian gods in 202, 213–214, 216
as paradoxographer 200–201, 226–227, 231–232, 235–241
philosophical orientation of 201 n.6
and the pious brothers of Catania 200–201, 230–231, 235–241
and Pliny (the Elder) 10–11, 234–235, 240–241
poetic and/or mythological explanations of Aetna rejected by 202, 209–210, 216, 238–239
poetics of novelty in 202, 226, 239
and Seneca 201 n.6, 204, 206–209, 214–216, 224–225
volcanic aesthetics in 200–201, 225–227, 239
volcanic theory of 201–202, 222–223
See also Aetna, Mount; Index Locorum
analogy. *See* similes and analogies
Antigonus of Carystus 227 n.72, 228–229, 232 n.89, 233 n.91, 237
Apollonius of Myndus 121–122, 125–126
Aratus 54–55, 57 n.5, 64, 72 n.45, 97 n.99, 171–174, 177–178
Aratean tradition 64–66, 171–174, 176–178
See also Index Locorum
Archimedes 82–83, 219 n.51
Aristophanes 26–27, 40–41; *see also* Index Locorum
Aristotle 47 n.74, 49 n.83, 53–54, 103, 111, 120 n.43, 121 n.48, 189 n.104
inconceivable pleasures in 44–45, 48–49, 205 n.16
nature as demiurge in 44–45, 215 n.42
wonder and philosophy in 43–46, 125 n.56, 228–229, 248–249
See also pseudo-Aristotle; Plato; Index Locorum
pseudo-Aristotle
On the Universe 231
On Wondrous Things Heard 229, 230–231, 238
See also Index Locorum
Artemidorus (of Parium?) 120–122, 125–126
ascent 13 n.4, 24–26, 34, 42, 48, 79, 109 n.28, 154–155
deflationary use of 26–27, 39 n.60
Nero's ascent in Lucan 155–159
See also flight of the mind; view from above
assimilation to god (*homoiōsis theōi*) 42, 48, 63–64, 70–72, 79–81, 110–111, 204
astrology
and empire 91–99
as *telos* of human intellectual development 95–96, 208–209

astrology (*Continued*)
criticism of 80, 81 n.70, 84, 206–209, 243, 256
kinds of 58–59
principles of 58–59, 60–61
real-life astrologers 73–74, 83–85, 88, 121 n.47
relationship to philosophy 2–3
See also Augustus; Lucan; Manilius
Astronomica. *See* Manilius
astronomy. *See Aetna* poet; astrology; constellations; cosmos; Lucan; Manilius; planets; Seneca; stars
Augustine 95
Augustus
and astrological/astronomical symbolism 56–57, 63–64, 65 n.27, 91, 93–94
Augustan culture 29, 37, 64–66, 68, 80, 84 n.77, 112–113
Augustan Principate 8–10, 91–94, 128–129, 243
"horologium" of 93–94
Octavian 143, 191
and the Porticus Vipsania 93–94

Bellerophon 56–57, 71–74, 80–81
Burke, Edmund 5–8, 16, 18–19, 47, 75 n.55, 147, 160 n.43

Cacus. *See* Hercules
Caesar 63–64, 120 n.43, 152 n.25
Caesar's calendar 190 n.108
See also Lucan
Callimachus 15, 39–41, 68 n.35, 155 n.31, 188 n.103, 189–190, 202, 227–228; *see also* Index Locorum
causes
causal analysis vs. paradoxography 226–241, 244–246, 252–253, 256
nature's causes in Hellenistic philosophy 45–50, 228–229, 231–232
nature's causes in Plato and Aristotle 42–45
nature's causes in the rhetorical tradition 52–55, 150, 229–230
Celsus 179–180
Chrysippus 48–49, 106 n.20; *see also* Stoicism
Cicero 25–27, 49–50, 52–54, 62 n.16, 64–66, 84, 88, 154–155, 214–215; *see also* Index Locorum
Civil War. *See* Lucan
comets 119–129
constellations 66–67, 71–72, 89–90, 168–169; *see also* astrology; Manilius; stars
cosmos
collapse of 37–38, 62–63, 98–99, 108, 130 n.70, 135–136, 145, 158–159, 163 n.48, 169 n.59, 173, 186–187
cosmic spectatorship 25–26, 34, 48–50, 50–55, 63–64, 128–129, 140, 153–159, 165–166, 174–178, 203–204
divinity of 17–18, 48–49, 58, 132–133, 214–216, 240
See also Lucan; Lucretius; Manilius; nature; Seneca; stars; Stoicism
Ctesebius 219 n.52, 220–221

De rerum natura. *See* Lucretius
Demetrius 32 n.42, 53–54
Democritus 16–17, 45–46, 62 n.16, 121–122, 167 n.54; *see also* Index Locorum

earthquakes 12–13, 37–38, 130, 133–134, 162, 164–167, 170–171, 201–202, 210–214, 217; *see also* sublimity; subterranean wind theory
ekpurōsis. *See* Stoicism
Empedocles
in Horace 39–40, 57–58, 253 n.15
Love and Strife 21, 57–58, 102, 212 n.30, 252–253
in Lucretius 32–34, 224–225
and the sublime at Rome 9 n.14, 54–55
Enceladus 27 n.34, 202, 210, 213–214
Endymion 248–251, 256–257
Epicureanism
ataraxia 21–22, 29, 37–38, 46–47, 131–132, 134, 151–152
atoms and/or void 14–15, 20, 33–34, 37–38, 57–58, 62 n.19, 93, 114, 131–132, 152–153, 165–166, 173–174, 204 n.11
meteorology in 46–47
multiple explanations 46–47, 61–62, 150–151
and *nil admirari* tradition 46–47
See also Epicurus; Lucan; Lucretius; Stoicism
Epicurus 46–50; *see also* Lucretius; Index Locorum
Eratosthenes 82–83
eros 3–4, 15, 19–21, 41, 50–51, 90, 141–142, 191–192, 229–230, 248–249, 252
Eudoxus of Cnidus 190–191, 194 n.114

falling. *See* sublime
Flavian period 1–2, 10–11, 199–200, 210, 234, 240–241, 251, 254–255

flight of the mind 13, 24–27, 34, 48, 81–82, 104, 106 n.18, 153–159, 196–198, 247–248; *see also* ascent; view from above
freedom and slavery 24, 51–52, 77–78, 80–81, 251

Galen 49 n.82, 215 n.42
genre 54–55, 82–83
Germanicus 19 n.18, 64–66, 146 n.2, 243
Giants and/or Gigantomachy 23–24, 32–33, 72 n.46, 202, 213–214
god. *See Aetna* poet; assimilation to god; cosmos; Manilius; nature; Plato; Seneca; Stoicism
greatness of soul/thought (defined) 23–24

Hades 39
  Aidoneus 166
  Dis 213–214
  See also underworld
Hermogenes 53–54, 150, 169, 230 n.84; *see also* Index Locorum
Hercules
  and Cacus 27–29, 37–38, 162–163
  in Lucretius 29 n.39
Herodotus 32 n.44, 163–164, 189–190, 196
Heron of Alexandria 219–221, 223 n.62
Hesiod 12–13, 48, 72–73, 210–211; *see also* Index Locorum
Hipparchus 248–251, 256–257
Homer 19–20, 33–34, 48, 56–57, 82, 152–153, 177–178, 196 n.120, 227
  Homeric exegesis, ancient 54–55, 230 n.82
  See also Index Locorum
homoiōsis theōi. *See* assimilation to god
Horace 8–9, 27 n.32, 32 n.45, 39–40, 45–46, 73, 83 n.76, 189–190; *see also* Index Locorum

imperium. *See Aetna* poet; astrology; Lucan; Manilius; Pliny (the Elder); Seneca; scientific sublime; sublimity
infinity 6, 48, 83, 91–99, 124–126, 189–190, 234–235; *see also* Manilius
initiation
  as model for study of nature 12–13, 17–18, 127–128, 235
inspiration 13, 16–17, 52 n.89, 156–157, 160–162, 224–225
iuuat 64–65, 67, 128–129, 154 n.29, 156–157, 184–185, 206–207, 214–215, 236, 248–249

Kant, Immanuel 5–8, 16, 18–19, 47, 61 n.15, 75 n.55, 82 n.73, 108 n.25, 147, 160 n.43, 246, 251–252
katabasis 12–13

leaping. *See* sublime
lightning and/or thunderbolts
  as paradigms of sublimity 3–5, 7, 19, 35, 53–54, 167–168
  being thunderstruck 35, 71, 167–168, 175
  theories of 167–168
Longinus
  decentering from the study of the sublime 5
  enslavement as metaphor in 51–52, 77–78
  flawed sublimity in 31–32, 50, 69–70, 79–80, 124, 254–255
  greatness of soul in 4–5, 23 n.23, 24, 32 n.42
  investigation of nature and cosmos in 4–5, 50–52, 197, 229–230, 247–248
  and paradoxographical tradition 229–230, 240
  will to sublimity of 51–53
  See also Aetna poet; Lucan; Manilius; Pliny (the Elder); Seneca; Index Locorum
Lucan (*Civil War*)
  Acoreus in 150 n.18, 187–198
  Amyclas in 171–174
  and Aratean tradition 171–174, 176–178
  ascent in 153–159
  Bardi in 152–153
  Caesar in 3–4, 10, 27 n.33, 145, 147–149, 151–153, 162–163, 167–168, 170–174, 177–178, 180, 184–185, 187–198
  Cato in 148 n.12, 152 n.23, 168–169, 178–187
  cosmic dissolution in 145, 158–159, 163 n.48, 169 n.59, 173, 186–187
  dark modes of sublimity in 27 n.33, 146–147, 160–169, 176–178, 184–186
  and didactic poetry 171 n.63, 175 n.73, 181–186, 188 n.103
  Druids in 152–154
  earthquakes in 162, 164–166, 170–171
  and Epicureanism 193–194
  Erictho and Thessalian witches in 160–161, 164–169, 184–185, 195
  humor and the sublime in 154–155, 157–158, 174, 178–179, 181–182, 185–187, 196
  and Longinus 147, 148 n.10, 160–161, 166–168, 172, 197

Lucan (*Civil War*) (*Continued*)
and Lucretius 147–148, 150–153, 159, 164–168, 170–171, 176, 181–182, 186–187, 191–192, 197
and Manilius 145, 146 n.2, 148, 150, 156–157, 159–160, 166–169, 174–178, 182 n.89
meteorology in 146, 170–174, 187–188, 195
and Nero 155–159, 188, 196
the Nile in 187–198
Phaethon in 155–159, 162–163, 173–175, 184–185, 192
Phemonoe in 160–164, 166–169, 184–185, 195
plague in 179–184
Pompey in 148 n.12, 153–155, 157–159, 171, 174–178, 185–186, 196–197
and Seneca 145–148, 150–152, 154, 156–159, 160, 166–171, 174–175, 177–178, 187–198
snakes in 159, 178–187
stars and planets in 149–150, 155–159, 164–165, 168–169, 171–178, 181, 184–185, 190–195
and Stoicism 169 n.59, 178 n.80, 179 n.82
sublimity of *imperium* in 145, 173–174, 177–178, 190–192, 195
tides in 149–153
See also Pliny (the Elder); Index Locorum
Lucilius (Iunior) 101–102, 142–144, 224–225
Lucretius (*De rerum natura*)
and Cicero 49–50, 52–53
compression model of knowledge in 95
earthquakes in 21, 37–38, 166, 170–171
decentering from the study of the sublime 8–9
Empedocles in 32–34, 39–40
Epicurus's battle with *religio* in 23–24, 28, 170
Epicurus's flight in 23–25, 35, 51, 97, 117, 159, 247–248, 250
honeyed cup in 76
Horace's deflationary use of Lucretian sublime 39–40
and *nil admirari* tradition 46–47
the Nile in 190 n.107
and paradoxographical tradition 190 n.107, 229, 231–232
Ovid's Phaethon as commentary on 32–33, 36–37
Phaethon in 34–35, 36 n.53, 68
plague in 38 n.57, 181–182, 221
Sicily in 229
thunderbolts in 3–4, 167–168
Virgil's Hercules and Cacus as commentary on 28–29
Virgil's Silenus as commentary on 14–15
uoluptas and *horror* in 20–22, 29, 47, 49–50, 114, 165–167, 204, 208
See also *Aetna* poet; Epicureanism; Lucan; Manilius; Pliny (the Elder); Seneca; Index Locorum

madness 4–5, 17 n.14, 28, 39–40, 134, 137–139, 162 n.47, 206–207, 225 n.67, 246–247
magic and magicians 81–83, 166–167, 178–179, 185–187
magnanimitas, magnitudo animi. See greatness of soul/thought
Manilius (*Astronomica*)
accumulation vs. compression models of knowledge in 94–97
aesthetics of the infinite 56–57, 66, 69–70, 79, 91
and Aratean tradition 64–66
and arctic sublime 89–90
as magician 81–83
as planet 75–76, 88–89
as transgressive 56–57, 70, 72–75, 81–83
assimilation to god in 63–66, 70–72, 79–81
astrologer as daredevil in 56–57, 67–74, 82–83, 126–127
astrology and *imperium* in 56–57, 65–66, 91–99
astrology as supreme intellectual pursuit in 95–96, 208–209
celestial chariot ride in 67–74, 97
cosmos as Roman in 63–64, 92–96
descent of god into man in 79
flawed sublimity in 69–70, 79–80
god as subject of 58, 71, 74–75, 78–79, 81–83
insufficient information to cast a horoscope in 58–59, 91
and Longinus 57 n.4, 61 n.13, 69–70, 77–80
and Lucretius 57–58, 62–63, 68, 76, 82–83, 93
map making in 93–94
mathematics and numbers in 9–10, 56–58, 74–91, 94–97, 110–111, 204, 252–253
the Milky Way in 59–70, 74 n.53, 92, 98–99, 126–127, 153 n.27, 160, 211 n.29
and Ovid 65 n.27, 68, 156–157

Phaethon in 56–57, 61–64, 63 n.21, 67–74, 80–81, 90 n.82, 91 n.84, 98
and Pythagoreanism 81–82
and Stoicism 57–58, 69 n.37, 79 n.64, 80–81
sweet and/or pretty poems rejected by 75–76, 88–89
tightrope walker in 72–74
See also *Aetna* poet; Lucan; Pliny (the Elder); Seneca; Index Locorum
Martial 32 n.45, 243–244
mathematical poetry 82–83; *see also* Manilius
medicine 2–3, 59 n.9, 132 n.75, 146, 179–180, 186–187
Medusa 89 n.81, 181, 182 n.89
megalophrosunē. *See* greatness of soul/thought
megalopsuchia. *See* greatness of soul/thought
meteorology 2–3, 46–47, 103, 110–112, 119–120, 123; *see also* Epicureanism; Lucan; Seneca; sublimity
the Milky Way 2–3, 6–7, 25–26, 59–66
in Aratus and Aratean tradition 64–66
See also Manilius
Mount Aetna. *See* Aetna
multiple explanations. *See* Epicureanism
music of the spheres 25–26, 70–71
mystery cults 213 n.35; *see also* initiation

nature
natura artifex (nature the demiurge) 44–45, 200–201, 210, 214–216, 220, 222–223, 225, 232–235, 237–241
natura artifex (nature the demiurge) and paradoxography 232–235, 240–241, 247–248, 254, 256
Nature (personified) 255–257
nature and/or cosmos equated with god 17–18, 48–49, 58, 132–133, 214–215, 240
moral value of the study of 21–22, 48–50, 53–54, 100–102, 103 n.13, 137, 142–147, 199–200, 206–207, 238–239, 247–248, 251–252, 256–257
Natural History. *See* the Pliny (the Elder)
natural philosophy 2–3, 14–15, 28, 103 n.12, 146, 148 n.13, 150–151, 171–172, 195, 203, 231–232
Natural Questions. *See* Seneca
Nero
as Phaethon 112–113, 155–159, 200–201, 225
depravity and/or destabilizing influence of Neronian Rome 10, 100–102, 128–129, 137, 149, 159, 199–200, 243, 251
sent an expedition to find source of the Nile 116–118, 187–188, 191 n.109
and water organs 218–219
See also Aetna poet; Lucan; Seneca
Nicander 182, 185–187
Nigidius Figulus 81–82, 193 n.113
nil admirari tradition 45–50, 195, 254–255
the Nile
cause(s) of flood and location of source as paradigmatically unknowable 189–190
See also Lucan; Lucretius; Nero; Seneca

ocean/Ocean 4–5, 7, 149, 152–164, 190 n.107, 196, 229–230
Octavian. *See* Augustus
Orpheus 14–15, 74 n.51, 200–201
Ovid 36–37, 54 n.94, 68, 71–72, 108–109, 134 n.85, 156–159, 162 n.46, 182 n.90, 189–190, 202 n.9, 204; *see also* Index Locorum

paradoxography 195, 200–201, 226–241, 245–248, 254–257; *see also* Aetna; *Aetna* poet; causes; Longinus; Lucretius; nature; Pliny (the Elder); scientific sublime; Seneca; Valerius Maximus
Parmenides 9 n.14, 12–13, 15–17, 24–25, 24 n.25, 37–38, 40–41, 54–55
Pericles 4, 52–53, 181–182
Petronius 137, 218–219, 247 n.7, 251
Phaethon (paradigm of sublime subject, student of nature) 34–37
possible decline in appeal of after Nero's demise 156 n.34, 200–201, 225
See also Lucan; Lucretius; Manilius; Nero; Seneca
physics 2–3, 48–49; *see also* Seneca
plague. *See* Lucan; Lucretius
planets
as cause of Nile's flood 192–193
comets as extra-zodiacal planets 119–129
order of 149 n.15, 248
planetary theory 121–122, 124–125, 127–128, 194–195
See also stars
Plato 25–27, 38–39, 49–50, 52–53, 98, 108 n.26, 162 n.47
and assimilation to god 42, 48, 61, 63–64, 204–205
and mathematics 81–82

Plato (*Continued*)
wonder and philosophy in 41–43, 61
See also Aristotle; assimilation to god; Index Locorum
Pliny (the Elder) (*Natural History*)
and *Aetna* poet 10–11, 234–235, 240–241
alienation from scientific sublime in 244–251
grandeur of nature in 251–252, 256
and *imperium* 94–95, 243–244
and Longinus 246–248, 251, 254–255
and Lucan 245–246, 250–251
and Lucretius 247–248, 250–251, 256–257
and Manilius 94–95, 235
and paradoxographical tradition 233–235, 245–248, 254–257
and Seneca 101, 114, 131–132, 134
and Stoicism 232 n.90, 244–245
will to sublimity in 254
See also Index Locorum
Pliny (the Younger) 30–32, 73, 124
Pompey. *See* Lucan
Ptolemy, Claudius 13–15, 24–25, 60 n.12, 81–82, 249 n.9
Pythagoras 39 n.59, 45–46, 49 n.83, 51 n.85
in Ovid 151 n.21, 156–157
Pythagoreans 120 n.43
Pythagoreanism 81–82

rainbows 41–42, 60–61, 102
recusatio 54–55, 206–207
rivers 4–5, 7, 113, 115–117, 161, 187–198, 190 n.107
rivers of fire 209, 229–230

Salmoneus 56–57, 70–74, 80–81
scientific sublime, the
alienation from 244–251
anachronism of "science," "scientist," "scientific" in ancient contexts 2–3
ancient theoretical approaches to 40–55
as antidote to moral decline and excesses of empire 137–144
defined 1, 12–15, 242–243
deflationary use of 26–27, 37–40, 154–155, 185–186
and empire 3–4, 9–11, 29, 37, 91–99, 112–113, 137, 142–144, 173–174, 190–192, 195, 216–226, 242–244, 248–251
impotence and/or breakdown of 169–198, 242–243
multiplicity of 3–4, 29, 37
and *nil admirari* tradition 45–50
tension with paradoxography 228–229, 231–232, 236–241, 244–248, 254–257
weaponization of 196
Seneca (*Natural Questions*)
comets in 119–129
consolatio in 108, 130, 133, 136–137
cooperative vs. solitary study in 112–119, 141–144
cosmic viewpoint in 103–104, 130
degradation of the sublime in 10, 100–101, 137–144
flood in 134–137
Hostius Quadra in 139–142
imperium and the sublime in 142–144, 243–244
leaping into the abyss in 108–110
and Longinus 124, 131 n.74, 134 n.85, 137 n.91, 190 n.107, 240, 251
and Lucilius (Iunior) 101–102, 142–144, 224–225
and Manilius 102, 110–113, 120 n.43, 124, 126–129, 143–144, 147 n.6, 206–209, 235
meteorology's symbolic value in 101–102, 112
and Neronian Rome 10, 100–102, 127–129, 137
and Nero 107 n.23, 112–113, 116–118, 188, 191 n.109
the Nile in 115–118, 142–144, 188–190
order of books of 103
paradoxography in 143 n.101, 190 n.107, 229, 231–232, 235
Phaethon in 101–102, 106–113, 115–116, 135 n.86, 141–142, 156–157, 250–251
praemeditatio malorum in 136–137
proficiens in 101–102, 108–109, 130, 135, 141
the inexperienced in 133–135, 138–139
sage in 100–102, 108–109, 128–130, 133–134, 138–140
spiritus and/or *aër* in 130–133, 214–215
study of nature as aporetic but all-embracing and ennobling in 100–102, 109–111, 125–129
study of nature as initiation into the mysteries in 127–128, 235

study of nature as interactive spectacle in 118–119
study of nature as study of god in 110–111, 132–133, 214–215
subject as meteorology vs. physics 103
violence of nature in 130–133
See also *Aetna* poet; Lucan; Pliny (the Elder); Index Locorum
Sextians 81–82
similes and analogies 3–4, 59–60, 64–66, 93, 118–119, 127–128, 158–159, 162, 167–168, 179–180, 216–223, 226
slavery. *See* freedom and slavery
Socrates 26–27, 39, 41–45, 52–54, 61, 205
spectacle, sublime (defined) 22–23
stars
as mundane 208
as source of anxiety 46
as source of sublimity 4–5, 13–14, 43–44, 51–52, 56–57, 129, 229–230, 234–235
sublimity of stars as out of reach 214–215, 248–251, 255–257
vs. planets 120–121, 124–125, 127–129
See also *Aetna* poet; Aratus; astrology; comets; constellations; cosmos; Lucan; Manilius; Milky Way; planets
Stoicism
conflagration (*ekpurōsis*) 98
contrasted with Epicureanism 193–194
deluge not caused by moral decline 136 n.88
elemental volatility of Stoic world 130 n.72
god in 48–49, 132–133, 214–215
and *natura artifex* 214–215
and *nil admirari* tradition 46–47, 228–229
and paradoxography 228–229, 231–232
physics as theology in 48–49
praemeditatio malorum 136–137
sage as elevated/sublime 49 n.81, 133–134
sage vs. fool 141
sage vs. *proficiens* 101–102
spiritus and/or *aër* 130–133, 214–215
sympathy 80–81, 253 n.15
tension 132
See also *Aetna* poet; Chrysippus; Lucan; Manilius; Pliny (the Elder); Seneca; Zeno
storms 30, 64–65, 171–174
Strabo
and *nil admirari* tradition 46
on climbing Aetna 237
on secrecy enhancing grandeur 213 n.35
See also Index Locorum
sublime, the
anxiety about achieving 137, 171–178, 186, 246–251
arctic sublime 89–90
aristocratic or elite connotations of 16–17, 23–24, 32, 49–50, 52–53, 98–99, 243
as *consolatio* 108, 130, 133, 136–137, 174–175, 177–178
as discourse vs. experience 1, 8, 16–17
beauty and 25, 44–45, 75 n.55, 137–138, 230–231, 239 n.104
criticism and mockery of 26–27, 38–40, 137–144, 149–155, 157–158, 174, 178–179, 181–182, 185–187, 196, 203–209
decentering Longinus from study of 8
decentering Lucretius from study of 9
defined 16–18
dramatizing 19–20, 22
dual logic of (defined) 16–18
encoding (defined) 19–22
falling as paradigm of 29–40
leaping as paradigm of 37–40
mechanical sublime 216–226
multiplicity of 3–4, 29, 37
Phaethon (paradigm of sublime subject, student of nature) 34–37
poverty/wealth and 31–32, 128–129, 172, 251–252
proptreptic use of 21–22, 61
rhetorical tradition and 50–55
sublime subject (defined) 23–28
tightrope walker (paradigm of sublime subject) 30–31
See also Empedocles; Longinus; Lucan; Lucretius; Manilius; scientific sublime; Seneca; Stoicism; sublimity
sublimity
dark modes of 27–29, 62–63, 108, 146–147, 160–169, 176–178, 184–186, 200, 209–210, 247–248
flawed sublimity 30–34, 50, 69–70, 79–80, 124, 254–255
historical origins of discourse on 16–17
of *imperium*/political power 3–4, 9–11, 29, 37, 63–64, 91–99, 112–113, 137, 142–144, 173–174, 190–192, 195, 216–226, 242–244, 248–251

sublimity (*Continued*)
will to sublimity 22–23, 27–28, 39–40, 52–53, 58–59, 179–180, 236 n.94, 254
See also Aetna; lightning and/or thunderbolts; Longinus; Lucan; Manilius; Pliny (the Elder); scientific sublime; stars; sublime
subterranean wind theory (of volcanism, seismic activity) 162, 170 n.62, 201–202; *see also* wind

Thales 38–39, 189–190, 206–207
thunderbolts. *See* lightning and/or thunderbolts
tightrope walker 30–31, 72–74
Typhos/Typhoeus 27 n.34, 162, 209–210, 229–230

underworld 20, 46–47, 114, 166, 228–229, 238; *see also* Hades; katabasis

Valerius Maximus 232–234
Vesuvius 199 n.1, 223 n.63, 243–244
view from above 24–27, 34, 48, 81–82; *see also* ascent; flight of the mind
Virgil
Aeneid and/or epic tradition 27–29, 37–38, 62–63, 70, 82–83, 95–96, 134, 152–153, 171–172, 173 n.68, 177
attribution of *Aetna* to 199 n.1
Eclogues 14–15
Georgics and/or didactic tradition 156, 181–182, 184 n.93, 185–187
See also Lucretius; Index Locorum
Vitruvius 22–23, 219–221
void. *See* Epicureanism
volcanoes. *See* Aetna; Vesuvius; subterranean wind theory

wind 22–23, 132 n.75, 149, 162, 167–168, 170–171, 207, 210–211, 220; *see also* subterranean wind theory

Zeno 46 n.73, 48–49, 204 n.11, 214–215

